Springer Tracts in Mechanical Engineering

Springer Tracts in Mechanical Engineering (STME) publishes the latest developments in Mechanical Engineering - quickly, informally and with high quality. The intent is to cover all the main branches of mechanical engineering, both theoretical and applied, including:

- Engineering Design
- Machinery and Machine Elements
- Mechanical Structures and Stress Analysis
- Automotive Engineering
- Engine Technology
- Aerospace Technology and Astronautics
- Nanotechnology and Microengineering
- Control, Robotics, Mechatronics
- MEMS
- Theoretical and Applied Mechanics
- Dynamical Systems, Control
- Fluids Mechanics
- Engineering Thermodynamics, Heat and Mass Transfer
- Manufacturing
- Precision Engineering, Instrumentation, Measurement
- Materials Engineering
- Tribology and Surface Technology

Within the scope of the series are monographs, professional books or graduate textbooks, edited volumes as well as outstanding PhD theses and books purposely devoted to support education in mechanical engineering at graduate and post-graduate levels.

Indexed by SCOPUS, zbMATH, SCImago.

Please check our Lecture Notes in Mechanical Engineering at https://link.springer.com/bookseries/11236 if you are interested in conference proceedings.

To submit a proposal or for further inquiries, please contact the Springer Editor **in your region**:

Ms. Ella Zhang (China)
Email: ella.zhang@springernature.com
Priya Vyas (India)
Email: priya.vyas@springer.com
Dr. Leontina Di Cecco (All other countries)
Email: leontina.dicecco@springer.com

All books published in the series are submitted for consideration in Web of Science.

Siqiang Wang · Shunying Ji

Computational Mechanics of Arbitrarily Shaped Granular Materials

Springer

Siqiang Wang
Department of Engineering Mechanics
Dalian University of Technology
Dalian, Liaoning, China

Shunying Ji
Department of Engineering Mechanics
Dalian University of Technology
Dalian, Liaoning, China

ISSN 2195-9862 ISSN 2195-9870 (electronic)
Springer Tracts in Mechanical Engineering
ISBN 978-981-99-9926-2 ISBN 978-981-99-9927-9 (eBook)
https://doi.org/10.1007/978-981-99-9927-9

This Springer imprint is published by the registered company Springer Nature Singapore Pte Ltd.
The registered company address is: 152 Beach Road, #21-01/04 Gateway East, Singapore 189721, Singapore

Preface

Granular materials are composed of a large number of discrete particles of different shapes and sizes, which are manifested as sand and sea ice in nature, soil and concrete in engineering structures, and ores and grains in industrial production, etc. They are one of the most widely occurring materials in natural environments, engineering applications, and daily life. Due to the morphological complexity, non-homogeneity, and randomness of granular materials at the microscale, they also exhibit complex mechanical behaviors at the macroscale. In order to effectively solve the mechanical problems of granular materials in different engineering fields, Cundall and Strack established the discrete element method (DEM) in the 1970s. However, the method still has many challenges in real morphology construction, efficient search algorithm, multi-contact force model, bond-fracture criterion between arbitrarily shaped elements, and high-performance parallel computing.

The discrete element method initially used two-dimensional discs or three-dimensional spheres to represent solid particles and was characterized by computational simplicity and efficient performance. However, spherical and irregular particles have significant differences in macroscopic and microscopic mechanical properties, while the multiple collisions, low flowability, and interlocking of irregular particles significantly affect the mechanical behaviors of granular materials. In recent years, discrete element methods for arbitrarily shaped particles have been rapidly developed. These methods can not only construct convex-shaped particles, such as spheres, ellipsoids, and cylinders, but also arbitrarily shaped particles with concave surface properties or internal cavities, such as gravel, sand, and rings. It is worth noting that superquadric equations, polyhedrons, and dilated polyhedrons are only suitable for simulating convex particle materials due to the limitation of contact algorithms. A single contact point between convex-shaped particles has difficulty reflecting the mechanical properties of multiple contact points between concave-shaped particles.

In order to reasonably describe the geometrical properties of arbitrarily shaped particles, the combined particle method has been developed. This method combines different numbers of spheres, ellipsoids, cylinders, superquadric elements, convex polyhedrons, and other basic elements to construct arbitrarily shaped particles. Although the combined particle method has been a significant improvement over the traditional convex-shaped particle model in describing complex particle morphology, it is an approximate approach and introduces new errors at the single particle level. Another class of discrete element methods for describing arbitrarily shaped granular materials includes Fourier series forms, non-uniform rational spline, spherical harmonic functions, energy conservation theory, level set algorithms, and directed distance field methods. These methods significantly improve the computational accuracy of DEM simulations and can be used to construct convex and concave particles. It is worth noting that the contact detection between arbitrarily shaped particles is more complicated than that of spherical particles. The ratio of the search time to the overall consumption time increases significantly, which also limits the number of particles and the computation time for the DEM simulation. Therefore, fast and accurate contact algorithms for arbitrarily shaped particles are still difficult challenges in the current research of discrete element methods. Nevertheless, research on discrete element methods for arbitrarily shaped granular materials has made great progress in both contact algorithms and software development.

For the software development of discrete elemental methods for arbitrarily shaped granular materials, a variety of software has been developed and commercialized, including: the software GDEM developed by the team of Prof. Shihai Li and Prof. Chun Feng from the Institute of Mechanics of the Chinese Academy of Sciences, the software DEMms developed by the team of Prof. Wei Ge and Prof. Limin Wang from the Institute of Process Engineering of the Chinese Academy of Sciences, the software DEMSLab developed by Prof. Yongzhi Zhao's team at Zhejiang University, the software CoSim developed by Prof. Wenjie Xu's team at Tsinghua University, the software SudoSim developed by Prof. Jidong Zhao's team at the Hong Kong University of Science and Technology, the software MatDEM developed by Prof. Chun Liu's team at Nanjing University, the software AgriDEM developed by Prof. Jianqun Yu's team at Jilin University, SDEM software developed by Prof. Shunying Ji's team at Dalian University of Technology, and so on. For the development of foreign software, a variety of software and open-source DEM calculation programs have been successfully commercialized and guided industrial productions.

This book introduces discrete element methods for arbitrarily shaped granular materials and is divided into seven chapters. Chapter 1 is the introduction, which mainly introduces the theoretical construction of arbitrarily shaped particles and high-performance discrete element methods. Chapters 2–5 introduce the discrete element methods for the superquadric model, multi-superquadric and poly-superquadric models, smooth polyhedron model, and level set model, respectively. Chapter 6 introduces high-performance GPU parallel algorithms for arbitrarily shaped granular materials and the development of analytical software for discrete element methods. Chapter 7 introduces the flow, mixing and segregation characteristics of arbitrarily shaped granular materials in silos and rotating drums.

The research in this book is supported by the National Key Research and Development Program of China (Grant Nos. 2021YFA1500302, 2018YFA0605902), the National Natural Science Foundation of China (Grant Nos. 41576179, 51639004, U20A20327, 42176241, 52101300, 12102083, 12202095), the Dalian Scientific

Innovation Foundation of China (Grant No. 2022RQ004), the China Postdoctoral Science Foundation (Grant No. 2022M710587), and other projects. It is also supported by the State Key Laboratory of Structural Analysis, Optimization and CAE Software for Industrial Equipment of Dalian University of Technology.

I am very grateful for the guidance of Prof. Hayley Shen and Prof. Hung Tao Shen of Clarkson University, USA. Under the guidance of the two professors, the authors began to engage in the research of discrete element methods in 2002, and applied them to different engineering fields. I would also like to pay special tribute to the research work of doctoral students Shanshan Sun, Shaocheng Di, Shuai Shao, Lu Liu, Xue Long, Yongjun Li, Shuailin Wang, Shuai Kong, Biyao Zhai, Shaomin Liang, and Xu Li, who graduated from the Group of Computational Granular Mechanics of Dalian University of Technology. The cooperation with Dalian Dizao Science and Technology Co., Ltd. on the discrete element software SDEM also enriches the achievement of this book.

Dalian, China
September 2023

Siqiang Wang
Shunying Ji

Contents

Chapter 1
Introduction

Granular materials are widely found in nature or industrial production, and are complex systems composed of a large number of discrete solid particles (Kou et al. 2017; Clerc et al. 2008). Granular materials have the special mechanical properties of solids or fluids, and solid–liquid-like transformation phenomena occur under certain conditions. The energy of granular systems can be dissipated rapidly by the friction and damping between particles, and the contact force and direction can be adjusted by the particle rearrangement. The local load can be extended in space and then form a stable granular system. The Discrete Element Method (DEM), proposed by Cundall and Strack, has become an important tool for modeling the mechanical properties of granular materials and solving related engineering problems (Cundall and Strack 1979). This method was first developed using regular elements of 2D disks or 3D spheres, and is characterized by its simplicity and high computational efficiency. However, complex systems consisting of irregular particles are common in nature or industry (Cleary 2000). Spherical and irregular particles differ significantly in their macro–micro mechanical properties, while multiple collisions, low flowability, and interlocking between irregular particles significantly affect the mechanical response of granular materials. Although the spherical particles considering the rolling friction model can be used to reflect the dynamic behavior of the irregular particles, it is difficult for this model to effectively capture the certain characteristics of granular materials, such as void fraction and coordination number (Xie et al. 2019). More importantly, the macro–micro mechanical properties obtained from spherical granular materials cannot be simply extended to irregular granular materials (Lu et al. 2015). Therefore, the development of theoretical constructions and contact models for arbitrarily shaped particles remains a crucial issue in the current discrete element method (Kafashan et al. 2019).

S. Wang and S. Ji, *Computational Mechanics of Arbitrarily Shaped Granular Materials*, Springer Tracts in Mechanical Engineering,
https://doi.org/10.1007/978-981-99-9927-9_1

In recent years, discrete element methods for irregular particles have been rapidly developed, including ellipsoid models based on quadratic equations (Zhou et al. 2011), superquadric models based on continue function representations, polyhedral models based on geometric topology (Govender et al. 2014), and dilated polyhedral models based on the Minkowski sum algorithm. These DEM models are usually suitable for constructing convex particles, and the contact detection between particles is single-point contact, which makes it difficult to accurately reflect the motion behaviors of concave granular materials. In order to characterize the geometrical properties of concave particles, the combined particle model has been developed. This method combines different numbers of basic elements such as spheres, ellipsoids, superquadrics, cylinders, and polyhedrons to construct arbitrary particle shapes (Liu and Zhao 2020). Considering the complexity of the contact detection between concave particles, its contact problem is usually transformed into a contact problem between several basic elements. The disadvantages of the combined particle method are that its computational efficiency decreases significantly with the increase in the number of basic elements, and the computational accuracy depends on the contact pattern between particles. Moreover, the novel discrete element methods for describing concave granular materials include Fourier series forms based on functional envelopes (Lai et al. 2020), spherical harmonic functions (Garboczi and Bullard 2017), energy conservation theories based on discretized surfaces (Feng 2021a, b, c), and level set algorithms (Vlahinić et al. 2017). Such methods significantly improve the computational accuracy of DEM simulations and can be used to construct arbitrary particle shapes. It is worth noting that the contact detection between irregular particles is more complicated than that for spherical particles. Meanwhile, the ratio of the search time to the overall computation time increases significantly, which limits the number of irregular particles and makes it difficult to accurately reflect the kinematics of large-scale irregular granular materials (Govender et al. 2015a, b).

In order to meet the large-scale computational needs of discrete element methods in real engineering problems, parallel computing methods usually adopt Central Processing Unit (CPU) parallelism or Graphics Processing Unit (GPU) parallelism. Among them, CPU parallel methods include the Open Multi-Processing (OpenMP) method based on the shared-memory approach, the Message Passing Interface (MPI) method based on the distributed memory model (Yan and Regueiro 2018), and the MPI-OpenMP method, which is a mixture of the above two algorithms (Liu et al. 2014a). The above methods use the CPU as the computational core, and each processor cooperates with each other to achieve the purpose of accelerating the solution speed. The disadvantage of CPU parallelism is that its computational efficiency usually depends on the number of computational cores, which leads to an increase in the cost of computer hardware and is susceptible to limitations such as load imbalance. Another approach is GPU parallelism, which has a parallel architecture consisting of thousands of smaller and more efficient cores. Compared to CPU parallelism, GPUs are inexpensive and suitable for operations with high arithmetic densities and simple logic branches, which provides a reliable hardware basis for large-scale numerical simulations of irregular granular materials (Chen and Matuttis 2013).

Granular materials are widely used in many fields, such as food transportation, mineral engineering, pharmaceutical processing, and geohazard prediction. Their flow states are affected by boundary conditions, material properties, particle shapes, and other factors, resulting in different flow patterns. During gravity-driven granular flow, the particles are rearranged to form an arch structure and a network of force chains with different strengths, which changes the granular flow state from continuous to intermittent flow. In addition, particle shape significantly affects the flow state of granular materials. Compared with spherical particles, the surface sharpness and aspect ratio of the irregular particles restrict the relative motion of the particles and enhance the shear strength of the granular material, and the flow state of the particles is changed from a continuous flow to a blocking state (Toson and Khinast 2017). Therefore, the study of the flow characteristics of irregular particles has an important reference for the application of granular materials in industrial production and design.

1.1 Discrete Element Methods for Arbitrarily Shaped Particles

In recent years, the discrete element method has become an important tool for analyzing the macro–micro flow properties of granular materials. With the gradual improvement of the simulation accuracy, more studies have focused on the real morphology of particles, which has led to the further development of the contact theory and search algorithm for irregular particles.

1.1.1 Arbitrarily Shaped DEM Model Based on Functional Representations

Ellipsoid modeling is a common method for constructing irregular particles, which can be used to construct oblate or prolate ellipsoidal particles with different aspect ratios and has the advantages of numerical stability and computational efficiency, as shown in Fig. 1.1. Lin and Ng (1995) developed a numerical computation program "ELLIPSE3D" for ellipsoids. The program uses two contact methods, including the geometric potential algorithm and the common normal method. Among them, the geometric potential algorithm aims to find the point on the particle surface, and this contact point has the smallest geometric potential. The common normal method aims to find two points on the particle surface, and the line between the two points is parallel to the surface normal of the two particles, respectively. For a two-dimensional ellipsoid or three-dimensional ellipsoid, different contact detection methods have different definitions of contact points due to mathematical considerations and result in different overlaps between particles. However, the overlap between

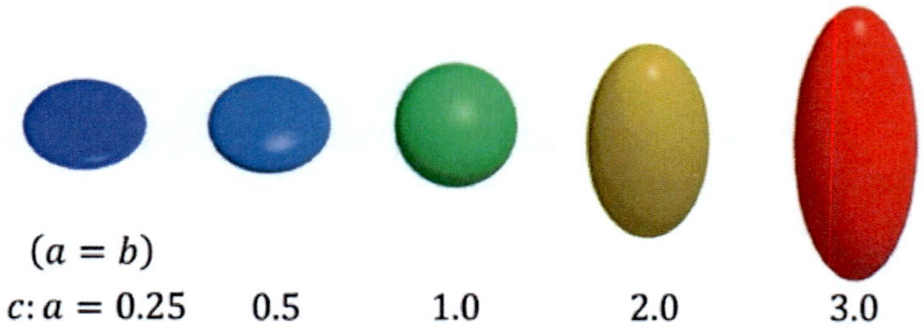

Fig. 1.1 Ellipsoid models with different aspect ratios

particles in DEM simulations is usually less than one thousandth of the particle size. Therefore, the differences arising from these two approaches are negligible. Hopkins (2004) proposed the dilated ellipsoid based on a mathematical model, which equates the contact between two ellipsoids to the contact between several spheres within the dilated radius of each ellipsoid. This method simplifies the contact detection between convex particles, and the shortest distance between particles within the dilated distance can be obtained by calculating the gradient of one constrained surface relative to the other constrained surface. Zheng et al. (2013) developed a model of normal and tangential contact force between ellipsoidal particles based on contact mechanics and finite element method, which is a semi-theoretical model. Its simulation parameters can be determined directly from material properties. During the collision of two viscoelastic ellipsoids, the numerical results of contact time, maximum contact force, rebound velocity, and rebound angular velocity of the ellipsoid particles are in good agreement with the numerical results obtained by the finite element method.

The superquadric model is an extension of the quadratic equation, and the ellipsoidal, cylindrical, and cubic particles with different aspect ratios and surface sharpness can be obtained by varying the five parameters of the function (Chen et al. 2020). The superquadric equation was first used by Barr to characterize the shape of non-regular particles in 1981 (Barr 1981). The equation was subsequently applied to computer graphics and can represent complex geometries. Williams and Pentland (1992) further applied the superquadric equation to computer-aided design and DEM simulations. This equation was combined with multibody dynamics, which provides higher advantages in terms of stability, time step, and response speed. Han et al. (2006) developed the superquadric equation for two-dimensional non-regular particles. He used an adaptive algorithm to construct convex polygonal elements for 2D superquadric equations and converted the contact detection between two superquadric particles into the contact detection between two polygonal particles. A linear search is performed on the two polygonal elements to determine the contact point and overlap region of the polygons, and this method is stable and robust in calculating the two-dimensional superquadric particles. Wellmann et al. (2008) proposed the common normal concept to calculate the overlap and contact normal between the superquadric particles, and the contact detection between the particles is transformed into a two-dimensional unconstrained optimization problem. Meanwhile, Newton's iterative method and Levenberg–Marquardt method (LM) are used to calculate the contact points between particles, which have high numerical accuracy. Portal et al. (2010) proposed an implicit solution algorithm for superquadric particles, which transforms the contact detection between particles into a nonlinear convex optimization problem. The objective function is the shortest distance between two particle surfaces, and the constraint equations are the implicit superquadric equations. The optimization problem is solved by using the interior point method and sequential quadratic programming method.

In addition, Lu et al. (2012) used both continuous function representation (CFR) and discrete function representation (DFR) for modeling the superquadric particles. For the discrete function representation, an adaptive discrete metathesis method is used to distribute discrete points on the particle surface, and a contact reference list is established to speed up the contact detection between particles. For the continuous function representation, a Newton iterative method is used to numerically calculate the contact points between two particles. The computational efficiency of the superquadric particles represented by discrete functions decreases with the increase in the number of surface points. When the particles have a sufficient number of surface points, the superquadric particles represented by discrete functions have the same numerical results as the superquadric particles represented by continuous functions. Podlozhnyuk et al. (2016) used the open-source DEM code "LIGGGHTS" which can generate superquadric particles of different morphologies, as shown in Fig. 1.2. Before calculating the inter-particle forces, the number of potential contact pairs is reduced, and the computational efficiency is improved by the bounding sphere and oriented bounding box methods. For inter-particle contact detection, the midway point method is used to calculate the contact points and directions between the particles, and the validity and robustness of the superquadric model are verified by comparing with the theoretical, experimental, and finite element results. Kildashti et al. (2020) established a generalized geometric force model for superquadric particles, which calculates the contact points between particles based on the modified common normal method. The radius of curvature at the contact points is calculated based on the local principal curvatures and principal directions, and the numerical results of superquadric particles are in good agreement with those obtained by the finite element method.

Although ellipsoid and superquadric equations can be used to construct different particle shapes, these shapes are centrosymmetric and limit the further engineering applications of both models. To address this limitation, the poly-ellipsoid model and poly-superquadric model have been developed. Peters et al. (2009) developed the poly-ellipsoid model for the centrosymmetric geometry of ellipsoid particles. The model stitches together eight one-eighth ellipsoids to form a geometrically asymmetric poly-ellipsoidal particle with different aspect ratios. The contact detection between particles is simplified to that between ellipsoids by using a geometric potential algorithm, and the quadrant positions of the contact points are used to determine whether the particles are in contact or not. Zhang et al. (2018a) used computed tomography images to construct poly-ellipsoid particles with smooth surfaces and calculated the overlap between ellipsoids based on a geometric potential algorithm.

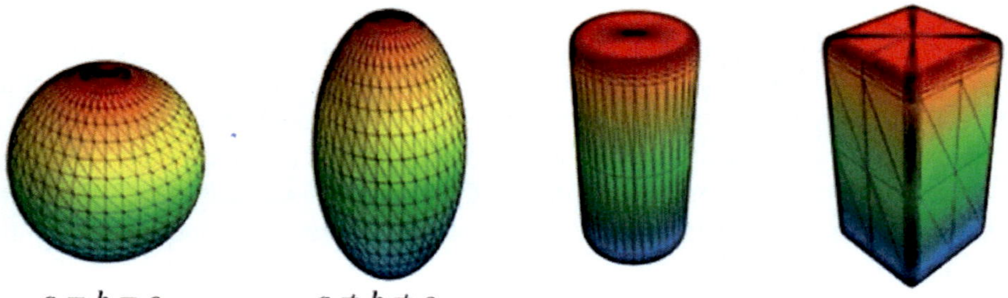

Fig. 1.2 Differently shaped particles constructed by superquadric models (Podlozhnyuk et al. 2016)

Prior to accurate contact detection, the prejudgment bounding sphere and bounding box methods for poly-ellipsoidal particles are proposed. By introducing the prejudgment technique, its running speed for sharp ellipsoids is 13 times faster than that of the traditional search method, while the running speed for non-sharp ellipsoids is 58 times faster than that of the traditional search method. Zhao and Zhao (2019) developed a poly-superquadric model for the geometric limitation of the center symmetry of the superquadric particles, as shown in Fig. 1.3. This model involves stitching together one-eighth of eight superquadric elements to form a convex particle shape that can be described with different elongation rates, sharpness, and geometric asymmetry. Eight superquadric equations and eight control equations are linked, and new functional equations are formed with the surface normal as the independent variable and the Cartesian coordinates as the dependent variable. A novel optimization method based on hybrid LM (Levenberg–Marquardt) and GJK (Gilbert–Johnson–Keerthi) methods is used to calculate the overlap between poly-superquadric particles. This model shows good stability, efficiency, and robustness in numerical simulations of particle stacking and triaxial compression.

1.1.2 Arbitrarily Shaped DEM Model Based on Geometric Topology

Polyhedral and dilated polyhedral models are universal methods for describing irregular particles based on geometric topology. Feng and Owen (2004) proposed an energy-based generalized contact model for the contact detection between two-dimensional polygons, where the contact force and normal direction are uniquely determined. When two polygons intersect, the line between the two intersection points is defined as the tangent direction. The direction perpendicular to this line is the contact normal direction, and the midpoint of the line is the contact point. Moreover, this model can be applied to arbitrary convex polygonal particles as a unified contact theory, and it has faster computational efficiency and numerical stability.

Fig. 1.3 Differently shaped particles constructed by poly-superquadric models (Zhao and Zhao 2019)

Nezami et al. (2006) used the common plane method to calculate the overlap between three-dimensional polyhedral particles, and the perpendicular plane of the shortest distance between the two particles is defined as the common plane, as shown in Fig. 1.4. The computational efficiency of finding the common plane is improved by introducing the shortest link method, which is 17 times faster than the traditional common plane method. Fraige et al. (2008) developed a multi-point contact model and used it to calculate the interaction between cubic particles. During particle collision, this model has more accurate contact characteristics than the single-point contact model, and the numerical results are closer to the experimental results. Gui et al. (2016) developed a new soft-sphere-imbedded pseudo-hard-particle model (SIPHPM) for polyhedral particles. This model is bounded by a series of softballs which are adjusted around the equilibrium position according to position, orientation, and particle shape. The advantage of this approach is that it does not need to deal with the forces between the soft spheres, and it also simplifies the contact detection between polyhedrons by employing a simple contact detection between spheres. Feng and Tan (2019) introduced the extended polytope algorithm (EPA) to the calculation of the forces between polyhedral particles. The method is combined with the Gilbert–Johnson–Keerthi (GJK) algorithm to determine the position of the point within the Minkowski difference region and the minimum distance from the boundary on the basis of the Minkowski difference operation. The shortest distance and direction from the boundary to the origin is the overlap and normal direction between the two polyhedrons. When the EPA algorithm is used to calculate the overlap between polyhedral particles, only the coordinates of the corner points of

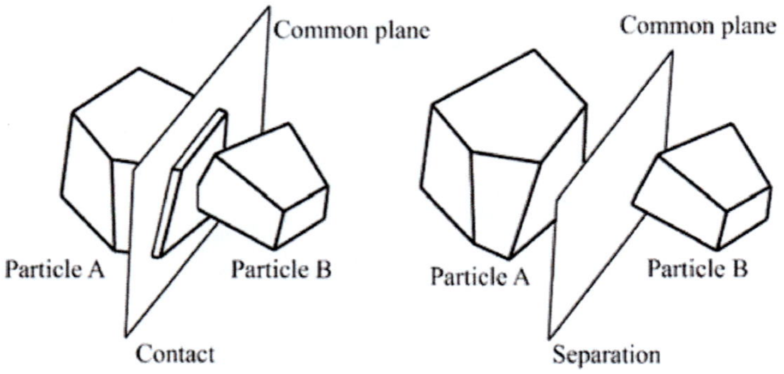

Fig. 1.4 Polyhedron models based on the common plane method (Nezami et al. 2006)

the particles are needed, and the geometric information of the edges and faces is not required. Therefore, this algorithm has remarkable simplicity and high computational efficiency.

Liu and Ji (2018) developed a dilated polyhedron model based on the Minkowski sum algorithm, as shown in Fig. 1.5. This model is formed by extending a spherical radius outward from the polyhedron, and the contact modes between polyhedrons include the plane, cylinder, and sphere contacts. This method avoids singular results at corner points and prismatic edges, which improves the computational efficiency of DEM simulations. Furthermore, Liu and Ji (2020) established an iterative solution algorithm based on second-order expansion functions for the contact problem between dilated polyhedral particles. This model approximates the outer contour of the dilated polyhedron with the weighted summation of the second-order expansion function, and the contact detection between the dilated polyhedral particles is transformed into an optimization problem of finding the shortest distance between the polyhedrons. Subsequently, the Newton iterative algorithm is used to calculate the quadratic system of nonlinear equations, and the approximate contact points can be obtained. Based on the location of the approximate contact point, the possible contact modes are determined by the iterative calculation. In DEM simulations, the numerical results are in general agreement with the experimental results in simulating the ballasted shear stresses.

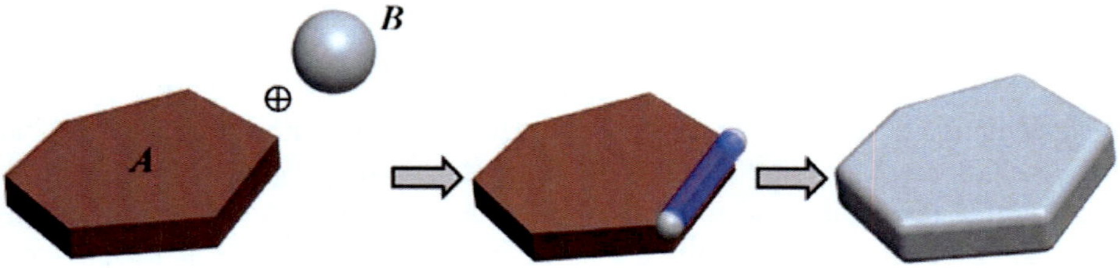

Fig. 1.5 Particles constructed by dilated polyhedron models (Liu and Ji 2018)

1.1.3 Arbitrarily Shaped DEM Model Based on Combined Particle Approach

Although ellipsoids, superquadric equations, poly-ellipsoids, poly-superquadric equations, polyhedrons, and dilated polyhedral models can describe different particle shapes, these shapes are limited by contact theory to calculate only convex-shaped particles. In order to overcome this limitation, the combined particle model has received widespread attention. This model combines different numbers of convex elements such as spheres, ellipsoids, superquadric particles, cylinders, and polyhedrons to construct realistic particle shapes. Considering the complexity of the contact detection between the combined particles, its contact theory is usually transformed into a contact problem between several basic elements. Vu-Quoc et al. (2000) used the combined sphere method to construct ellipsoidal particles similar to soybeans and proposed a new contact model of elastic–plastic frictional contact to calculate the normal and tangential forces between particles on the basis of the traditional contact mechanics of Mindlin and Deresiewicz. This model was used to simulate the granular flow process in an inclined groove, and the numerical results were in good agreement with the experimental results. Kruggel-Emden et al. (2008) used the combined sphere method to construct a sphere and analyze the rebound characteristics of the sphere impacting a flat wall. The numerical results of the combined sphere method were dependent on the particle arrangement. Thus, the combined sphere method has limitations when used to approximate a sphere, and its generalization to other arbitrary shapes will face difficulties. Although the combined sphere method has been a significant improvement over the sphere model in describing complex particle morphology, it is an approximation-based method that may introduce new errors, at least at the single particle level. To address this issue, Ji and Shen (2006) applied different contact force models to simulate the flow characteristics of combined spheres and found that the deviations introduced at the single particle level may offset or compensate each other in granular systems. In non-regular granular systems, the shape of each particle differs from the real shape, and the collision deviation caused by a single particle may be negligible in the whole granular system. Li et al. (2015) used the combined sphere method to approximate the real particle shape, as shown in Fig. 1.6. Three solutions based on modified heuristic algorithms are proposed for the combined sphere method, including the solid-filling scheme, the surface-covering scheme, and the triangular surface-covering scheme. Generally, the solid-filling scheme approximates complex particle shapes using fewer spheres and has better numerical accuracy, while the surface-covering scheme consumes less computational time in representing partial particle shapes. The triangular surface coverage scheme is composed of less than a few thousand triangles, which approximates the particle shape in a short time and uses a smaller number of spherical particles.

In addition to the combined model with the sphere as the basic element, combined models for other shapes of particles have been developed. Liu and Zhao (2020) proposed a combined superquadric model similar to the combined sphere method.

Fig. 1.6 Differently shaped particles constructed by combined sphere models (Li et al. 2015)

This method describes an arbitrarily shaped particle with a combination of several superquadric elements, while the superquadric particles interact with each other. Three tablet-shaped particles are constructed using the combined superquadric method, and the packing heights and the dynamic angle of repose of the particles in the rotating cylinder are in good agreement with the experimental results. Meanwhile, the computational efficiency of the combined superquadric model is lower compared with the traditional sphere model, and it depends on the aspect ratio, blockiness parameter, and number of basic superquadric elements. Rakotonirina et al. (2019) composed several convex polyhedrons with a certain overlap and formed a combined polyhedral particle. The contact detection between combined polyhedral particles is simplified to the contact detection between polyhedrons, and the GJK (Gilbert–Johnson–Keerthi) algorithm is used to calculate the overlap and normal direction between convex polyhedral elements. In contrast to the composite sphere model, this model avoids the artificial roughness that arises when approximating the particle shapes and preserves the geometrical features of faces, edges, and vertices.

Kidokoro et al. (2015) combined two hemispheres and a cylindrical particle and formed a spherocylinder particle. The contact detection between spherocylinder particles was categorized into four contact modes: (1) parallel contact between two cylinders; (2) cross contact between two cylinders; (3) contact between two hemispheres; and (4) contact between a hemisphere and a cylinder. In order to examine the accuracy of this model, the flow process of spherocylinders in vibrational transport and inclined grooves is compared with the experimental results, and the numerical results are basically consistent with the experimental results. Meng et al. (2018) composed a number of spherocylinder particles with a certain overlap and formed a combined spherocylinder particle, as shown in Fig. 1.7. The packing characteristics of the combined spherocylinder particles are analyzed by analytical models and relaxation algorithms, and the Monte Carlo method is used to optimize the packing structure and eliminate the locally ordered structure. The results show that the concave particles have higher coordination numbers than the convex particles, and there are more interlocking and entanglement effects in the concave particles, which leads to the formation of stable packing structures.

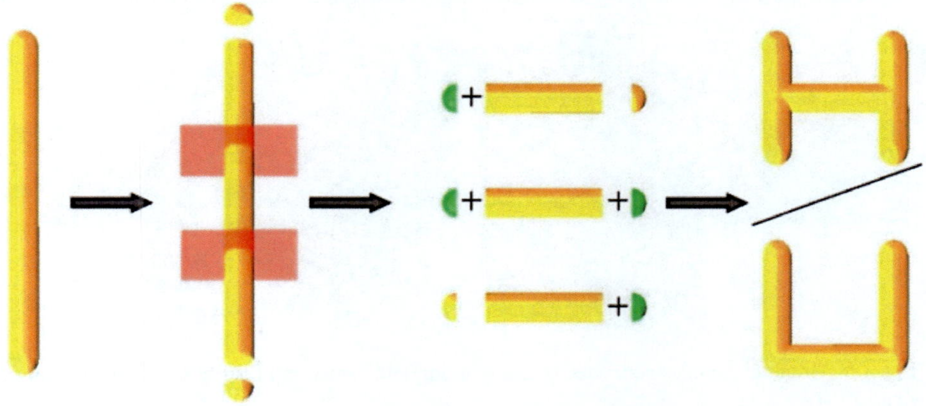

Fig. 1.7 Concave particles constructed by composed spherocylinder models (Meng et al. 2018)

1.1.4 Novel Discrete Element Methods for Arbitrarily Shaped Particles

It is worth noting that the combined particle method overcomes the shape limitations such as geometrical symmetry and strict convexity of the basic elements. Meanwhile, this method can simplify the complex contact into the contact detection between the basic elements. This reduces the complexity of numerical calculations and improves the stability of DEM simulations. However, the computational efficiency of this method decreases significantly with the increase in the number of basic elements, and its computational accuracy depends on the contact patterns between particles. In order to truly reflect the kinematic properties of arbitrarily shaped particles, novel non-regular construction methods have attracted widespread attention. Li et al. (2019) proposed a model of heart-shaped particles with concave surfaces. The heart-shaped particles use the lattice method to calculate the contact points and do not need to invoke the Newton iterative algorithm and the corresponding Jacobi inverse matrix. This contact algorithm does not diverge during the solution process, thus ensuring the numerical stability of DEM simulations. The time taken to calculate the contact points between superquadric particles by the grid method is less than that of the traditional Newton iterative algorithm, thus demonstrating the high computational efficiency of this method.

Mollon and Zhao (2014) combined the random field theory with a particle generation algorithm based on the Fourier function, so as to generate a realistic particle model with a complex shape and controllable accuracy. This algorithm consists of five steps: (1) execution of Voronoi subdivision with constraints within a polyhedral cell with a given volume fraction and a number of particles; (2) calculation of the corrected covariance matrix and eigenvalues from the correlation matrix and Fourier spectrum of each Fourier series; (3) a normalized radius is used to represent each Voronoi cell, which is projected into the eigenspace of the covariance matrix along with the correlation coefficients, and these coefficients are normalized to obtain the

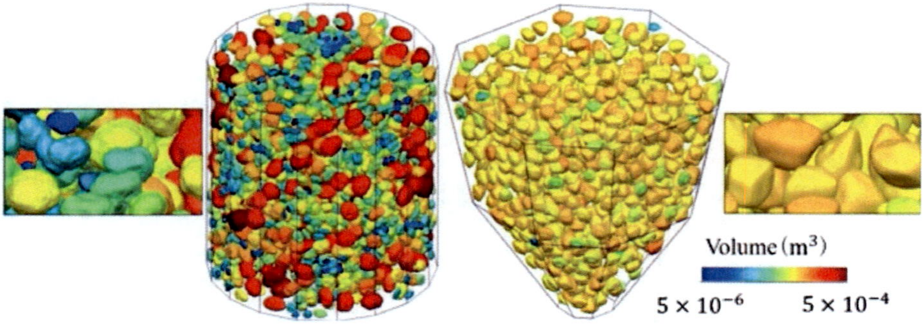

Fig. 1.8 Granular samples generated by the random field theory and Fourier function (Mollon and Zhao 2014)

fitting coefficients in units of variance; (4) calculation of the particle radius in each sample as well as fitting and stochastic feature parameters; (5) the generated particles are stored in their respective cells and these particles are scaled according to volume fractions or constraints. Using the above steps, particle packing structures with different aspect ratios, surface sharpness, sphericity, size distributions, volume fractions, and other key features can be obtained, as shown in Fig. 1.8.

Kawamoto et al. (2016) proposed an arbitrary morphological discrete element method based on level set functions. This method combines X-ray computed tomography (XRCT) to represent the real particle shape in pixel form and reconstructs the image using a discretized level set function, which can capture the complex morphology of real particles. The level set function is a scalar implicit function, and the value of the function represents the distance from a spatial point to the particle surface. For a two-dimensional particle, the outer surface of the particle is represented as a contour line, as shown in Fig. 1.9a. Meanwhile, more contour lines are added inside or outside the particle contour. These contour lines represent the distance or height from the particle surface, which is positive outside the particle and negative inside the particle. The grid is then superimposed over the space containing the particle contours. Figure 1.9b shows the height values for each grid node, which are the values of the discretized level set function. The value of the level set function for each grid node can be calculated by interpolating the surrounding grid nodes. If the functional value of a node is less than zero, the point is located inside the particle. Otherwise, the point is located outside the particle. The original particle surface is reconstructed by interpolating the points, as shown in Fig. 1.9c. Furthermore, the discrete element method based on the level set theory is used to simulate the triaxial compression process of real-shaped granular materials. The numerical results of stress–strain and volume–strain agreed well with the experimental results, which confirms the effectiveness of the level set algorithm.

Su and Yan (2018) used the micro X-ray computed tomography (μXCT) technique to visualize real particles in three dimensions at high resolution and used the spherical harmonic function for particle reconstruction. A spherical harmonic function contains both real and imaginary parts and contains several order derivatives. The influence of

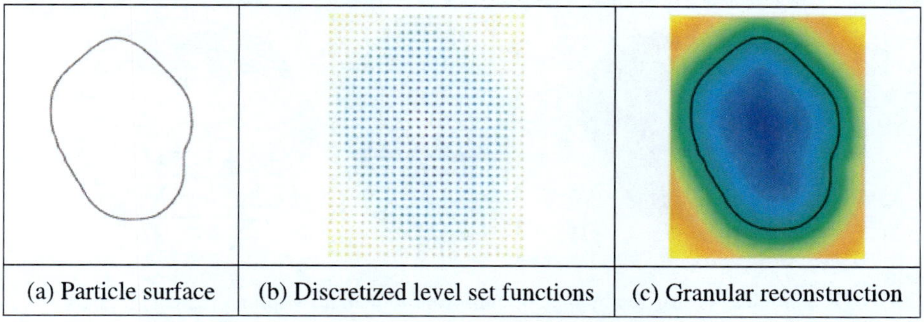

Fig. 1.9 Arbitrarily shaped particles constructed by the level set functions (Kawamoto et al. 2016)

the coefficients of the spherical harmonic function and the degree of mesh partitioning on determining the particle size and shape is investigated, and the correspondence between different shapes and descriptive parameters is established. Meanwhile, a probabilistic method considering the coefficients of the spherical harmonic function is proposed based on principal component analysis (PCA) and empirical cumulative distribution function (ECDF) to reconstruct the real particles in three dimensions. In addition, the shape of real particles can be accurately described when the maximum power of the spherical harmonic function is 10, as shown in Fig. 1.10.

Liu et al. (2020c) described arbitrary particle shapes based on non-uniform rational basis splines (NURBS). NURBS determines arbitrary particle shapes by parameters such as control points, node vectors, weights, and function orders. The arbitrary particle shape is transformed into the particle shape described by NURBS, as shown in Fig. 1.11. This model is more suitable for constructing particle shapes with smooth surfaces since the spline is continuous and higher order derivable in NURBS. With the same number of sampling points, the particle shape described by NURBS has higher numerical accuracy, and its spherical particles described by 20 control points are basically the same as the traditional spherical particles. Moreover, the model saves 80% of the computational time when using 500 sampling points compared to a discrete element method based on facet descriptions. Thus, this model is suitable for

Fig. 1.10 Granular samples generated based on spherical harmonic functions (Su and Yan 2018)

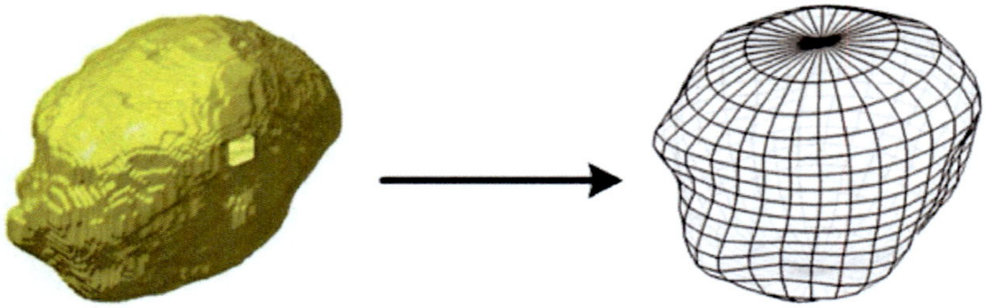

Fig. 1.11 Particles constructed by the non-uniform rational basis splines (Liu et al. 2020c)

large-scale numerical simulations of non-regular particles. In previous studies, the NURBS method has been coupled with the finite element method and successfully used to analyze deformable particles (Espath et al. 2015).

Feng (2021a, b) developed a polyhedron algorithm based on the energy conservation principle for arbitrary morphological particles in nature or industrial applications. This method envelopes the surface of arbitrary morphological particles with several triangular elements, which form a polyhedral model. The total energy of the isolated system remains constant in a given reference frame, and it is assumed to be conserved over time. In conventional DEM simulations, the total energy of granular materials is always conserved if the damping, plastic deformation, and heat transfer modes of the system are neglected. The contact normal of a particle is defined as the gradient of the energy function. When the energy function of a pair of particles is given, the normal direction, the contact point, and the force will be calculated automatically without introducing additional assumptions. When a linear contact energy function is used for the contact volume, the contact points between the two neighboring particles can be determined and used to calculate the contact force, which significantly improves the computational efficiency and applicability of the energy-conserving contact model. Figure 1.12 shows the simulation of the mixing and packing process of banana-shaped particles and pear-shaped particles using the energy-conserving contact theory. It can be found that the method has good numerical stability in DEM simulations and applies to granular materials with arbitrary morphology.

1.2 High Performance Computation of Discrete Element Methods

Considering the large number of granular materials in real engineering applications and natures, traditional discrete element methods have low computational efficiency, which limits the further engineering applications. NVIDIA company proposed the Compute Unified Device Architecture (CUDA) in 2007, which aimed to utilize the general-purpose computing power of GPUs. The programming language is a CUDA C language, and thousands of smaller and more efficient cores are used to achieve

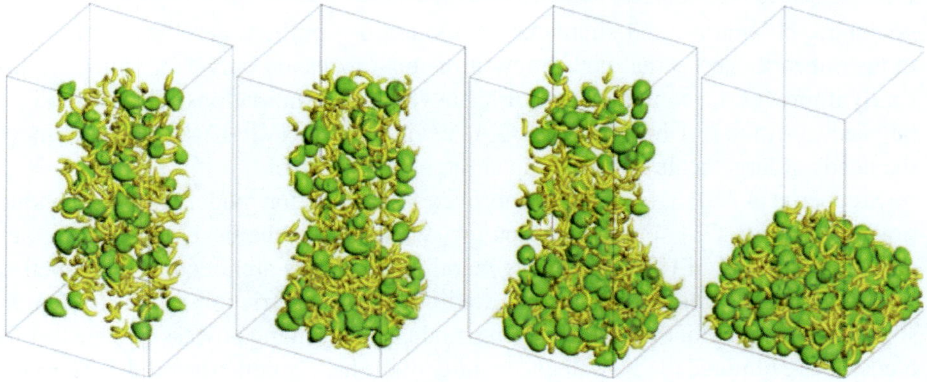

Fig. 1.12 Packing processes of arbitrarily shaped particles simulated using an energy-conserving contact theory (Feng 2021a)

powerful parallel computing capabilities. The GPU parallel approach was first used for spherical particles in DEM simulations since GPU parallelism is suitable for operations with high arithmetic density and simple logic branches.

1.2.1 GPU Parallel Algorithm for Spherical Particles

Nishiura and Sakaguchi (2011) proposed a parallel vector algorithm in a shared-memory system that overcame the load imbalance in a distributed memory system. In order to avoid the defect of simultaneous memory accesses when building a list of neighboring pairs, a neighbor list is constructed by reordering the particle indices. This approach avoids redundant memory contention when superimposing contact forces on each particle by creating a reference list containing contact pair indices. Xu et al. (2011) improved the computational efficiency of GPU parallelism by ignoring the superposition effect of tangential forces between spherical particles. The regional decomposition method was used, and more than 200 GPUs were applied to simulate the industrial-scale granular materials with an acceleration ratio of 11 times. Zheng et al. (2012) developed a GPU parallel framework based on a homogeneous grid algorithm, and the grids are divided into a wide phase and a narrow phase and unit-parallelized separately. This approach significantly improved the computational efficiency of the discrete elemental method for simulating dense concrete flow. The GPU parallel method had an acceleration ratio of about 73 times compared with CPU computation. Yue et al. (2015) established a CPU-GPU heterogeneous architecture, which optimizes the storage structure and adopts shared memory to avoid access conflicts and maximize the GPU memory bandwidth frequency. The acceleration ratio of this parallel method is 19.6 times when the number of spherical particles is 20,000. Govender et al. (2015a, b) developed the spherical GPU parallel algorithm "BLAZEDEM", which transforms concave geometric boundaries into convex geometric boundaries and simulates the motion of 4 million particles for 1 s in an industrial mill. The actual consumption of this process is only 1 h, which significantly improves the computational efficiency of DEM simulations. This method can be used to simulate 1 billion at the GPU device of the NVIDIA K40, which meets the needs of large-scale computation at the industrial level.

Steuben et al. (2016) introduced improved sliding friction models and heat conduction models into GPU parallel computing, which led to the development of a new methodology "GPGPU" based on a generalized software architecture. This method was qualitatively compared with the traditional sliding friction model to demonstrate the reliability of the current sliding model, and the icing process of wind turbine blades was examined to illustrate the validity of the heat conduction model. Tian et al. (2017) used a multi-GPU parallel algorithm to improve the computational efficiency of the discrete element method. The one-dimensional region decomposition method is used to realize distributed computing, and a dynamic load balancing strategy is applied to ensure the efficiency of each GPU, as shown in Fig. 1.13. The method used the asynchronous communication to reduce the time cost of communication and can simulate large-scale granular materials on the order of 100 million. He et al. (2018) developed a GPU parallelization technique for multi-grid search in the large-size-distributed granular materials, which achieves the parallelism of a single GPU. This three-level parallel and efficient computational approach optimized the associated memory layout and also simulated complex motion behaviors at the particle scale, including elastic-plastic contact and deformation, fluid-solid coupling, etc., which significantly improved the computational efficiency and the reliability prediction of compaction behaviors in simulating the granular compaction process.

In addition, the GPU parallel-based spherical discrete element method is widely used in many fields, such as chemical engineering, cold zone engineering, geotechnical engineering, and food engineering. Ren et al. (2013) used a GPU parallel algorithm to simulate the mixing and flow process of spherical particles in a 7.5 L mixer. When the fill rate is in the range of 50–60%, the mixer has the highest operating efficiency and productivity, and the granular material has the best mixing rate and lowest

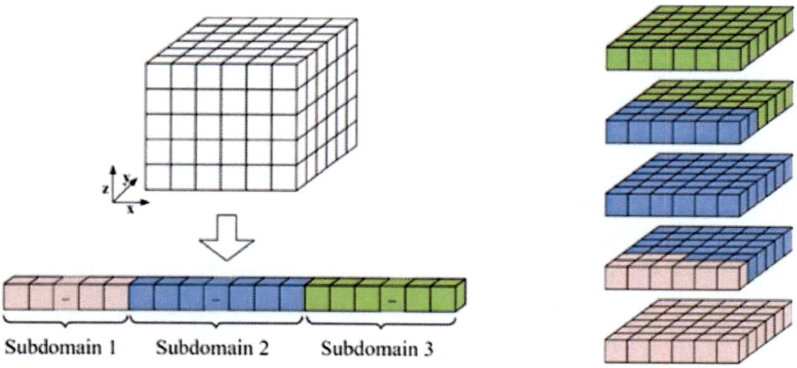

Fig. 1.13 GPU parallel algorithm based on the one-dimensional domain decomposition (Tian et al. 2017)

energy consumption at a rotation speed of 30 rpm. Seo et al. (2014) applied the GPU parallel algorithm to a system containing two-component particles and simulated the mixing behaviors of spherical particles during the mixing process. The deviation between the numerical and experimental results was less than 15%. Hazeghian and Soroush (2015) used a GPU parallel algorithm to simulate the fault rupture process of soil particles. The results showed that the angle between the direction of maximum principal strain and the fault varied along the fault rupture, while the pinch angle increased with the increase in soil mobility. Faults in dense sand layers produced greater slopes than in loose sand layers, which can cause more severe damage to neighboring structures. Yu et al. (2015) used a GPU parallel algorithm to simulate the mixing behaviors of spherical granular materials in a rotating drum. Alternating baffles are used to guide the particles in an axial motion, which has a superimposed effect on the dispersion and convection of the particles and improves the mixing efficiency of granular materials in the axial direction.

Hazeghian and Soroush (2015) used a GPU parallel algorithm to study the microscopic mechanism of the shear zone of dense sandstone in fractured layers. The results showed that the shear zone on the macroscopic scale underwent localized torsion or bending, which was related to the softening behavior of the granular material, the creation of more voids, the high rotational velocity of the particles, and the rapid energy dissipation in the granular bed. The sliding between particles dominated the changes in the shear zone, and the rolling mechanism also played a crucial role in the energy dissipation in the shear zone. Peng et al. (2016) used a GPU parallel algorithm to simulate the gravity-driven flow process of spherical granular materials in a silo. The silo can be divided into nine regions, including the loose accumulation region, flow region, shear layer, flow transition region, flow contraction region, stagnation region, vertical flow region, centripetal flow region, and free-fall region. Meanwhile, the particles have different flow rates and accelerations in different flow regions. Peng et al. (2018) further analyzed the critical transition region that existed between the stable flow region and the unstable flow region, and the increase in the void ratio of the granular material in this region might lead to the unstable flow of the particles in the silo. Zheng et al. (2018) extended the GPU parallel algorithm for spherical particles to the DEM-FEM coupled simulation, as shown in Fig. 1.14. More than 15 times acceleration ratio was achieved in the numerical calculation of the interaction between pneumatic tires and sand particles. The numerical results of the total traction force, traction bar pull, and driving resistance were in good agreement with the experimental results, which confirmed the validity of the current DEM-FEM coupling algorithm based on GPU parallelism. Long et al. (2019) used GPU parallel algorithms to simulate the sea ice under uniaxial compression and three-point bending, and the spherical bond-fragmentation model was introduced into the DEM simulations. The effects of micro-scale parameters such as particle size, sample size, bond strength, and inter-particle friction coefficient on the uniaxial compressive and bending strengths of sea ice at the macroscopic scale were analyzed, and the correspondence between the micro-parameters and the macroscopic mechanical strengths of sea ices are established.

Fig. 1.14 Interaction between a tire and sand particles based on GPU parallel algorithms (Zheng et al. 2018)

1.2.2 GPU Parallel Algorithm for Non-spherical Particles

Due to the fact that contact detection between irregular particles is more complicated than that of spherical particles, the consumption time of irregular particles is usually several or tens of times that of spherical particles, which limits the further application of irregular discrete element methods in industries. Thus, GPU parallel algorithms provide an important tool for large-scale numerical simulation of irregular particles.

Gan et al. (2016) introduced GPU parallel algorithms into the ellipsoidal discrete element method and simulated the flow characteristics of irregular granular materials in a blast furnace, a screw conveyor, and a rotating drum. When the aspect ratio of particles deviated from 1, a single GPU had a faster running speed than a single CPU, which indicated that the GPU parallel algorithm is more suitable for non-regular granular materials. Meanwhile, the MPI-based multi-GPU parallel algorithm has a faster running speed than the single GPU parallelism, i.e., the 32 GPUs parallelism is 18 times faster than single GPU parallelism. This method can also be used to simulate tens of millions of irregular granular materials. Gan et al. (2019) further used a multi-GPU approach based on MPI to analyze the influence of the particle size distribution, particle shape, and rotating speed on the energy dissipation of non-regular granular materials. The results showed that particles with rough surfaces had greater energy dissipation than smooth particles, and granular materials with certain particle size distributions had more significant cushioning performance under impact.

In addition, GPU parallel methods for polyhedral and dilated polyhedral models have been developed. Zhang et al. (2013) developed a GPU parallel algorithm for 2D triangular particles, which was based on the DEM-FEM coupling method proposed by Antonio Munjiza. Then, a dynamic domain decomposition technique was introduced, which increased the computational speed by 80 times. Govender et al. (2014) developed the GPU parallel code "BLAZEDEM-GPU" for convex polyhedral particles, which fully utilized the task-level parallelism on the GPU device and provided a real-time view of the motion of the polyhedral particles. Meanwhile, the code optimized the contact detection of convex polyhedral particles based on the split-plane approach, which significantly reduced the memory consumption on the GPU device and enabled the large-scale numerical computation of 34 million polyhedrons on a single NVIDIA K6000 GPU (Govender et al. 2015a, b). Govender et al. (2019) further developed a GPU parallel algorithm for concave polyhedrons, which was based on the contact detection between convex polyhedrons. This method was validated by comparing the numerical and experimental results of the flow process of convex and concave polyhedrons in the hopper. The results showed that the flow process of convex polyhedrons was intermittent, while concave polyhedrons were more likely to form arch structures during the flow process. Liu et al. (2020a, b) introduced a bond-fracture model with polyhedral elements based on the BLAZEDEM-GPU code, which can simulate the derivation, extension, and fracture processes of microscopic cracks. The results of uniaxial compression of Brazilian disks and limestone were in good agreement with the experimental results, indicating that this method can be used to model the mechanical properties of granular materials at both microscopic and macroscopic scales. Lubbe et al. (2020) improved the computational efficiency of the traditional BLAZEDEM-GPU code by considering the size of the computational domain, particle number, particle density, particle size, and particle shape, and the consumption time was reduced by 20%.

GPU parallel algorithms are also widely used in DEM simulations of arbitrarily shaped particles. Longmore et al. (2013) constructed sand particles using the combined sphere method in a GPU parallel architecture, which reduced the consumption of texture memory while increasing the memory space used for storing the particles. This enabled rapid visualization of the granular material and accurate simulation of the elastic hysteresis and strain loss of the sands during the compression and unloading processes, the flow behavior of particles in the hopper, and the formation of dynamic angles of repose. Liu et al. (2020a, b) introduced the bonding-fracture model of spherical particles into the GPU parallel algorithm and simulated the crushing process of large-scale ballast under external loads. The results showed that the spherical bonding-fragmentation model based on GPU parallelism can better simulate the evolution of the rock damage process as well as the emergence and expansion of cracks, and its numerical results were in good agreement with the experimental results. Boehling et al. (2016) used GPU parallel algorithms to simulate the process of spraying and coating of 1 million tablet-shaped particles inside a rotating drum. Each tablet-shaped particle consisted of 8 spheres, and the number of real spheres in the DEM simulation was 8 million. Numerical results showed that the number of nozzles and spray rate had a significant effect on the coating coefficient of the tablet particles, while the rotor load and rotational speed had a negligible effect on the coating coefficient. In addition, decreasing the spray rate and increasing the rotational speed also resulted in the optimal industrial parameters. Kureck et al. (2019) developed a tablet-shaped model with biconvex planes and utilized GPU parallel algorithms to simulate the coating and encapsulation process of 2 million tablet-shaped particles in a rotating drum, as shown in Fig. 1.15. Compared with the tablet-shaped particles constructed by combined sphere models, this method had higher computational accuracy and faster running efficiency, and the corresponding numerical results were in better agreement with the experimental results.

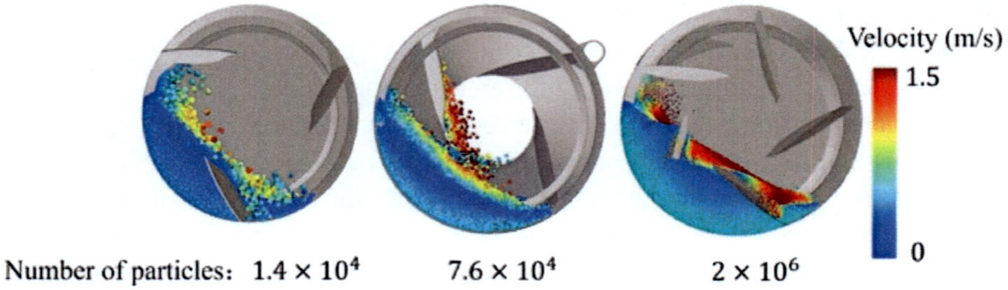

Fig. 1.15 Different number of tablet-shaped particles in a rotating drum (Kureck et al. 2019)

1.3 DEM Analysis of the Flow Characteristics of Granular Materials

Granular materials are complex systems composed of a large number of solid particles, which are widely present in nature and industrial productions. The discrete element method (DEM) is an important tool for simulating the macroscopic and microscopic characteristics of granular materials, and spherical particles were initially used to represent solid particles. However, irregular granular materials are widely found in nature and industrial productions. Compared with spherical particles, irregular particles have lower fluidity and stronger interlocking properties. The particle shape significantly affects the contact mode between the particles, thereby changing the flow state of granular materials. More importantly, the conclusions drawn from spherical granular materials cannot simply applied to irregular granular materials.

1.3.1 Flow Properties of Granular Materials in Silos

The flow process of granular materials has been widely studied by experiments and DEM simulations; these studies focus more on the effects of particle size, shape, material parameters, orifice diameter, hopper angle on the mass flow rate, particle velocity, and the force chain structure. Magalhães et al. (2016) used a discrete element method to study the flow of spherical granular materials in a conical hopper. It was found that the velocity of particles located in the center of the hopper was higher than the velocity of particles near the side walls for the same height, as shown in Fig. 1.16. Local mass flow index (LMFI) was used to quantify the different flow states, and this velocity distribution showed a power relationship with the distance from the hopper apex. Wan et al. (2018) investigated the effect of the orifice shape on the flow rate of granular materials. As the orifice changed from sharp to smooth, the flow rate of particles increased. The smooth orifice reduced the volatility of the instantaneous flow rate, while the hopper diameter and bottom thickness had no effect on the granular flow rate.

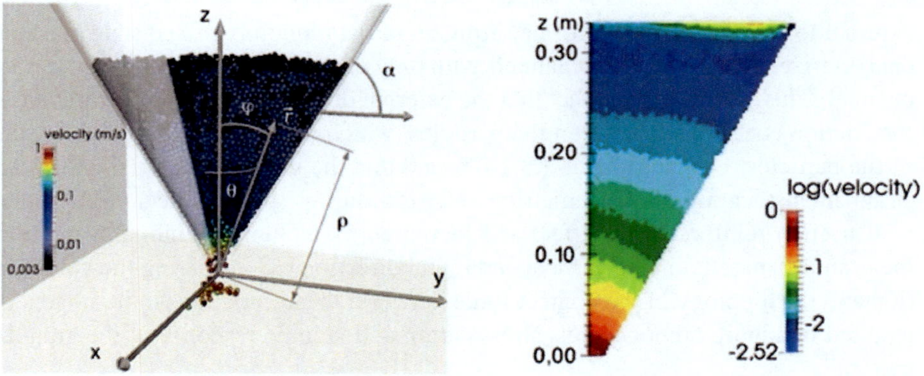

Fig. 1.16 Velocity distribution of granular materials in a cone silo (Magalhães et al. 2016)

Zhou and Sun (2013) found that the shear rate of the granular material was enhanced in the front and back regions during the flow process. This was mainly due to the fact that the particles in the front and back regions were driven by collisions during the slope flow, which in turn drove the granular flow. In the middle region, particle friction drove the granular flow. As the flow state changed from friction-driven to collision-driven, the progressively larger fluctuations led to a decrease in the volume fraction and an increase in the flowability of the granular material. Zhang et al. (2014) found that wall friction had a significant effect on the flow rate of particles when the height of the granular bed was high enough. When the friction coefficient between the particle and the wall was increased to 0.5, the flow rate of particles decreased to 30% of what it would be without friction. This effect was closely related to the Janssen effect (Zhang et al. 2014). Verbucheln et al. (2015) obtained a more homogeneous mass flow of particles by adding a threaded structure to the inner wall of the pipe, which eliminated the need for the external driving energy. Collisions between particles in a pipe with a threaded structure were more homogeneous, and there were fewer fluctuations in the solids fraction of the particles as they flowed through the narrow pipe. Meanwhile, the mass flow rate of granular materials in a threaded pipe can be accurately predicted by introducing parameters as a function of pipe diameter and thread wavelength based on the traditional Beverloo equation.

In addition, granular materials under external pressure show many unique mechanical properties. External energy is continuously input into granular materials through boundary conditions, such as vibration and shear, which in turn changes the macroscopic structure and microscopic mechanical response of granular materials. Liu and Nagel (1998) proposed a flow transition phase diagram for loose granular materials, which described the correspondence between the particle temperature, load, and density, and the flow and blockage states of granular materials. When the particle temperature was increased or an external pressure load was applied, the granular material changed from the blocked state to the flow state. Liu et al. (2008) found that granular materials exhibited significant structural transformations with increasing external loads. They are transformed from an initial randomly mixed state to a state characterized by independent channels with only one type of particle present in each channel. This was due to the fact that the external driving force indirectly increased the friction coefficients between the particles, which in turn changed the flow state of the particles. Bernhardt et al. (2016) found that the granular material underwent strain intensification and volumetric expansion during straight shear, while edge-wall friction reinforced this effect and increased the volumetric inhomogeneity of the granular material in the vertical and radial directions. Considering the void fraction and the topology of the contact force network, these variations in homogeneity resulted in a more homogeneous stress–strain and volume response of the granular material.

Asadzadeh and Soroush (2016) found that the direction of the strong contact force was gradually rotated toward the direction of the principal stresses during straight shear, while the network of the strong contact force chains was uniformly distributed in the central region of the granular material and away from the vertical boundary. In addition, the internal stresses in the granular material induced anisotropy during the straight shear process and remained almost unchanged after the peak shear stress was reached. Peng et al. (2021) designed a suspended silo and eliminated the Janssen effect, which loaded the top pressure directly to the orifice at the bottom of the silo. The flow rate of particles increased linearly with the increase in external pressure, and larger silo diameters with the same orifice size have less pressure effect on the granular material. This indicated the existence of characteristic ratios of granular materials with respect to silo diameter, orifice diameter, and flow rate, which consequently changed the granular materials from a pressure-driven forced flow to a conventional Beverloo flow. Ji (2013) investigated the solid-liquid-like phase transition process in the shear flow of spherical particles, and the mechanical behavior of the granular material at the macroscopic scale was dominated by the inter-particle contact force at the microscopic scale. The diagram of the solid-liquid phase transition was developed by analyzing the motion behavior of particles at different volume fractions and shear rates, and the evolution of the distribution of the contact forces between particles during the phase transition was also determined.

Spherical particles are more prone to sliding and rotating compared to irregular particles, which leads to premature yielding of the microstructure of the granular materials through rolling failure modes and faster mass flow rate of granular materials in the hopper (Cleary and Sawley 2002). In recent years, the focus of research on the flow properties of granular materials has gradually shifted from spherical particles to irregular particles. Höhner et al. (2012) simulated the unloading process of granular materials in a hopper using spherical particles, composite spherical particles, and polyhedral particles. Particles with greater sharpness had a lower flow rate and higher residual volume in a flat-bottomed hopper. Meanwhile, the angularity of the particle surface hindered the movement of the particles, which formed an arch structure above the orifice. Besides, polyhedral particles constructed using the combined sphere method have a smoother surface than polyhedral particles with multiple sharp vertices and edges, and this shape difference results in combinatorial sphere particles having less resistance to flow than polyhedral particles. Therefore, different particle shapes approximated by different non-regular discrete element methods may produce different simulation results. Zeng et al. (2017) analyzed the velocity patterns and contact force fluctuation of ellipsoidal particles in a silo. Based on the magnitude and frequency of the flow fluctuations, the whole flow phase is categorized into three types, including the initial state, the violent state, and the slight state. At the top of the silo, the disappearance of inter-particle contact force occurs only in the stage of violent fluctuation. There exists a one-to-one correspondence between the velocity fluctuation and the disappearance of contact force, i.e., the time of occurrence of the disappearance of contact force is the peak time of the velocity fluctuation. Meanwhile, the disappearance of the velocity fluctuation and contact force is more obvious for particles in a hopper with a smaller angle compared to a larger hopper angle.

Liu et al. (2014b) simulated the flow process of ellipsoidal particles in a hopper. As the particle shape gradually deviated from spherical particles, the mixing region near the side walls decreased while the stagnation region at the bottom angle increased, as shown in Fig. 1.17. Spherical particles had the fastest flow rate, while the flow rate of ellipsoids decreased as the aspect ratio deviated from 1. Fraige et al. (2008) compared the flow characteristics of spherical and cubic particles. The results showed that spheres had faster flow rates than cubic particles, and a large number of cubic particles remained in the hopper at the end of the discharging process. Cleary and Sawley (2002) and Soltanbeigi et al. (2018) simulated the flow process of superquadric particles in a hopper. The surface sharpness of the particles increased the flow resistance and decreased the flow rate, while it had essentially no effect on the flow pattern. Besides, the aspect ratio not only reduced the flow rate of particles, but also affected the flow pattern transition of the granular materials. The flow of elongated particles no longer exhibited free flow, but more like a continuum of deformation and fragmentation. Yielding of the particles in the flow process led to the fragmentation of the microstructure and the creation of a large number of voids. These voids were supported by arching stresses, which transformed the flow pattern of the particles into a funnel flow.

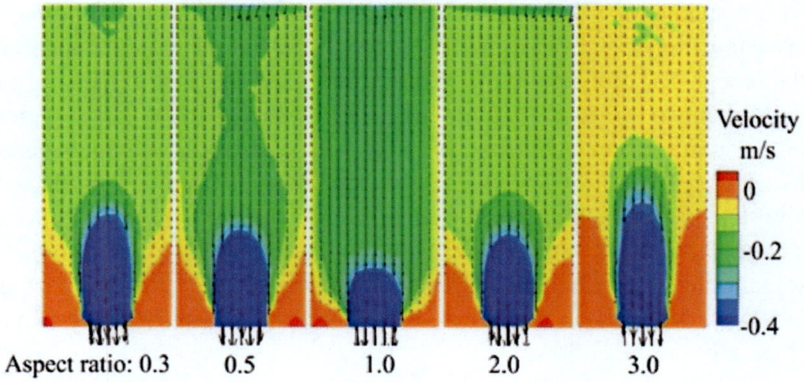

Fig. 1.17 Effect of aspect ratios on the velocity distribution of ellipsoidal particles in a cylindrical silo (Liu et al. 2014b)

In industrial production and applications, the flow pattern of granular materials is critical to silo or hopper design and processing. Generally, there are two types of flow patterns for granular materials in a silo. The first is where all particles have a uniform velocity and flow out of the orifice at the same time, and this flow is defined as mass flow. The second is that the granular material has a non-uniform velocity distribution, i.e., the velocity of particles located in the center of the hopper is higher than that of particles near the sidewalls, which is defined as funnel flow. It is worth noting that the particle shape and silo structure have a superimposed effect on the flow pattern of granular materials. Zhang et al. (2018b) used a discrete element method to simulate the flow process of ellipsoidal particles in a conical silo and to investigate the critical transition height between the mass flow and the funnel flow in the silo. In the mass flow, the void fraction of the granular material is small, and the granular orientation is close to horizontal. In the funnel flow, the void fraction of the granular materials becomes larger, and the granular orientation is gradually vertical. In addition, particle velocity, void fraction, and particle orientation varied with the vertical height of the granular material, and the truncation position is defined as the critical height for the flow pattern transition. Meanwhile, the critical height increased with increasing hopper angle or silo diameter, while it decreased with increasing orifice diameter.

González-Montellano et al. (2011) found that the friction between the combined spherical particles and the silo wall and the hopper angle significantly affected the flow pattern of the granular material. The mass flow index (MFI) was used to quantify the inhomogeneity of the velocity distribution of particles in the silo. MFI < 0.3 indicates that the particles are in the funnel flow. Meanwhile, the MFI decreases with increasing particle friction or decreasing hopper height. Li et al. (2016) changed the flow pattern of granular materials by reducing particle–wall and particle–particle friction and thus inducing the formation of a stable mass flow of granular materials. Compared with particle–particle friction, particle–wall friction played a crucial role in the flow pattern transition of granular materials. As the friction coefficient between the particle and the wall increased, the flow pattern of the granular material changed from mass flow to funnel flow with the presence of a transition region. Han et al. (2019) investigated the discharging process of ellipsoidal particles and found that there is a flow transition from mass flow to funnel flow of the granular material at the initial moment. The direction of the principal stress acting on the ellipsoidal particles in the silo changed from axial to horizontal, and the change in the particle orientation was reflected in the flow pattern transition at the macroscopic level. Govender et al. (2018) simulated the flow process of polyhedral particles in a rectangular silo, and the flow state of polyhedral particles changed from funnel flow to mass flow as the hopper angle increased. Jin et al. (2010) investigated the flow patterns of spherical, ellipsoidal, and hexahedral particles in a silo. The results showed that the cubic particles were basically in mass flow, while the spherical and ellipsoidal particles were basically in funnel flow.

1.3.2 *Mixing and Segregation Properties of Granular Materials in Rotating Drums*

Rotating drums are widely used in industrial production and processing, such as separation, mixing, drying, and coating processes. Numerical simulation of granular materials in rotating drums is a crucial tool for studying the flow characteristics of particles, and the mechanistic studies of their mixing and segregation characteristics are valuable references for the application of granular materials in industries.

The flow and mixing characteristics of granular materials in horizontally rotating drums have been investigated mainly by discussing the effects of different rotational speeds, directions, sizes, filling fractions, elastic modulus, and inter-particle friction coefficients on the mixing indices, durations, and particle temperatures (Chand et al. 2012; Gui and Fan 2015; Xiao et al. 2017). Chou et al. (2016) investigated the effect of granular friction on the mixing characteristics of mono-dispersed and bi-dispersed granular materials in the horizontal rotating drum. In both types of granular materials, the dynamic angle of repose increased with increasing particle friction or rotational speed. The flow behaviors of particles were more complex in bi-disperse granular materials compared to mono-dispersed granular materials, and the regions dominated by frictional effects and the regions dominated by potential energy can be determined from the angle of repose. Halidan et al. (2018) simulated the mixing process of granular materials in two- and four-bladed mixers, as shown in Fig. 1.18. The mixing rate of the granular material increased with increasing impeller velocity, and it decreased beyond the critical speed. The mixing of particles is faster in a four-blade mixer compared to the two-blade mixer. As the filling level of particles increased, the flow state of particles changed from sliding flow to circulating flow and then to cascading flow.

Particle shape has a significant effect on the mixing characteristics of granular materials inside a rotating drum and has been widely studied (Gui et al. 2018). Gui et al. (2017) used the SIPHPM method to construct polygonal particles and used the discrete element method to simulate the mixing process of irregular particles inside a horizontal rotating drum. Quadrilateral, hexagonal, and triangular particles had high mixing degrees, while pentagonal, heptagonal, octagonal, and decagonal particles had low mixing degrees. At the same rotational speed, the kinetic energy of particles increased with the number of polygons, which is mainly due to the lower energy dissipation of the particles close to the circle. Höhner et al. (2014)

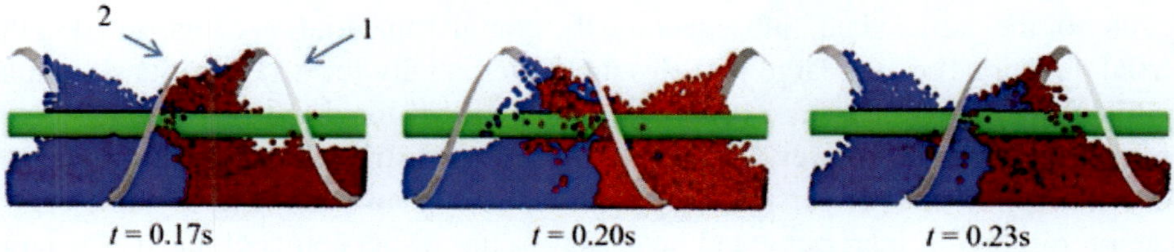

Fig. 1.18 Axial mixing processes of granular materials in a two-bladed mixer (Halidan et al. 2018)

compared the mixing behaviors of spheres, polyhedrons, and smooth polyhedrons inside a rotating drum. Spheres had the fastest mixing rate compared to polyhedrons and smooth polyhedrons. Polyhedral particles with sharp corners enhanced the interlocking between particles and flow resistance, and thus smooth polyhedrons had a faster mixing rate than polyhedral particles. Cleary (2013) used superquadric equations to simulate the mixing process of spherical and irregular particles in a plow mixer. Compared to spherical particles, superquadric particles had sharp edges and planes, which limited the relative rotation and sliding between particles. Meanwhile, the sharp vertices enhanced the shear strength of the granular materials and reduced the flow and mixing rate.

Ma and Zhao (2017) used superquadric equations to construct ellipsoidal particles with different aspect ratios and studied the mixing behaviors of irregular granular materials in a horizontal rotating drum. The mixing degree of particles decreased with increasing the rotational speed of the drum. The high rotational speed limited the diffusion behavior of particles in the axial direction, while the aspect ratio of particles had no effect on the axial diffusion. Meanwhile, the long axis of the ellipsoids is more inclined to be perpendicular to the axis of the drum with the continuous rotation of the drum. Ma and Zhao (2018) further investigated the mixing characteristics of cylindrical particles with different aspect ratios inside a rotating drum. Spherical particles had continuous flow behaviors, while cylindrical particles had intermittent flow characteristics, as shown in Fig. 1.19. The mixing rate of the particles increased with decreasing rotational speed and aspect ratio, and the direction of the main axis of the cylindrical particles was parallel to the sidewalls of the cylinder. Lu et al. (2014) used superquadric equations to construct the irregular particle morphology and prevented the particles from sliding close to the wall of the drum by increasing the surface roughness of the drum. Compared with the smooth wall, the rough wall can reduce the dynamic angle of repose of particles, the relative thickness of the flow layer, the shear rate, and the whole temperature of the granular materials. Meanwhile, the effects of rough walls on the dynamic angle of repose of the granular material and the relative thickness of the flow layer became more pronounced at higher rotational speeds and lower gravitational accelerations.

Due to the influence of different parameters (e.g., size, density, shape, etc.), the granular materials exhibited segregation behaviors in the axial or radial direction inside the horizontal rotor. Yang et al. (2017) investigated the radial and axial segregation characteristics in a horizontal drum due to the difference in particle sizes. The differences of the particle size led to the radial segregation, i.e., large particles were located at the periphery of the particle bed while small particles were located at the center of the particle bed. Subsequently, the granular materials are segregated in the axial direction, i.e., the large particles are preferentially located near the sidewalls, while the small particles are located in the middle of the drum. Yang et al. (2018) further discussed the time evolution of the frequency distribution and spatio-temporal distribution of axial dispersion coefficients in binary mixtures. Rapid radial segregation resulted in increasing and decreasing axial dispersion coefficients for large and small particles in the active and passive zones, respectively, while the frequency distribution of axial segregation coefficients for all particles followed the normal

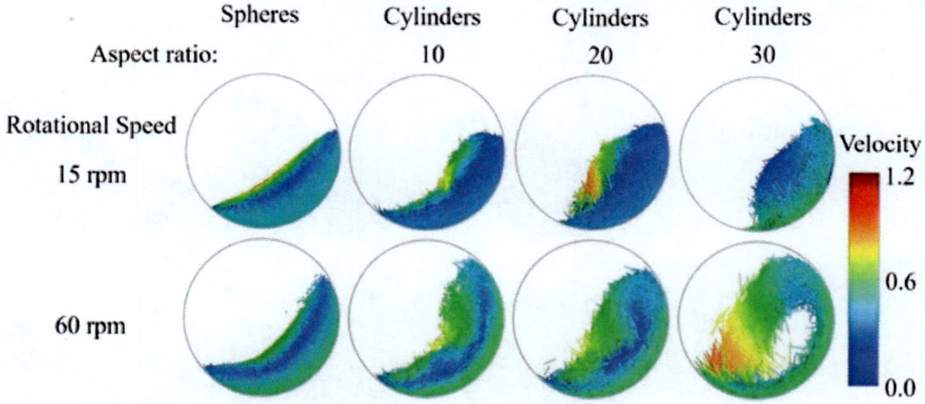

Fig. 1.19 Velocity distribution of spheres and cylindrical particles with different aspect ratios in a rotating drum (Ma and Zhao 2018)

distribution. Cui et al. (2014) discussed different axial segregation phenomena due to sidewall effects and inter-particle forces, and these two factors were also influenced by the rotational speed of the drum and sidewall friction. When wall friction is low, small particles are close to the side walls in the drum with lower rotational speeds, whereas the axial segregation pattern of small particles is random in the drum with higher rotational speeds.

Xu et al. (2010) found that larger particle size or density ratios facilitated the formation of segregation patterns of granular materials, whereas the penetration effect was limited and segregation was difficult to occur in ternary or polydispersed granular materials. Xu et al. (2021) also observed that the granular materials exhibited segregation patterns in binary mixtures due to the competition between inter-particle forces and wall effects. Besides, different filling fractions and rotational speeds caused opposite axial segregation phenomena. One is that the small particles are located in the center of the drum, and the large particles are near the sidewall. The other is that the small particles are close to the sidewall, and the large particles are located in the middle of the drum. Arntz et al. (2014) found that the particle radius, density, and mass had a significant effect on the mixing and segregation characteristics of the binary mixtures and had essentially no effect on the flow state. The rotational speed of the drum determined the flow state of particles, which consequently determined the mixing degree of granular materials.

It is worth noting that particle shape has a significant effect on the segregation characteristics of granular materials. Maione et al. (2017) used wooden cylindrical particles and steel spheres to form a binary mixture and investigated the axial segregation behaviors of the binary mixture experimentally and through DEM simulations. The wood particles were close to the sidewall of the drum, while the steel spheres were located in the center of the drum. This was because the difference in particle density led to axial segregation of the particles. He et al. (2019) investigated the

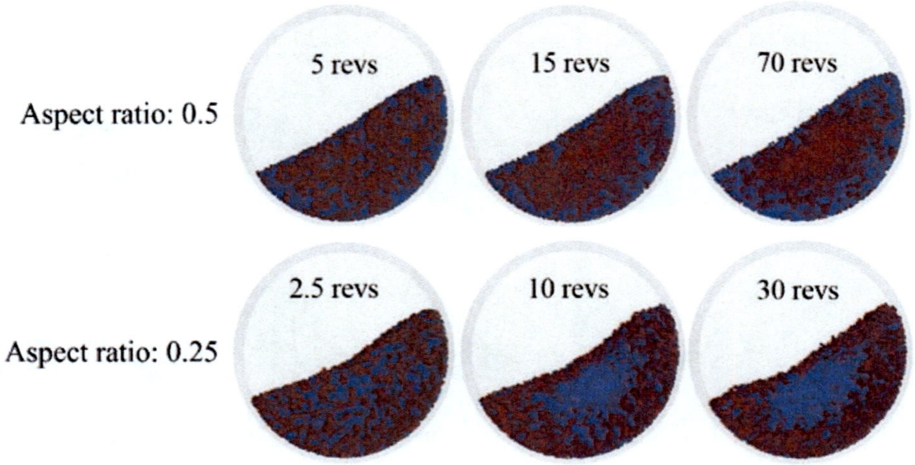

Fig. 1.20 Radial segregation processes of spheres and ellipsoid particles with different aspect ratios in a rotating drum (ellipsoids are marked in red, and spheres are marked in blue) (He et al. 2019)

segregation behavior of binary mixtures composed of spherical and ellipsoidal particles inside a horizontal rotating drum. For binary mixtures consisting of ellipsoids and spheres with an aspect ratio of 0.5, ellipsoids tended to be distributed in the center of the granular bed, while spheres were more likely to be distributed in the periphery of the granular bed. Reverse segregation occurs when the difference in shape of the particles in the mixture increases, i.e., the spheres are located at the center of the granular bed while the ellipsoids are located at the periphery of the granular bed, as shown in Fig. 1.20. This was mainly due to the shape difference leading to the formation of more voids, where the spheres entered the center of the granular bed by penetration.

Pereira et al. (2011) investigated the radial segregation characteristics of cylindrical and cubic particles, and these segregation behaviors were mainly caused by the buoyancy or penetration of the particles. The radial segregation pattern of long cylindrical particles was not a streaky shape but a crescent-shaped pattern. For thin cylinders (the diameter is much larger than the length), a streaky segregation pattern can be observed within the granular bed. Besides, cubic particles usually showed thinner and longer streaky segregation patterns. Yang et al. (2020) investigated the segregation process of cylindrical particles inside a horizontal rotating drum by increasing the ratio of length to diameter of the particles. The small cylinders were located at the center of the granular bed, while the large cylinders were located at the periphery of the granular bed. The aspect ratio of particles increased the dynamic angle of repose of the granular bed, which resulted in a larger total kinetic energy and corresponding vibration amplitude, as well as increased the flow velocity in the active region.

1.4 Summary

In recent years, the numerical simulation has been developed from the initial sphere to ellipsoid, superquadric equation, polyhedron, dilated polyhedron, combined particle model, heart-shaped particle, spherical harmonic function, random field theory, Fourier series form, level set method, non-uniform rational spline, energy conservation theory, and many other non-regular particle models. Different construction methods and contact theories have unique advantages as well as certain shortcomings for the flexibility of the algorithm, numerical stability, and computational efficiency. Therefore, the development of a more stable and efficient non-regular discrete element method is still an important challenge in the numerical simulation. Moreover, different GPU parallel algorithms are suitable for different non-regular discrete element methods, and the complex non-regular granular contact theory significantly affects the computational efficiency of GPU parallelism. Thus, GPU parallel algorithms still need to be further improved for the computational needs of large-scale non-regular granular materials in practical engineering fields.

Furthermore, the particle shape and boundary conditions have a superimposed effect on the flow properties of granular materials. Meanwhile, the shape differences constructed by different irregular discrete element methods may lead to different motion behaviors of granular materials. Therefore, the effects of different irregular discrete element methods on the flow characteristics of granular materials and the potential mechanisms need to be further investigated in future studies.

References

Arntz MMHD, Beeftink HH, Den Otter WK, Briels WJ, Boom RM (2014) Segregation of granular particles by mass, radius, and density in a horizontal rotating drum. AIChE J 60:50–59

Asadzadeh M, Soroush A (2016) Fundamental investigation of constant stress simple shear test using DEM. Powder Technol 292:129–139

Barr AH (1981) Superquadrics and angle-preserving transformations. IEEE Comput Graph Appl 1:11–23

Bernhardt ML, Biscontin G, O'Sullivan C (2016) Experimental validation study of 3D direct simple shear DEM simulations. Soils Found 56:336–347

Boehling P, Toschkoff G, Knop K, Kleinebudde P, Just S, Funke A, Rehbaum H, Khinast JG (2016) Analysis of large-scale tablet coating: modeling, simulation and experiments. Eur J Pharm Sci 90:14–24

Chand R, Khaskheli MA, Qadir A, Ge B, Shi Q (2012) Discrete particle simulation of radial segregation in horizontally rotating drum: effects of drum-length and non-rotating end-plates. Phys A 391:4590–4596

Chen J, Matuttis HG (2013) Optimization and OpenMP parallelization of a discrete element code for convex polyhedra on multi-core machines. Int J Mod Phys C 24:1350001

Chen H, Zhao S, Zhou X (2020) DEM investigation of angle of repose for super-ellipsoidal particles. Particuology 50:53–66

Chou SH, Hu HJ, Hsiau SS (2016) Investigation of friction effect on granular dynamic behavior in a rotating drum. Adv Powder Technol 27:1912–1921

Cleary PW (2000) DEM simulation of industrial particle flows: case studies of dragline excavators, mixing in tumblers and centrifugal mills. Powder Technol 109:83–104
Cleary PW (2013) Particulate mixing in a plough share mixer using DEM with realistic shaped particles. Powder Technol 248:103–120
Cleary PW, Sawley ML (2002) DEM modelling of industrial granular flows: 3D case studies and the effect of particle shape on hopper discharge. Appl Math Model 26:89–111
Clerc MG, Cordero P, Dunstan J, Huff K, Mujica N, Risso D, Varas G (2008) Liquid–solid-like transition in quasi-one-dimensional driven granular media. Nat Phys 4:249–254
Cui Z, Zhao Y, Chen Y, Liu X, Hua Z, Zhou C, Zheng J (2014) Transition of axial segregation patterns in a long rotating drum. Particuology 13:128–133
Cundall PA, Strack ODL (1979) A discrete numerical model for granular assemblies. Géotechnique 29:47–65
Espath LFR, Braun AL, Awruch AM, Dalcin LD (2015) A NURBS-based finite element model applied to geometrically nonlinear elastodynamics using a corotational approach. Int J Numer Meth Eng 102:1839–1868
Feng YT (2021a) A generic energy-conserving discrete element modeling strategy for concave particles represented by surface triangular meshes. Int J Numer Meth Eng 122:2581–2597
Feng YT (2021b) An energy-conserving contact theory for discrete element modelling of arbitrarily shaped particles: contact volume based model and computational issues. Comput Methods Appl Mech Eng 373:113493
Feng YT (2021c) An energy-conserving contact theory for discrete element modelling of arbitrarily shaped particles: basic framework and general contact model. Comput Methods Appl Mech Eng 373:113454
Feng YT, Owen DRJ (2004) A 2D polygon/polygon contact model: algorithmic aspects. Eng Comput 21:265–277
Feng YT, Tan Y (2019) On Minkowski difference-based contact detection in discrete/discontinuous modelling of convex polygons/polyhedra. Eng Comput 37:54–72
Fraige FY, Langston PA, Chen GZ (2008) Distinct element modelling of cubic particle packing and flow. Powder Technol 186:224–240
Gan JQ, Zhou ZY, Yu AB (2016) A GPU-based DEM approach for modelling of particulate systems. Powder Technol 301:1172–1182
Gan J, Evans T, Yu A (2019) Impact energy dissipation study in a simulated ship loading process. Powder Technol 354:476–484
Garboczi EJ, Bullard JW (2017) 3D analytical mathematical models of random star-shape particles via a combination of X-ray computed microtomography and spherical harmonic analysis. Adv Powder Technol 28:325–339
González-Montellano C, Ayuga F, Ooi JY (2011) Discrete element modelling of grain flow in a planar silo: influence of simulation parameters. Granul Matter 13:149–158
Govender N, Wilke DK, Kok S, Els R (2014) Development of a convex polyhedral discrete element simulation framework for NVIDIA Kepler based GPUs. J Comput Appl Math 270:386–400
Govender N, Wilke DN, Kok S (2015a) Collision detection of convex polyhedra on the NVIDIA GPU architecture for the discrete element method. Appl Math Comput 267:810–829
Govender N, Rajamani RK, Kok S, Wilke DN (2015b) Discrete element simulation of mill charge in 3D using the BLAZE-DEM GPU framework. Miner Eng 79:152–168
Govender N, Wilke DN, Pizette P, Abriak N-E (2018) A study of shape non-uniformity and poly-dispersity in hopper discharge of spherical and polyhedral particle systems using the Blaze-DEM GPU code. Appl Math Comput 319:318–336
Govender N, Wilke DN, Wu C-Y, Tuzun U, Kureck H (2019) A numerical investigation into the effect of angular particle shape on blast furnace burden topography and percolation using a GPU solved discrete element model. Chem Eng Sci 204:9–26
Gui N, Fan J (2015) Numerical study of heat conduction of granular particles in rotating wavy drums. Int J Heat Mass Transfer 84:740–751

Gui N, Yang X, Jiang S, Tu J (2016) A soft-sphere-imbedded pseudo-hard-particle model for simulation of discharge flow of brick particles. AIChE J 62:3562–3574
Gui N, Yang X, Tu J, Jiang S (2017) Numerical simulation and analysis of mixing of polygonal particles in 2D rotating drums by SIPHPM method. Powder Technol 318:248–262
Gui N, Yang X, Tu J, Jiang S (2018) Numerical study of the motion behaviour of three-dimensional cubic particle in a thin drum. Adv Powder Technol 29:426–437
Halidan M, Chandratilleke GR, Dong KJ, Yu AB (2018) Mixing performance of ribbon mixers: effects of operational parameters. Powder Technol 325:92–106
Han K, Feng YT, Owen DRJ (2006) Polygon-based contact resolution for superquadrics. Int J Numer Meth Eng 66:485–501
Han Y, Jia F, Li G, Liu H, Li J, Chen P (2019) Numerical analysis of flow pattern transition in a conical silo with ellipsoid particles. Adv Powder Technol 30:1870–1881
Hazeghian M, Soroush A (2015) DEM simulation of reverse faulting through sands with the aid of GPU computing. Comput Geotech 66:253–263
He Y, Evans TJ, Yu AB, Yang RY (2018) A GPU-based DEM for modelling large scale powder compaction with wide size distributions. Powder Technol 333:219–228
He SY, Gan JQ, Pinson D, Zhou ZY (2019) Particle shape-induced radial segregation of binary mixtures in a rotating drum. Powder Technol 341:157–166
Höhner D, Wirtz S, Scherer V (2012) A numerical study on the influence of particle shape on hopper discharge within the polyhedral and multi-sphere discrete element method. Powder Technol 226:16–28
Höhner D, Wirtz S, Scherer V (2014) A study on the influence of particle shape and shape approximation on particle mechanics in a rotating drum using the discrete element method. Powder Technol 253:256–265
Hopkins MA (2004) Discrete element modeling with dilated particles. Eng Comput 21:422–430
Ji S (2013) Probability analysis of contact forces in quasi-solid-liquid phase transition of granular shear flow. Sci China Phys Mech Astron 56:395–403
Ji S, Shen HH (2006) Effect of contact force models on granular flow dynamics. J Eng Mech-ASCE 132:1252–1259
Jin B, Tao H, Zhong W (2010) Flow behaviors of non-spherical granules in rectangular hopper. Chin J Chem Eng 18:931–939
Kafashan J, Wiącek J, Abd Rahman N, Gan J (2019) Two-dimensional particle shapes modelling for DEM simulations in engineering: a review. Granul Matter 21:80
Kawamoto R, Andò E, Viggiani G, Andrade JE (2016) Level set discrete element method for three-dimensional computations with triaxial case study. J Mech Phys Solids 91:1–13
Kidokoro T, Arai R, Saeki M (2015) Investigation of dynamics simulation of granular particles using spherocylinder model. Granul Matter 17:743–751
Kildashti K, Dong K, Samali B (2020) An accurate geometric contact force model for super-quadric particles. Comput Methods Appl Mech Eng 360:112774
Kou B, Cao Y, Li J, Xia C, Li Z, Dong H, Zhang A, Zhang J, Kob W, Wang Y (2017) Granular materials flow like complex fluids. Nature 551:360–363
Kruggel-Emden H, Rickelt S, Wirtz S, Scherer V (2008) A study on the validity of the multi-sphere discrete element method. Powder Technol 188:153–165
Kureck H, Govender N, Siegmann E, Boehling P, Radeke C, Khinast JG (2019) Industrial scale simulations of tablet coating using GPU based DEM: a validation study. Chem Eng Sci 202:462–480
Lai Z, Chen Q, Huang L (2020) Fourier series-based discrete element method for computational mechanics of irregular-shaped particles. Comput Methods Appl Mech Eng 362:112873
Li CQ, Xu WJ, Meng QS (2015) Multi-sphere approximation of real particles for DEM simulation based on a modified greedy heuristic algorithm. Powder Technol 286:478–487
Li Y, Gui N, Yang X, Tu J, Jiang S (2016) Effect of friction on pebble flow pattern in pebble bed reactor. Ann Nucl Energy 94:32–43

Li C, Peng Y, Zhang P, Zhao C (2019) The contact detection for heart-shaped particles. Powder Technol 346:85–96

Lin X, Ng TT (1995) Contact detection algorithms for three-dimensional ellipsoids in discrete element modelling. Int J Numer Anal Meth Geomech 19:653–659

Liu L, Ji S (2018) Ice load on floating structure simulated with dilated polyhedral discrete element method in broken ice field. Appl Ocean Res 75:53–65

Liu L, Ji S (2020) A new contact detection method for arbitrary dilated polyhedra with potential function in discrete element method. Int J Numer Meth Eng 121:5742–5765

Liu AJ, Nagel SR (1998) Jamming is not just cool any more. Nature 396:21–22

Liu Z, Zhao Y (2020) Multi-super-ellipsoid model for non-spherical particles in DEM simulation. Powder Technol 361:190–202

Liu X, Qi H, Ge W, Li J (2008) Pattern formation in particle systems driven by color field. Particuology 6:515–520

Liu H, Tafti DK, Li T (2014a) Hybrid parallelism in MFIX CFD-DEM using OpenMP. Powder Technol 259:22–29

Liu SD, Zhou ZY, Zou RP, Pinson D, Yu AB (2014b) Flow characteristics and discharge rate of ellipsoidal particles in a flat bottom hopper. Powder Technol 253:70–79

Liu GY, Xu WJ, Govender N, Wilke DN (2020a) A cohesive fracture model for discrete element method based on polyhedral blocks. Powder Technol 359:190–204

Liu GY, Xu WJ, Sun QC, Govender N (2020b) Study on the particle breakage of ballast based on a GPU accelerated discrete element method. Geosci Front 11:461–471

Liu S, Chen F, Ge W, Ricoux P (2020c) NURBS-based DEM for non-spherical particles. Particuology 49:65–76

Long X, Ji S, Wang Y (2019) Validation of microparameters in discrete element modeling of sea ice failure process. Part Sci Technol 37:550–559

Longmore J-P, Marais P, Kuttel MM (2013) Towards realistic and interactive sand simulation: a GPU-based framework. Powder Technol 235:983–1000

Lu G, Third JR, Müller CR (2012) Critical assessment of two approaches for evaluating contacts between super-quadric shaped particles in DEM simulations. Chem Eng Sci 78:226–235

Lu G, Third JR, Müller CR (2014) Effect of wall rougheners on cross-sectional flow characteristics for non-spherical particles in a horizontal rotating cylinder. Particuology 12:44–53

Lu G, Third JR, Müller CR (2015) Discrete element models for non-spherical particle systems: from theoretical developments to applications. Chem Eng Sci 127:425–465

Lubbe R, Xu W-J, Wilke DN, Pizette P, Govender N (2020) Analysis of parallel spatial partitioning algorithms for GPU based DEM. Comput Geotech 125:103708

Ma H, Zhao Y (2017) Modelling of the flow of ellipsoidal particles in a horizontal rotating drum based on DEM simulation. Chem Eng Sci 172:636–651

Ma H, Zhao Y (2018) Investigating the flow of rod-like particles in a horizontal rotating drum using DEM simulation. Granul Matter 20:41

Magalhães FGR, Atman APF, Moreira JG, Herrmann HJ (2016) Analysis of the velocity field of granular hopper flow. Granul Matter 18:33

Maione R, Kiesgen De Richter S, Mauviel G, Wild G (2017) Axial segregation of a binary mixture in a rotating tumbler with non-spherical particles: experiments and DEM model validation. Powder Technol 306:120–129

Meng L, Wang C, Yao X (2018) Non-convex shape effects on the dense random packing properties of assembled rods. Phys A 490:212–221

Mollon G, Zhao J (2014) 3D generation of realistic granular samples based on random fields theory and Fourier shape descriptors. Comput Methods Appl Mech Eng 279:46–65

Nezami EG, Hashash YMA, Zhao D, Ghaboussi J (2006) Shortest link method for contact detection in discrete element method. Int J Numer Anal Meth Geomech 30:783–801

Nishiura D, Sakaguchi H (2011) Parallel-vector algorithms for particle simulations on shared-memory multiprocessors. J Comput Phys 230:1923–1938

Peng L, Xu J, Zhu Q, Li H, Ge W, Chen F, Ren X (2016) GPU-based discrete element simulation on flow regions of flat bottomed cylindrical hopper. Powder Technol 304:218–228

Peng L, Zou Z, Zhang L, Zhu Q, Li H (2018) GPU-based discrete element simulation on flow stability of flat-bottomed hopper. Chin J Chem Eng 26:43–52

Peng Z, Zhou J, Zhou J, Miao Y, Cheng L, Jiang Y, Hou M (2021) External pressure dependence of granular orifice flow: transition to Beverloo flow. Phys Fluids 33:043313

Pereira GG, Pucilowski S, Liffman K, Cleary PW (2011) Streak patterns in binary granular media in a rotating drum. Appl Math Model 35:1638–1646

Peters JF, Hopkins MA, Kala R, Wahl RE, Munijiza A (2009) A poly-ellipsoid particle for non-spherical discrete element method. Eng Comput 26:645–657

Podlozhnyuk A, Pirker S, Kloss C (2016) Efficient implementation of superquadric particles in discrete element method within an open-source framework. Comput Part Mech 4:101–118

Portal R, Dias J, De Sousa L (2010) Contact detection between convex superquadric surfaces. Arch Mech Eng 57:165–186

Rakotonirina AD, Delenne J-Y, Radjai F, Wachs A (2019) Grains3D, a flexible DEM approach for particles of arbitrary convex shape—part III: extension to non-convex particles modelled as glued convex particles. Comput Part Mech 6:55–84

Ren X, Xu J, Qi H, Cui L, Ge W, Li J (2013) GPU-based discrete element simulation on a tote blender for performance improvement. Powder Technol 239:348–357

Seo IS, Kim JH, Shin JH, Shin SW, Lee SH (2014) Particle behaviors of printing system using GPU-based discrete element method. J Mech Sci Technol 28:5083–5087

Soltanbeigi B, Podlozhnyuk A, Papanicolopulos S-A, Kloss C, Pieker S, Ooi JY (2018) DEM study of mechanical characteristics of multi-spherical and superquadric particles at micro and macro scales. Powder Technol 329:288–303

Steuben J, Mustoe G, Turner C (2016) Massively parallel discrete element method simulations on graphics processing units. J Comput Inf Sci Eng 16:031001

Su D, Yan WM (2018) 3D characterization of general-shape sand particles using microfocus X-ray computed tomography and spherical harmonic functions, and particle regeneration using multivariate random vector. Powder Technol 323:8–23

Tian Y, Zhang S, Lin P, Yang Q, Yang G, Yang L (2017) Implementing discrete element method for large-scale simulation of particles on multiple GPUs. Comput Chem Eng 104:231–240

Toson P, Khinast JG (2017) Impulse-based dynamics for studying quasi-static granular flows: application to hopper emptying of non-spherical particles. Powder Technol 313:353–360

Verbucheln F, Parteli EJ, Poschel T (2015) Helical inner-wall texture prevents jamming in granular pipe flows. Soft Matter 11:4295–4305

Vlahinić I, Kawamoto R, Andò E, Viggiani G, Andrade JE (2017) From computed tomography to mechanics of granular materials via level set bridge. Acta Geotech 12:85–95

Vu-Quoc L, Zhang X, Walton OR (2000) A 3-D discrete-element method for dry granular flows of ellipsoidal particles. Comput Methods Appl Mech Eng 187:483–528

Wan J, Wang F, Yang G, Zhang S, Wang M, Lin P, Yang L (2018) The influence of orifice shape on the flow rate: a DEM and experimental research in 3D hopper granular flows. Powder Technol 335:147–155

Wellmann C, Lillie C, Wriggers PA (2008) Contact detection algorithm for superellipsoids based on the common-normal concept. Eng Comput 25:432–442

Williams JR, Pentland AP (1992) Superquadrics and modal dynamics for discrete elements in interactive design. Eng Comput 9:115–127

Xiao X, Tan Y, Zhang H, Deng R, Jiang S (2017) Experimental and DEM studies on the particle mixing performance in rotating drums: effect of area ratio. Powder Technol 314:182–194

Xie C, Ma H, Zhao Y (2019) Investigation of modeling non-spherical particles by using spherical discrete element model with rolling friction. Eng Anal Bound Elem 105:207–220

Xu Y, Xu C, Zhou Z, Du J, Hu D (2010) 2D DEM simulation of particle mixing in rotating drum: a parametric study. Particuology 8:141–149

Xu J, Qi H, Fang X, Lu L, Ge W, Wang X, Xu M, Chen F, He X, Li J (2011) Quasi-real-time simulation of rotating drum using discrete element method with parallel GPU computing. Particuology 9:446–450

Xu Y, Gao X, Li T (2021) Numerical study of the bi-disperse particles segregation inside a spherical tumbler with discrete element method (DEM). Comput Math Appl 81:588–601

Yan B, Regueiro RA (2018) A comprehensive study of MPI parallelism in three-dimensional discrete element method (DEM) simulation of complex-shaped granular particles. Comput Part Mech 5:553–577

Yang S, Sun Y, Zhang L, Chew JW (2017) Numerical study on the axial segregation dynamics of a binary-size granular mixture in a three-dimensional rotating drum. Phys Fluids 29:103302

Yang S, Zhang L, Luo K, Chew JW (2018) DEM investigation of the axial dispersion behavior of a binary mixture in the rotating drum. Powder Technol 330:93–104

Yang S, Wang H, Wei Y, Hu J, Chew JW (2020) Flow dynamics of binary mixtures of non-spherical particles in the rolling-regime rotating drum. Powder Technol 361:930–942

Yu F, Zhou G, Xu J, Ge W (2015) Enhanced axial mixing of rotating drums with alternately arranged baffles. Powder Technol 286:276–287

Yue X, Zhang H, Ke C, Luo C, Shu S, Tan Y, Feng C (2015) A GPU-based discrete element modeling code and its application in die filling. Comput Fluids 110:235–244

Zeng Y, Jia F, Zhang Y, Meng X, Han Y, Wang H (2017) DEM study to determine the relationship between particle velocity fluctuations and contact force disappearance. Powder Technol 313:112–121

Zhang L, Quigley SF, Chan AHC (2013) A fast scalable implementation of the two-dimensional triangular discrete element method on a GPU platform. Adv Eng Softw 60–61:70–80

Zhang S, Lin P, Wang C-L, Tian Y, Wan J-F, Yang L (2014) Investigating the influence of wall frictions on hopper flows. Granul Matter 16:857–866

Zhang B, Regueiro R, Druckrey A, Alshibli K (2018a) Construction of poly-ellipsoidal grain shapes from SMT imaging on sand, and the development of a new DEM contact detection algorithm. Eng Comput 35:733–771

Zhang Y, Jia F, Zeng Y, Han Y, Xiao Y (2018b) DEM study in the critical height of flow mechanism transition in a conical silo. Powder Technol 331:98–106

Zhao S, Zhao J (2019) A poly-superellipsoid-based approach on particle morphology for DEM modeling of granular media. Int J Numer Anal Meth Geomech 43:2147–2169

Zheng J, An X, Huang M (2012) GPU-based parallel algorithm for particle contact detection and its application in self-compacting concrete flow simulations. Comput Struct 112–113:193–204

Zheng QJ, Zhou ZY, Yu AB (2013) Contact forces between viscoelastic ellipsoidal particles. Powder Technol 248:25–33

Zheng Z, Zang M, Chen S, Zeng H (2018) A GPU-based DEM-FEM computational framework for tire-sand interaction simulations. Comput Struct 209:74–92

Zhou GGD, Sun QC (2013) Three-dimensional numerical study on flow regimes of dry granular flows by DEM. Powder Technol 239:115–127

Zhou ZY, Zou RP, Pinson D, Yu AB (2011) Dynamic simulation of the packing of ellipsoidal particles. Ind Eng Chem Res 50:9787–9798

Chapter 2
Superquadric DEM Model Based on Functional Representations

The discrete element method (DEM), proposed by Cundall and Strack, has been shown to be a practical approach to studying various granular materials (Cundall and Strack 1979). In this approach, two-dimensional disks or three-dimensional spheres were initially employed because they enable simple calculation and efficient operation (Zhu et al. 2007, 2008). However, the granular systems commonly encountered in industry or nature comprise non-spherical grains. Although the particle shape strongly affects the dynamics of granular systems, such as packing (Gan et al. 2017), hopper discharge (Höhner et al. 2013), or conveying systems (Zhou et al. 2017), experimental measurements and numerical simulations have focused mainly on spherical particles. Meanwhile, the applicability of the conclusions drawn from spherical particle systems to non-spherical particle systems is doubtful (Lu et al. 2015). To model particles of various shapes reasonably, different methods, including a combined approach based on regular elements (Kruggel-Emden et al. 2008), dilated polyhedra based on Minkowski sum theory (Galindo-Torres and Pedroso 2010), arbitrary convex shapes based on orientation discretization (Dong et al. 2015), and superquadric elements based on continuous function representation (CFR) (Lu et al. 2012), have been developed. Contact detection between two superquadric elements has benefited from the development of approaches such as the common normal concept (Wellmann et al. 2008), the interior point algorithm (Chakraborty et al. 2008), and the geometric potential approach (Houlsby 2009). Compared to the common normal concept (Kildashti et al. 2018), the geometric potential approach has higher computational efficiency (Zhou et al. 2018). However, a disadvantage is that it may not meet better the definition of contact mechanics, i.e., the normal force is perpendicular to the surface of two overlapping particles. For the interaction with boundaries, Podlozhnyuk et al. (2016) presented an analytical solution for particle–wall contact; however, this algorithm is applicable only to the contact detection between a particle and a flat wall. For complex geometric boundaries, the corresponding functional form cannot be easily found. Therefore, a widespread and common approach is to

S. Wang and S. Ji, *Computational Mechanics of Arbitrarily Shaped Granular Materials*, Springer Tracts in Mechanical Engineering,
https://doi.org/10.1007/978-981-99-9927-9_2

use a standard triangular finite element surface mesh, and the corresponding interaction can be simplified as the contact detection between a superquadric particle and a triangular element. Cleary and Sawley (2002) simulated industrial granular flows by constructing a triangular finite element mesh.

This chapter introduces the contact detection between particles and between particles and triangular elements. Subsequently, the corresponding nonlinear contact force model considering the equivalent radius of curvature at a local contact point is applied for superquadric particles. Then, quantitative verification and validation studies are performed, including a flat wall impacted by a single cylinder with different blockiness parameters, the presence of the arch structure of cubes, and the dynamic hopper discharge of ellipsoids. Moreover, the effects of the particle shape and base angle on the discharge rate are analyzed to obtain a fundamental understanding of particle flow during the discharging process.

2.1 Construction of Superquadric Elements

In this section, the superquadric equation is introduced in detail. The superquadric equation, which allows the aspect ratio of particles and the smoothness of the surface to be changed efficiently, is a more common approach to describe particle shapes. Meanwhile, 80% of solid shapes can be represented by superquadrics (Williams and Pentland 1992), and other solid shapes can be derived from higher-dimensional hyper-quadrics (Zhong et al. 2016).

2.1.1 Superquadric Elements Based on Continuous Function Representation

The superquadric method was developed by expanding the quadric surface, which has been used extensively to describe particles with convex and concave shapes. The superquadric equation is given as (Barr 1981):

$$\left(\left|\frac{x}{a}\right|^{n_2} + \left|\frac{y}{b}\right|^{n_2}\right)^{n_1/n_2} + \left|\frac{z}{c}\right|^{n_1} - 1 = 0, \tag{2.1}$$

where a, b, and c are the half-lengths of the particles along the principal axes, and n_1 and n_2 control the blockiness of the cross-section of the particles. When $n_1 = n_2$, the above equation can be simplified to (Cleary 2004):

$$\left|\frac{x}{a}\right|^{n_2} + \left|\frac{y}{b}\right|^{n_2} + \left|\frac{z}{c}\right|^{n_2} - 1 = 0 \tag{2.2}$$

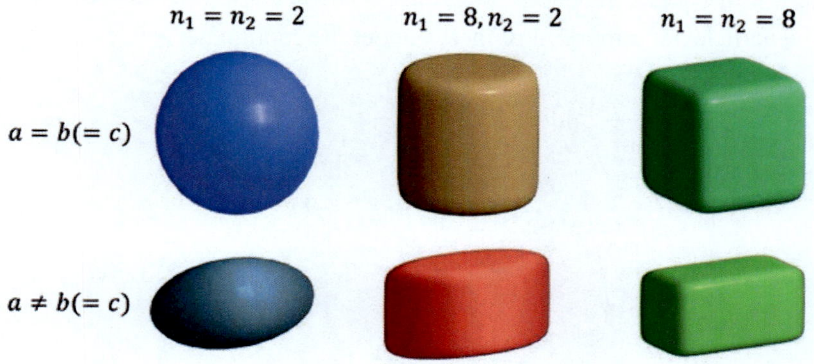

Fig. 2.1 Examples of three-dimensional superquadric elements

This simplified equation may not describe cylinder-shaped particles accurately. Therefore, we used Eq. (2.1) as the standard equation for the superquadric particles in the present work. An ellipsoid is obtained for $n_1 = n_2 = 2$, a cylinder-shaped element is obtained if $n_1 \gg 2$ and $n_2 = 2$, and a cubic shape is obtained if $n_1 = n_2 \gg 2$. For $n_1, n_2 \to \propto$, it is theoretically possible to describe real particles with sharp corners, but this model is usually limited by the search algorithm. Currently, the construction of non-spherical particles by the superquadric equation is usually given a reasonable range of n_1 and n_2, thereby ensuring the stability and high efficiency of the algorithm. Figure 2.1 shows the different particle shapes obtained by varying five parameters of the superquadric equation.

2.1.2 Calculation of Mass and Moment of Inertia

The closed space surrounded by the superquadric equation can constitute a non-spherical particle. The density of the granular material is denoted as ρ, and the mass can be obtained by integrating the mass of the particle, expressed as:

$$m = \iiint_V \mathrm{d}m = 8 \iiint_{V_1} \rho \mathrm{d}V$$

$$= 8 \int_0^c \int_0^{a\left[1-\left(-\frac{z}{c}\right)^{n_1}\right]^{\frac{1}{n_1}}} \int_0^{b\left\{\left[1-\left(\frac{z}{c}\right)^{n_1}\right]^{\frac{n_2}{n_1}}-\left(\frac{x}{a}\right)^{n_2}\right\}^{\frac{1}{n_2}}} \rho \mathrm{d}x\mathrm{d}y\mathrm{d}z. \tag{2.3}$$

The particle has a moment of inertia about the coordinate axis:

$$I_x = \iiint_V \left(y^2 + z^2\right) \mathrm{d}m$$

$$= 8 \int_0^c \int_0^{a\left[1-\left(-\frac{z}{c}\right)^{n1}\right]^{\frac{1}{n1}}} \int_0^{b\left\{\left[1-\left(\frac{z}{c}\right)^{n1}\right]^{\frac{n2}{n1}} - \left(\frac{x}{a}\right)^{n2}\right\}^{\frac{1}{n2}}} \rho\left(y^2 + z^2\right) \mathrm{d}x \mathrm{d}y \mathrm{d}z, \quad (2.4)$$

$$I_y = \iiint_V \left(x^2 + z^2\right) \mathrm{d}m$$

$$= 8 \int_0^c \int_0^{a\left[1-\left(-\frac{z}{c}\right)^{n1}\right]^{\frac{1}{n1}}} \int_0^{b\left\{\left[1-\left(\frac{z}{c}\right)^{n1}\right]^{\frac{n2}{n1}} - \left(\frac{x}{a}\right)^{n2}\right\}^{\frac{1}{n2}}} \rho\left(x^2 + z^2\right) \mathrm{d}x \mathrm{d}y \mathrm{d}z, \quad (2.5)$$

$$I_z = \iiint_V \left(x^2 + y^2\right) \mathrm{d}m$$

$$= 8 \int_0^c \int_0^{a\left[1-\left(-\frac{z}{c}\right)^{n1}\right]^{\frac{1}{n1}}} \int_0^{b\left\{\left[1-\left(\frac{z}{c}\right)^{n1}\right]^{\frac{n2}{n1}} - \left(\frac{x}{a}\right)^{n2}\right\}^{\frac{1}{n2}}} \rho\left(x^2 + y^2\right) \mathrm{d}x \mathrm{d}y \mathrm{d}z. \quad (2.6)$$

Due to the symmetry of the equations, the inertia of a superquadric particle is satisfied as:

$$I_{xy} = I_{xz} = I_{yx} = I_{yz} = I_{zx} = I_{zy} = 0 \quad (2.7)$$

Thus, the inertia tensor matrix form of a superquadric particle can be expressed as:

$$I = \begin{pmatrix} I_x & I_{xy} & I_{xz} \\ I_{yx} & I_y & I_{yz} \\ I_{zx} & I_{zy} & I_z \end{pmatrix} = \begin{pmatrix} I_x & 0 & 0 \\ 0 & I_y & 0 \\ 0 & 0 & I_z \end{pmatrix}. \quad (2.8)$$

2.2 Contact Algorithm and Contact Force Model

In this section, the contact algorithms between particles and between particles and structures are described in detail. The problem of contact between particles is transformed into solving a system of nonlinear equations, and the contact points between particles are calculated using the Newtonian iterative method. The particle-structure contact problem is transformed into a contact problem between particles and triangular elements, and the contact modes between the particles and the triangular elements include the face, edge, and vertex contacts.

2.2.1 Contact Detection Between Particles

Compared with the simple and efficient contact detection between spherical particles, the ratio of the time of contact detection between irregular particles to the entire simulation time will increase significantly, and the computational efficiency is mainly affected by the particle shape, boundary conditions, and search algorithm (Lu et al. 2012). Because of the complexity of contact detection between non-spherical particles and the orientation under different contact patterns, the bounding spheres and oriented bounding boxes (OBBs) have been well developed, as shown in Fig. 2.2. The bounding sphere is used as the first rough contact detection between two superquadric particles. If the distance between two particle centers is larger than the sum of their corresponding radii, they are not in contact. Otherwise, they are in contact. Moreover, OBB is the second rough contact detection. If the distance of two centers of particles projected onto the separation axis is larger than the sum of the projections of each box onto it, the boxes do not intersect each other.

Finally, the Newton–Raphson approach is used to calculate the overlap between the particles i and j, which is based on finding a "midway" point $\boldsymbol{X}_0$. The corresponding nonlinear equations are as follows (Podlozhnyuk et al. 2016):

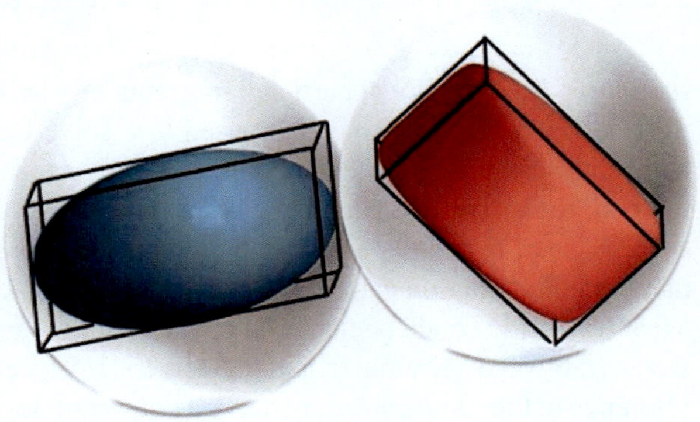

Fig. 2.2 Bounding spheres and oriented bounding boxes

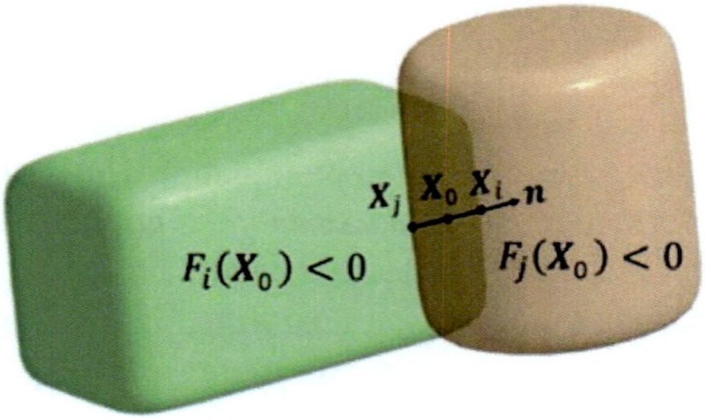

Fig. 2.3 Contact detection between superquadric particles

$$\begin{cases} \nabla F_i(\boldsymbol{X}) + \lambda^2 \nabla F_j(\boldsymbol{X}) = 0 \\ F_i(\boldsymbol{X}) - F_j(\boldsymbol{X}) = 0 \end{cases}, \tag{2.9}$$

where F_i and F_j are the superquadric equations of particle i and particle j in globally fixed coordinates, respectively, and λ^2 is the non-negative multiplier. Newton's iterative algorithm for this problem can be expressed as:

$$\begin{pmatrix} \nabla^2 F_i(\boldsymbol{X}) + \lambda^2 \nabla^2 F_j(\boldsymbol{X}) & 2\lambda \nabla F_j(\boldsymbol{X}) \\ \nabla F_i(\boldsymbol{X}) - \nabla F_j(\boldsymbol{X}) & 0 \end{pmatrix} \begin{pmatrix} \mathrm{d}\boldsymbol{X} \\ \mathrm{d}\lambda \end{pmatrix} = - \begin{pmatrix} \nabla F_i(\boldsymbol{X}) + \lambda^2 \nabla F_j(\boldsymbol{X}) \\ F_i(\boldsymbol{X}) - F_j(\boldsymbol{X}) \end{pmatrix}, \tag{2.10}$$

where $\boldsymbol{X} = \boldsymbol{X} + \mathrm{d}\boldsymbol{X}$ and $\lambda = \lambda + \mathrm{d}\lambda$. If $F_i(\boldsymbol{X}_0) < 0$ and $F_j(\boldsymbol{X}_0) < 0$ can be satisfied for point $\boldsymbol{X}_0$, the two particles are in contact, and the corresponding normal direction can be described as $\boldsymbol{n} = \nabla F_i(\mathbf{X}_0)/\|\nabla F_i(\mathbf{X}_0)\|$, as shown in Fig. 2.3. Then, given $\boldsymbol{X}_i = \boldsymbol{X}_0 + \alpha_0 \boldsymbol{n}$ and $\boldsymbol{X}_j = \boldsymbol{X}_0 + \beta_0 \boldsymbol{n}$, we can easily determine the unknown parameters α_0 and β_0 by Newton's method (Podlozhnyuk et al. 2016): $\alpha_0^{k+1} = \alpha_0^k - F_i\left(\boldsymbol{X}_i^k\right)/\left(\nabla F_i\left(\boldsymbol{X}_i^k\right) \cdot \boldsymbol{n}\right)$ and $\beta_0^{k+1} = \beta_0^k - F_j\left(\boldsymbol{X}_j^k\right)/\left(\nabla F_j\left(\boldsymbol{X}_j^k\right) \cdot \boldsymbol{n}\right)$, and the normal overlap is $\boldsymbol{\delta}_{\mathrm{n}} = \boldsymbol{X}_i - \boldsymbol{X}_j$.

2.2.2 Contact Detection Between Particles and Structures

In industry, non-spherical particles are frequently used; thus, a numerical calculation method (i.e., the DEM) has been used to resolve the interaction between particles and complicated geometric boundaries. Based on the advantages of the finite element method, the arbitrarily geometric boundary models can be discretized and meshed by a series of triangular elements with controllable accuracy. Therefore, the aforementioned interaction can be simplified to the interaction with a triangular element. Meanwhile, the contact detection between spheres and triangular elements has been well established (Kremmer and Favier 2001; Hu et al. 2013) and can provide a reference to the same contact patterns as superquadric particles, as shown in Fig. 2.4.

The vertices of a triangular element are denoted *A*, *B*, and *C*, where $\boldsymbol{x}_A$, $\boldsymbol{x}_B$, and $\boldsymbol{x}_C$ are three position vectors. Therefore, the three edge vectors are easily obtained and

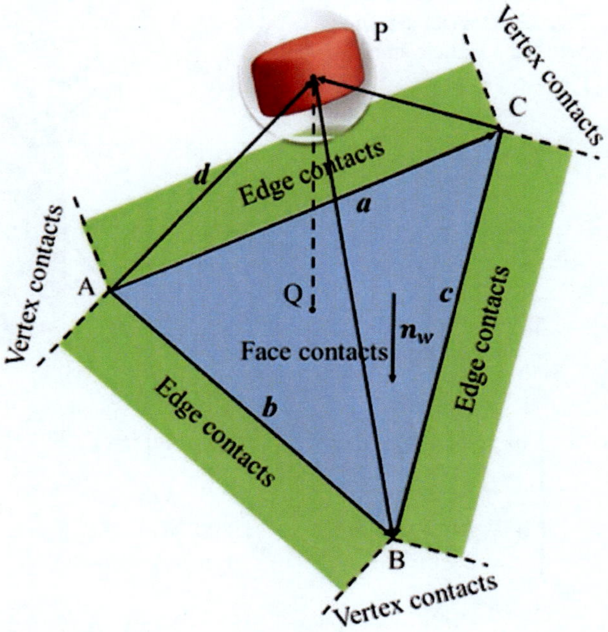

Fig. 2.4 Three contact modes between particles and triangular elements

denoted $\boldsymbol{a}$, $\boldsymbol{b}$, and $\boldsymbol{c}$, respectively. Meanwhile, the normal direction, $\boldsymbol{n}_w(n_x, n_y, n_z)$, is given by $\boldsymbol{n}_w = \boldsymbol{a} \times \boldsymbol{b}/|\boldsymbol{a} \times \boldsymbol{b}|$ and is directed outwards with respect to the particle. In addition, the centroid of particle and the projection point of the centroid onto the plane where the triangular element is located are denoted P and Q, respectively, and vector $\boldsymbol{d}$ directs from vertex A to point P. To determine if a particle is in contact with a triangular element, the contact between the particle and the plane where the triangle element is located must be detected first. The bounding sphere can be reintroduced as the first rough contact detection between a particle and a plane. The distance $|PQ|$ is expressed as $|PQ| = |\boldsymbol{d} \cdot \boldsymbol{n}_w|$. Therefore, the bounding sphere is in contact with the plane if $|PQ|$ is less than or equal to the radius R_0 of the sphere.

Secondly, the contact point $\boldsymbol{x}$ on the surface of the particle and the projection, $\boldsymbol{x}^*$, of the contact point onto the plane (Fig. 2.5) can be determined by the analytical solution of particle-plane contact, which has been described well previously (Podlozhnyuk et al. 2016). The final calculation results are as follows:

$$\begin{cases} \alpha_1 = \left(|bn_y|/|an_x|\right)^{1/(n_2-1)} \\ \gamma_1 = \left(1+\alpha_1^{n_2}\right)^{n_1/n_2-1} \\ \beta_1 = \left(\gamma_1|n_z c|/|n_x a|\right)^{1/(n_1-1)} \\ x = a/\left(\left(1+\alpha_1^{n_2}\right)^{n_1/n_2} + \beta_1^{n_1}\right)^{1/n_1} \operatorname{sign}(n_x) \\ y = \alpha_1 b|x|/a \operatorname{sign}(n_y) \\ z = \beta_1 c|x|/a \operatorname{sign}(n_z) \end{cases} \quad \text{if } n_x \neq 0, \tag{2.11}$$

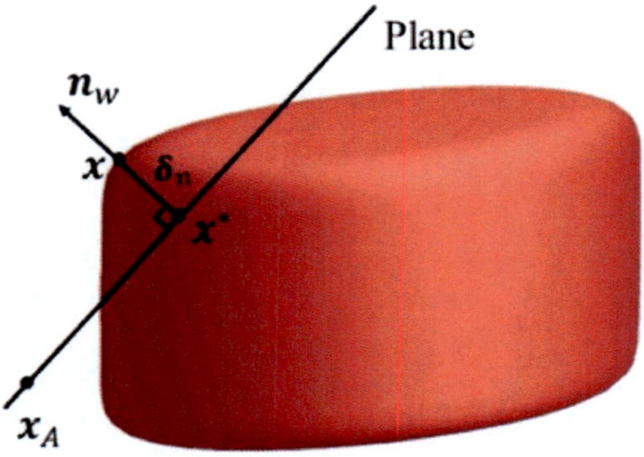

Fig. 2.5 Contact detection between a particle and a plane

$$\begin{cases} \omega = \left(\left(b|n_y|/n_1\right)^{n_1/(n_1-1)} + (c|n_z|/n_1)^{n_1/(n_1-1)} \right)^{1/n_1-1} \\ x = 0 \\ y = b\left(b|n_y|\omega/n_1\right)^{1/(n_1-1)} \operatorname{sign}(n_y) \\ z = c(c|n_z|\omega/n_1)^{1/(n_1-1)} \operatorname{sign}(n_z) \end{cases} \quad \text{if } n_x = 0, \tag{2.12}$$

$$\boldsymbol{x}^* = ((\boldsymbol{x}_A - \boldsymbol{x}) \cdot \boldsymbol{n}_w) \cdot \boldsymbol{n}_w + \boldsymbol{x}, \tag{2.13}$$

where a, b, c, n_1, and n_2 are the function parameters of a particle. Therefore, the particle is in contact with the plane when the value of $((\boldsymbol{x}_A - \boldsymbol{x}) \cdot \boldsymbol{n}_w)$ is less than or equal to zero.

Thirdly, to obtain three different contact modes, such as face, edge, and vertex contacts, it is necessary to determine whether projection point $\boldsymbol{x}^*$ is located in the triangular element. Point $\boldsymbol{x}^*$ in the triangular element can be expressed by α and β with respect to vertex A.

$$\boldsymbol{x}^* - \boldsymbol{x}_A = \alpha \boldsymbol{a} + \beta \boldsymbol{b}, \tag{2.14}$$

$$\alpha > 0 \quad \cap \quad \beta > 0 \quad \cap \quad \alpha + \beta < 1. \tag{2.15}$$

Equation (2.14) is then multiplied by vectors $\boldsymbol{a}$ and $\boldsymbol{b}$, respectively.

$$\left(\boldsymbol{x}^* - \boldsymbol{x}_A\right)\boldsymbol{a} = \alpha \boldsymbol{a} \cdot \boldsymbol{a} + \beta \boldsymbol{b} \cdot \boldsymbol{a}, \tag{2.16}$$

$$\left(\boldsymbol{x}^* - \boldsymbol{x}_A\right)\boldsymbol{b} = \alpha \boldsymbol{a} \cdot \boldsymbol{b} + \beta \boldsymbol{b} \cdot \boldsymbol{b}. \tag{2.17}$$

Here, the values of α and β can be calculated by Eqs. (2.16) and (2.17).

$$\alpha = \frac{(\boldsymbol{b} \cdot \boldsymbol{b})((\boldsymbol{x}^* - \boldsymbol{x}_A) \cdot \boldsymbol{a}) - (\boldsymbol{b} \cdot \boldsymbol{a})((\boldsymbol{x}^* - \boldsymbol{x}_A) \cdot \boldsymbol{b})}{(\boldsymbol{a} \cdot \boldsymbol{a})(\boldsymbol{b} \cdot \boldsymbol{b}) - (\boldsymbol{a} \cdot \boldsymbol{b})(\boldsymbol{b} \cdot \boldsymbol{a})}, \tag{2.18}$$

$$\beta = \frac{(\boldsymbol{a} \cdot \boldsymbol{a})((\boldsymbol{x}^* - \boldsymbol{x}_A) \cdot \boldsymbol{b}) - (\boldsymbol{a} \cdot \boldsymbol{b})((\boldsymbol{x}^* - \boldsymbol{x}_A) \cdot \boldsymbol{a})}{(\boldsymbol{a} \cdot \boldsymbol{a})(\boldsymbol{b} \cdot \boldsymbol{b}) - (\boldsymbol{a} \cdot \boldsymbol{b})(\boldsymbol{b} \cdot \boldsymbol{a})}. \tag{2.19}$$

If α and β satisfy Eq. (2.15) and the value of $F(\boldsymbol{x}^*)$ is less than or equal to zero, the particle is in face contact with the triangular element and the corresponding overlap $\boldsymbol{\delta}_n$ can be given by $\boldsymbol{\delta}_n = \boldsymbol{x}^* - \boldsymbol{x}$. Otherwise, edge contacts and vertex contacts will be detected.

Taking edge $\boldsymbol{b}(b_x, b_y, b_z)$ as an example (Fig. 2.6), the projection of centroid P on edge $\boldsymbol{b}$ is point Q. Points E and F are obtained by extending the distance of $\mp R_0$ from point Q in the direction of vector $\boldsymbol{b}$. In the same way, points A' and B' can also be obtained by extending the distance of $\mp R_0$ from vertices A and B in the direction of vector $\boldsymbol{b}$. This means that the particle may be in contact with the edge when the projection point Q is within the range of segments AA' and BB'. Then, the distance $|PQ|$ can be expressed by:

$$|PQ| = \left| \boldsymbol{d} - \frac{\boldsymbol{d} \cdot \boldsymbol{b}}{|\boldsymbol{b}|} \cdot \frac{\boldsymbol{b}}{|\boldsymbol{b}|} \right|. \tag{2.20}$$

A rough but necessary condition to prove edge contact is that the point Q satisfies Eq. (2.21):

$$|PQ| < R_0 \quad \cap \quad -R_0 < \frac{\boldsymbol{d} \cdot \boldsymbol{b}}{|\boldsymbol{b}|} < R_0. \tag{2.21}$$

Next, point $\boldsymbol{x}_{e1}$ with the minimum potential energy can be determined on segment EF. Here, segment EF can be expressed as a function of parameter t,

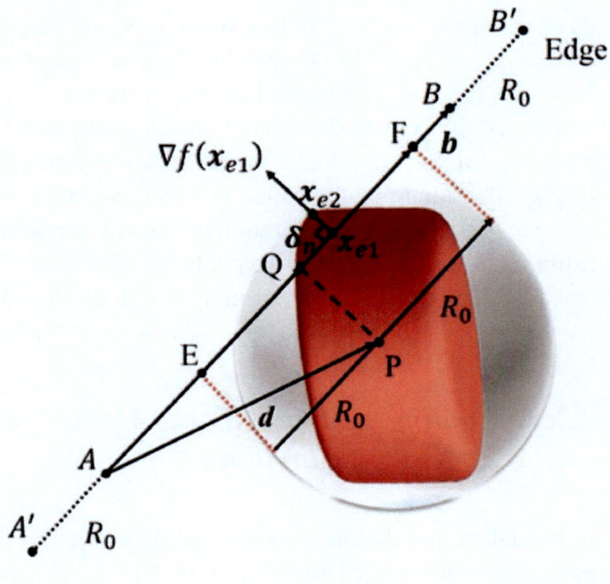

Fig. 2.6 Contact detection between a particle and an edge

$$\boldsymbol{x}'_E\left(x'_{Ex}, x'_{Ey}, x'_{Ez}\right) = \boldsymbol{x}_E + t\frac{\boldsymbol{b}}{|\boldsymbol{b}|}, \qquad t \in [0, |\boldsymbol{x}_F - \boldsymbol{x}_E|], \tag{2.22}$$

where $\boldsymbol{x}_E\left(x_{Ex}, x_{Ey}, x_{Ez}\right)$ and $\boldsymbol{x}_F\left(x_{Fx}, x_{Fy}, x_{Fz}\right)$ are the position coordinates of points E and F. The corresponding superquadric equation can be expressed as a function of parameter t, and the derivative of this equation with respect to t can be easily obtained:

$$f(t) = \left[\left|\frac{x'_{Ex}}{a}\right|^{n_2} + \left|\frac{x'_{Ey}}{b}\right|^{n_2}\right]^{n_1/n_2} + \left|\frac{x'_{Ez}}{c}\right|^{n_1} - 1, \tag{2.23}$$

$$\begin{aligned}\frac{\partial f}{\partial t} = n_1 \Bigg\{ & \left[\left|\frac{x'_{Ex}}{a}\right|^{n_2} + \left|\frac{x'_{Ey}}{b}\right|^{n_2}\right]^{\frac{n_1}{n_2}-1} \cdot \left[\frac{x_{Ex}}{a}\left|\frac{x'_{Ex}}{a}\right|^{n_2-1} \cdot \operatorname{sign}\left(x'_{Ex}\right)\right. \\ & \left. + \frac{x_{Ey}}{b}\left|\frac{x'_{Ex}}{a}\right|^{n_2-1} \cdot \operatorname{sign}\left(x'_{Ex}\right)\right] + \frac{x_{Ez}}{c}\left|\frac{x'_{Ez}}{c}\right|^{n_1-1} \cdot \operatorname{sign}\left(x'_{Ez}\right)\Bigg\}.\end{aligned} \tag{2.24}$$

Therefore, point $\boldsymbol{x}_{e1}$ is easily given by the dichotomy of the superquadric gradient function. If $f(\boldsymbol{x}_{e1}) < 0$, the particle and the triangular element are in edge contact. Another contact point, $\boldsymbol{x}_{e2}$, on the particle surface is expressed as $\boldsymbol{x}_{e2} = \boldsymbol{x}_{e1} + \alpha_e \nabla f(\boldsymbol{x}_{e1})$, and the unknown parameter α_e can be solved by Newton's method, i.e., $\alpha_e^{k+1} = \alpha_e^k - f\left(\boldsymbol{x}_{e2}^k\right)/\left(\nabla f\left(\boldsymbol{x}_{e2}^k\right) \cdot \nabla f(\boldsymbol{x}_{e1})\right)$. The corresponding overlap is $\boldsymbol{\delta}_n = \boldsymbol{x}_{e2} - \boldsymbol{x}_{e1}$. If the particle is not in contact with all three edges, vertex contacts will be detected.

Taking vertex $\boldsymbol{A}\left(x_{Ax}, x_{Ay}, x_{Az}\right)$ as an example, the particle and the triangular element are in contact if $f(\boldsymbol{x}_A) < 0$. In the same way as the method for calculating edge contacts, another contact point $\boldsymbol{x}_{A2}$ on the particle surface is expressed as $\boldsymbol{x}_{A2} = \boldsymbol{x}_A + \alpha_v \nabla f(\boldsymbol{x}_A)$, and the unknown parameter α_v can be obtained using Newton's method. The normal overlap is $\boldsymbol{\delta}_\mathrm{n} = \boldsymbol{x}_{A2} - \boldsymbol{x}_A$.

Finally, the contact determination rules are crucial for excluding invalid contacts when a particle is in contact with a series of triangular elements at the same time. It is worth noting that the contact of a superquadric particle with a single triangular element is significantly different from that of a spherical particle; however, the algorithm for determining effective contacts is the same. Therefore, this method can be extended to superquadrics. For more detailed calculations, refer to Hu et al. (2013).

2.2.3 Nonlinear Contact Force Model Considering Equivalent Curvature

In the DEM simulation of spherical elements, the nonlinear contact force model has been well established (Zhu et al. 2007, 2008); the ellipsoidal contact model,

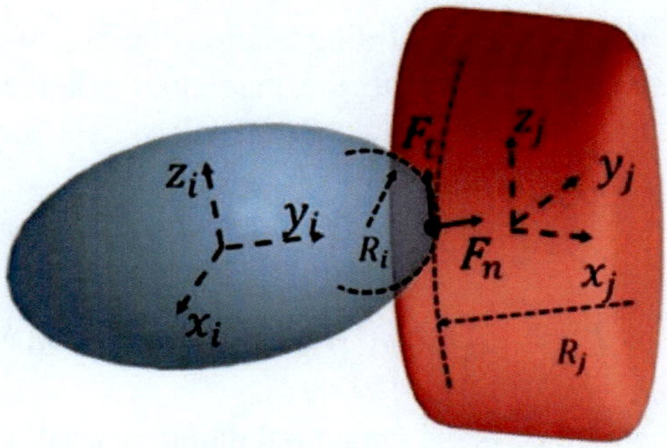

Fig. 2.7 Nonlinear force model considering the equivalent radius of curvature of the particles

which considers surface curvature, can also provide a good reference. Therefore, a corresponding nonlinear force model considering the equivalent radius of curvature is applied for superquadric elements, as shown in Fig. 2.7.

In this model, the normal force mainly includes the elastic force and the viscous force:

$$\boldsymbol{F}_{\mathrm{n}} = K_{\mathrm{n}}\boldsymbol{\delta}_{\mathrm{n}}^{3/2} + C_{\mathrm{n}}A\boldsymbol{v}_{\mathrm{n},ij}, \tag{2.25}$$

where C_{n} and $\boldsymbol{v}_{\mathrm{n},ij}$ are the normal damping coefficient and the normal relative speed, respectively, and K_{n} and A are the normal equivalent stiffness and the viscous parameter, respectively, which are given by:

$$\begin{aligned} K_{\mathrm{n}} &= 4/3E^*(R^*)^{1/2} \\ A &= \left(8m^*E^*\left(R^*\delta_{\mathrm{n}}\right)^{1/2}\right)^{1/2}, \end{aligned} \tag{2.26}$$

where $E^* = \frac{E}{2(1-\nu^2)}$, $m^* = \frac{m_i \cdot m_j}{m_i + m_j}$, and $R^* = \frac{R_i \cdot R_j}{R_i + R_j}$, and E and ν are the Young's modulus and Poisson's ratio, respectively. R is the equivalent radius of curvature at a contact point with a maximum value of ten times the radius of the volume-equivalent sphere, and this maximum is the same as the value used in the previous study (Podlozhnyuk et al. 2016). The parameter R is obtained using $R = 1/K_{\mathrm{mean}}$, where K_{mean} is the local mean curvature, given by (Goldman 2005):

$$K_{\mathrm{mean}} = \frac{\nabla F^{\mathrm{T}} \cdot \nabla^2 F \cdot \nabla F - |\nabla F|^2\left(\frac{\partial^2 F}{\partial x^2} + \frac{\partial^2 F}{\partial y^2} + \frac{\partial^2 F}{\partial z^2}\right)}{2|\nabla F|^3}. \tag{2.27}$$

The tangential contact force also includes the elastic force $\left(\boldsymbol{F}_{\mathrm{t}}^{\mathrm{e}}\right)$ and the viscous force $\left(\boldsymbol{F}_{\mathrm{t}}^{\mathrm{d}}\right)$ that considers the Mohr–Coulomb law. Here, $\boldsymbol{F}_{\mathrm{t}}^{\mathrm{e}}$ is given by:

$$\boldsymbol{F}_{\mathrm{t}}^{\mathrm{e}} = \mu_{\mathrm{s}}\left|K_{\mathrm{n}}\delta_{\mathrm{n}}^{3/2}\right|\left(1 - \left(1 - \min\left(\delta_{\mathrm{t}}, \delta_{\mathrm{t,max}}\right)/\delta_{\mathrm{t,max}}\right)^{3/2}\right), \tag{2.28}$$

where μ_s is the sliding friction coefficient, and δ_t is the total tangential displacement, which is added to the tangential relative displacement in each DEM step, given by $\delta_t = \delta_t + \boldsymbol{v}_{t,ij} \cdot dt$. Here, $\boldsymbol{v}_{t,ij}$ is the tangential relative speed, and $\delta_{t,\max} = \mu_s(2-\nu)/2(1-\nu) \cdot \delta_n$.

The tangential viscous force $\boldsymbol{F}_t^d$ is given by:

$$\boldsymbol{F}_t^d = C_t\left(6\mu_s m^*\left|K_n\delta_n^{3/2}\right|\left(1-\min\left(\delta_t, \delta_{t,\max}\right)/\delta_{t,\max}\right)^{1/2}/\delta_{t,\max}\right)^{1/2} \cdot \boldsymbol{v}_{t,ij}, \quad (2.29)$$

where C_t is the tangential damping coefficient.

When the particles rotate relative to each other, the moment $\left(\boldsymbol{M}_{r,ij}\right)$ caused by rolling friction can be expressed as:

$$\boldsymbol{M}_r = \mu_r R_i |\boldsymbol{F}_n| \hat{\boldsymbol{\omega}}_{ij}^n, \quad (2.30)$$

where μ_r is the rolling friction coefficient, and $\hat{\boldsymbol{\omega}}_{ij}^n$ is the normal relative rotating speed, given by $\hat{\boldsymbol{\omega}}_{ij}^n = \boldsymbol{\omega}_{ij}^n / \left|\boldsymbol{\omega}_{ij}^n\right|$.

2.3 Numerical Validation and Applications

In this section, quantitative verification and validation studies are performed, including a flat wall impacted by a single cylinder with different blockiness parameters, the presence of the arch structure of cubes, and the dynamic hopper discharge of ellipsoids. Moreover, the effects of the particle shape and base angle on the discharge rate are analyzed to obtain a fundamental understanding of particle flow during the discharging process.

2.3.1 Collision of Two Ellipsoidal Particles

To validate particle-boundary contact detection, the DEM simulations and theoretical results (Kodam et al. 2010) of a single cylinder impacting a flat wall are compared, as shown in Fig. 2.8. Although superquadric elements can construct cylinder-like particles, there is a difference between this shape approximation and a real cylinder. Figure 2.9a shows that six particles with different blockiness were modeled by changing the parameter n_1/n_2 ($n_2 = 2$). Meanwhile, gravity and friction are not considered during the impact process. The post-impact angular speed $\left(\omega_y^+\right)$ and translational speed $\left(V_z^+\right)$ for different impact angles are given by (Kodam et al. 2010):

$$\omega_y^+ = mV_z^-(1+\varepsilon)\cos(\phi_0+\theta_0)/\left(I_{yy} + mr^2\cos^2(\phi_0+\theta_0)\right), \quad (2.31)$$

$$V_z^+ = \omega_y^+ r \cos(\phi_0 + \theta_0) - \varepsilon V_z^-, \tag{2.32}$$

where m is the mass of the cylinder, V_z^- is the pre-impact translational speed of the cylinder, ε is the rebound coefficient at the contact point, ϕ_0 is the angle between the surface of the cylinder and the line connecting the contact point and the center of the cylinder, r is the distance between the contact point and the center of the particle, θ_0 is the angle between the cylinder surface and the wall, and I_{yy} is the mass moment of inertia about the y-axis. For each impact angle (θ_0), the initial pre-impact translational velocity $\left(V_z^-\right)$ and the rebound coefficient (ε) are constant at 1 m/s and 0.85, respectively.

For the various impact angles, the dimensionless rebound translational speed $\left(V_z^+/V_z^-\right)$ and angular speed $\left(r\omega_y^+/V_z^-\right)$ were calculated by DEM and compared with the analytical results (Kodam et al. 2010), as shown in Fig. 2.9b, c. As the blockiness parameter increases, the numerical results become closer to the theoretical results. When parameter n_1 is greater than eight, the numerical results do not change

Fig. 2.8 Scheme of cylinder–wall impact, in the xz plane

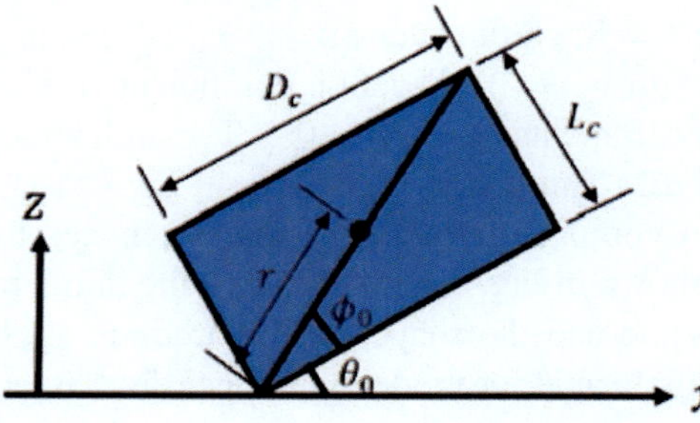

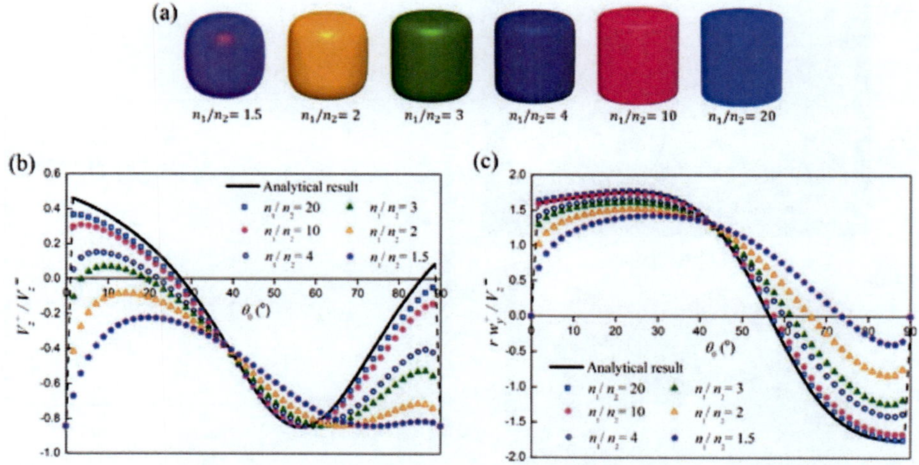

Fig. 2.9 Comparison between analytical results (Kodam et al. 2010) and DEM simulation results for cylinder–wall impact: **a** six particles with different blockiness, **b** the dimensionless post-impact translational velocity, V_z^+/V_z^-, and **c** the dimensionless post-impact angular velocity, $r\omega_y^+/V_z^-$

significantly with increasing blockiness. In addition, the point contact does not occur at the impact angles of 0 and 90° for the true cylinder. Therefore, for the theoretical solution, the translational rebound speed is only related to rebound coefficient ε, and the angular speed is zero. For the DEM simulations, the dimensionless translational rebound speed is 0.85, and the angular speed is zero. The results of the superquadric simulation agree well with the theoretical results. It is also shown that the blockiness has a significant influence on the macroscopic mechanical response of non-spherical particles.

2.3.2 Experimental Validation of Hopper Discharge

The observed discharge dynamics are typical of granular flow, and simulation using superquadric elements was carried out for validation. The first test is the presence of arch structure of cube-like particles under the same conditions (Fraige et al. 2008), i.e., particle side length $2a = 2b = 2c = 12.7\,\text{mm}$; blockiness parameter $n_1 = n_2 = 8$; particle density $\rho_p = 1200\,\text{kg/m}^3$; Young's modulus $E_p = 1\,\text{GPa}$; Poisson ratio $\nu_p = 0.3$; friction coefficient $\mu = 0.4$; and normal and tangential damping coefficients $c_n = c_s = 0.3$. The arch structures observed experimentally and in the DEM simulation are shown in Fig. 2.10a, b, where the orifice length was 80 mm. A comparison of the flow rate between the experimental and simulation results is shown in Fig. 2.10c. Although the initial packing structures of the cubes are not the same and the shapes of the arch are slightly different, the present algorithms are applicable for modeling hopper discharge and arching of non-spherical particles.

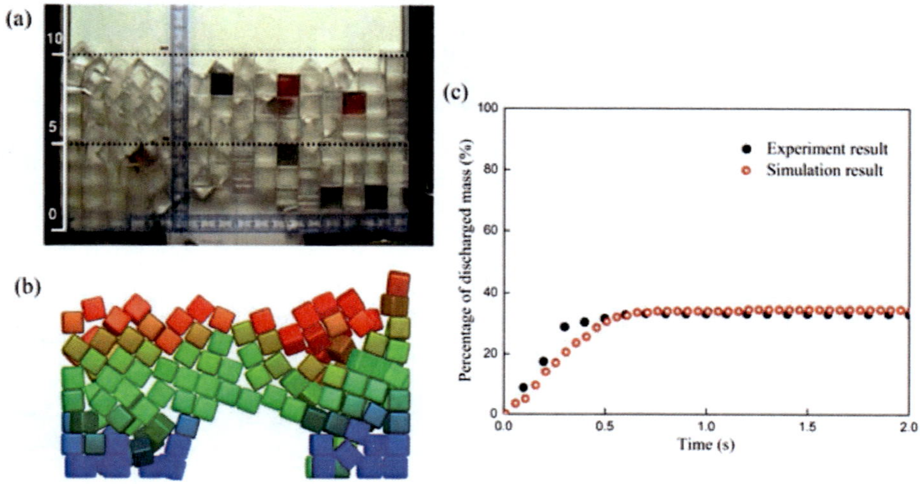

Fig. 2.10 Comparison between the experimental results (Fraige et al. 2008) and simulation results for cubes: **a** experimental arch structure, **b** DEM simulation of the arch structure, and **c** the percentage of discharged mass as a function of time

Table 2.1 Major computational parameters for the dynamic discharge of ellipsoids

Definition	Value	Definition	Value
Particle density (kg/m^3)	1338	Normal damping coefficient	0.3
Young's modulus (GPa)	1.0	Tangential damping coefficient	0.3
Poisson's ratio	0.29	Sliding friction coefficient	0.4
Time step (s)	5×10^{-6}		

The second test is the discharging of ellipsoids from a hopper. Real Smarties chocolate candies from a previous study (Liu et al. 2014) can be modeled by superquadric elements that have parameters of $2a = 13.56$ mm, $2b = 2c = 7.19$ mm, $n_1 = 2$, and $n_2 = 2$. The main calculation parameters are listed in Table 2.1. In the DEM simulation, the hopper was approximately 290 mm in length and 55 mm in width, and two orifice diameters were used: $D_0 = 54$ and 63 mm. A total of 10,500 particles with random initial positions and orientations were dropped into this container, and the orifice was not opened until all the particles had become motionless in the hopper.

Figure 2.11 shows the experimental (Liu et al. 2014) and simulation snapshots of hopper discharge at different moments. When the orifice was opened, a V-shaped flow pattern gradually appeared because of the difference in the flow rate of the particles on both sides and at the middle of the hopper. Finally, this pattern almost disappeared, and the remaining particles followed the shape of the slope. Figure 2.12 shows a comparison of the percentage of particles remaining in the hopper between the experimental (Liu et al. 2014) and simulation results. For the orifice having $D_0 = 54$ mm, good agreement was observed. For the orifice having $D_0 = 63$ mm, the percentage obtained from simulation is lower than that of experiment. The reason for this discrepancy may be that the friction coefficient and the damping coefficient have a more significant effect on the granular flow when the orifice becomes larger and the flow rate becomes faster. However, these parameters should better reflect the mechanical properties of the experimental material.

2.3.3 Analysis of the Flow Process of Granular Materials in a Silo

Knowledge of the effect of particle shape on the discharge rate is crucial for the reliable design and application of hoppers in industrial processes. Therefore, the aforementioned method was further tested for the simulation of hopper discharge. In these simulations, the hopper container consisted of an upper base and a lower hopper and measured $11D$ in length and $4D$ in width. Periodic boundary conditions were applied in width. The orifice was always $3.6D$ in length and $4D$ in width, and the base angles of $\theta = 15°, 30°, 45°, 60°$, and $75°$ were used for differently shaped particles to demonstrate the superposed effect of the hopper configuration. Meanwhile, the total

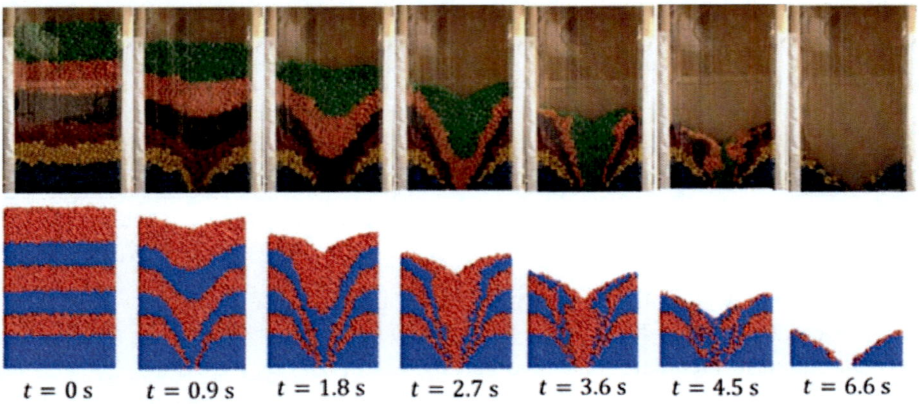

Fig. 2.11 Snapshots of the flow pattern between the experiment (Liu et al. 2014) (top) and simulation (bottom) at different moments

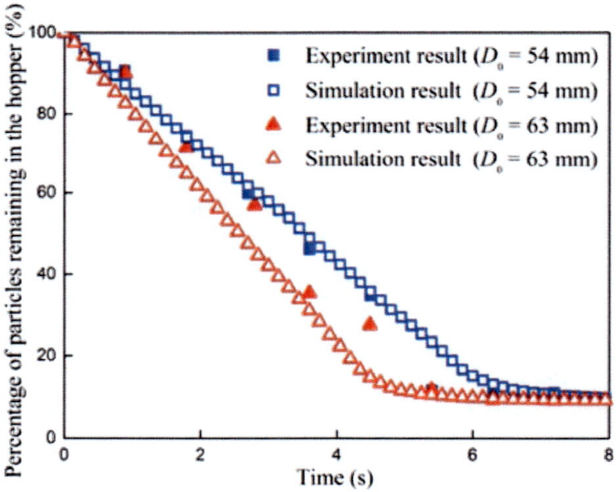

Fig. 2.12 Experimental (Liu et al. 2014) and simulated results of the percentage of particles remaining in the hopper when the orifice diameters are 54 and 63 mm

mass of the particles remained unchanged, approximately 0.1 kg. The superquadric elements have parameters of $D = 2b = 2c = 4\,\text{mm}$ and the aspect ratio given by $\alpha = 2a/D$. The shapes considered are ellipsoids, cylinders, and cubes with aspect ratios varying from 0.5 to 1.5, as shown in Fig. 2.13. Furthermore, the particle angularity increases with increasing blockiness parameter from 2 to 8. Given the geometric characteristics of the superquadric center symmetry, a quarter of the outline of the superquadric elements is shown in Fig. 2.14a. The shape of the particles can then be changed progressively between the spheres, cylinders, and cubes using a fixed aspect ratio of 1.0, as shown in Fig. 2.14b. The particle density, the coefficient of friction, and the DEM time step are 2500 kg/m^3, 0.1, and 1×10^{-6} s, respectively. The particles with random initial positions and orientations were dropped into the hopper to form a bed under gravity, and the orifice was not opened until all the particles had become motionless. In addition, it is necessary to phenomenologically observe the discharge process. Figure 2.15 shows a comparison of the discharging of spheres within a bed having $\theta = 30°$, ellipsoids within a bed having $\theta = 45°$, cylinders within a bed having $\theta = 60°$, and cubes within a bed having $\theta = 75°$ at 0, 0.5, and 1.0 s.

Figure 2.16 shows a comparison of the flow characteristics of the spheres (a), ellipsoids (b), cylinders (c), and cubes (d) within beds having $\theta = 15°$, 30°, 45°, 60°, and 75°. The percentage of discharged mass increased with time almost linearly for the different particle shapes and base angles, and the flow rate of the particles depended on both the particle shape and the base angle. Figure 2.17 shows the relationship between the average mass flow rate and the base angle for differently shaped particles. When the base angle was less than 45°, the discharge rate increased slightly with increasing base angle, and the particle shape appeared to be the primary

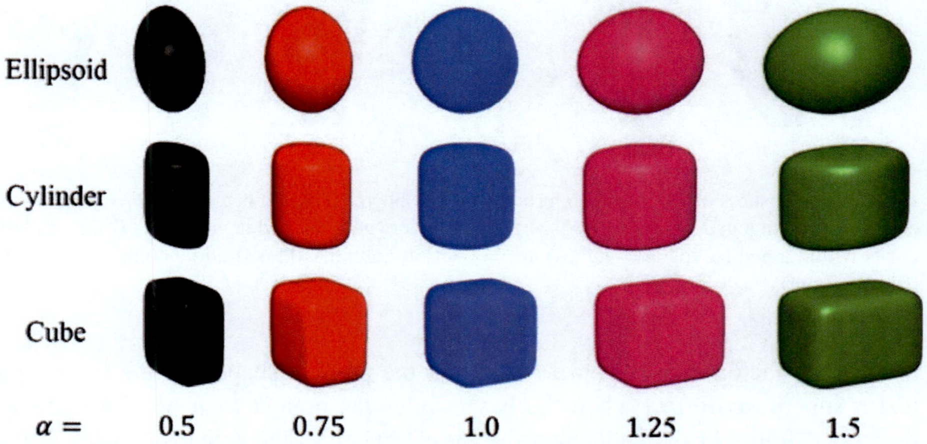

Fig. 2.13 Examples of superquadric particles with different aspect ratios

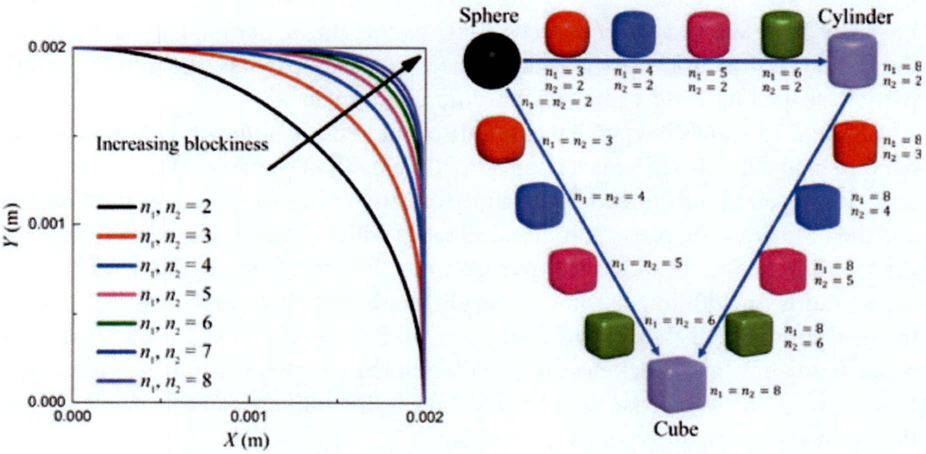

Fig. 2.14 Influence of blockiness parameters on the particle shapes

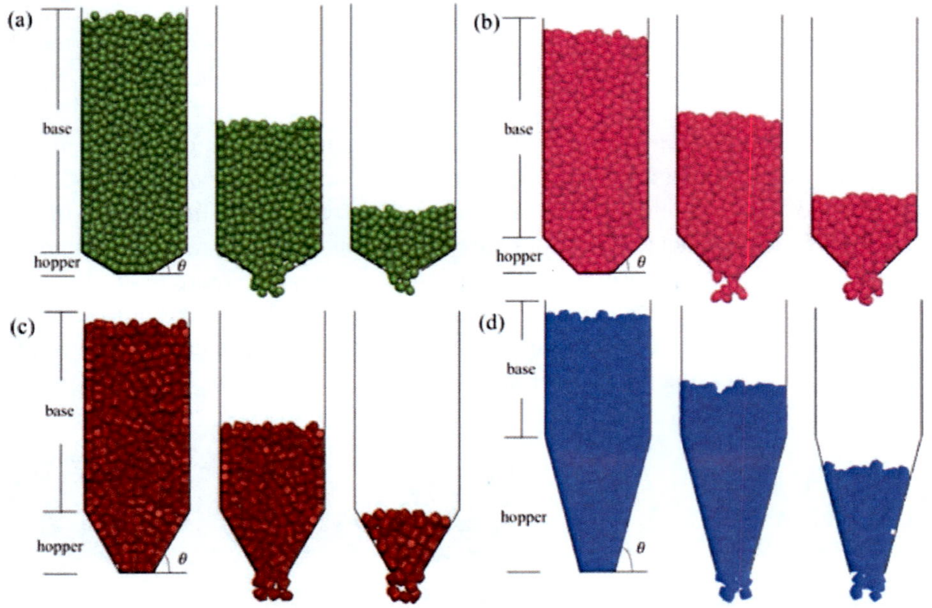

Fig. 2.15 Snapshots of the discharge process of the spheres within a bed having $\theta = 30°$ (**a**), the ellipsoids within a bed having $\theta = 45°$ (**b**), the cylinders within a bed having $\theta = 60°$ (**c**), and the cubes within a bed having $\theta = 75°$ (**d**) at $t = 0$ s (left column), 0.5 s (middle column), and 1.0 s (right column)

influencing factor. However, we found that the particle shape and the base angle had a superposed effect when the base angle was greater than 45°. In addition, the "flowability" of the cubes was the lowest, whereas the spheres had the fastest discharge rate for different base angles; the ellipsoids had a faster discharge rate than the cylinders. Meanwhile, the effect of the particle shape on the discharge rate increased with increasing base angle. This is mainly because when the base angle becomes large and the orifice is constant, the height of the hopper increases and the residence time of the particles in the hopper increases; thus, the influence of the particle shape on the flow rate becomes more pronounced.

Figure 2.18 shows the percentage of discharged mass of the ellipsoids (a), cylinders (b), and cubes (c) in beds having $\theta = 30°$ and 60°. The percentage of discharged mass increased almost linearly with time for different aspect ratios of the particles, and the discharge rate of the particles decreased with increasing the aspect ratio from 0.5 to 1.5. Figure 2.19 shows the average mass flow rate for particles with different aspect ratios. In addition, as shown, the ellipsoids were discharged fastest, whereas the discharge rate of the cubes was the lowest. For the same aspect ratio, the effect of the base angle and blockiness on the discharge rate is significant and cumulative. However, for the same base angle or blockiness, the influence of the aspect ratio on the discharge rate is of secondary importance.

Figure 2.20 shows a comparison of flow characteristics in the processes of changing from spheres to cylinders (a, b), cylinders to cubes (c, d), and spheres

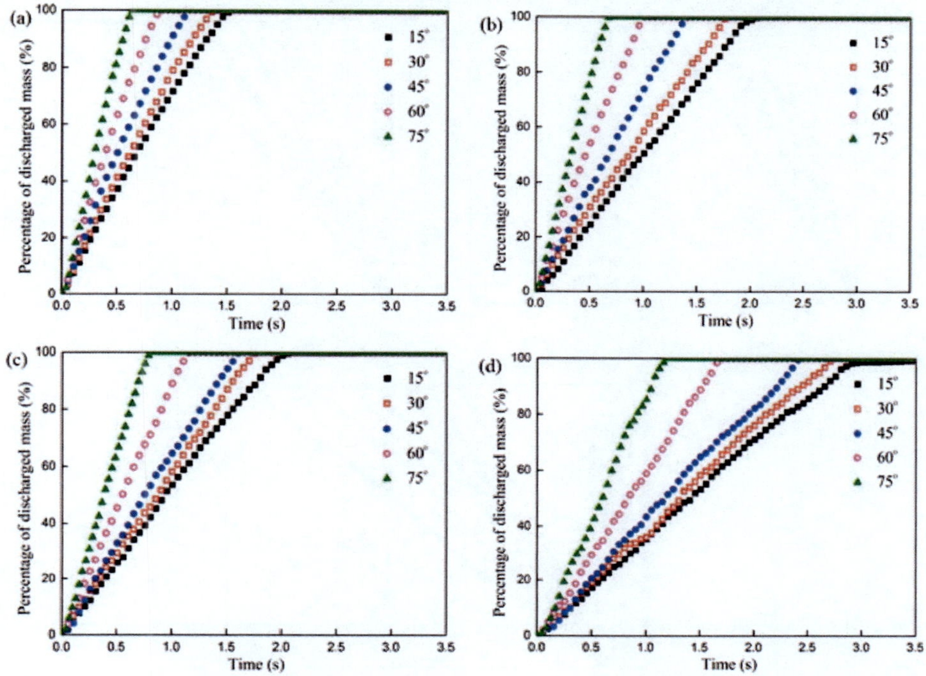

Fig. 2.16 Percentage of discharged mass as a function of time for the spheres (**a**), ellipsoids (**b**), cylinders (**c**), and cubes (**d**) within beds having different base angles

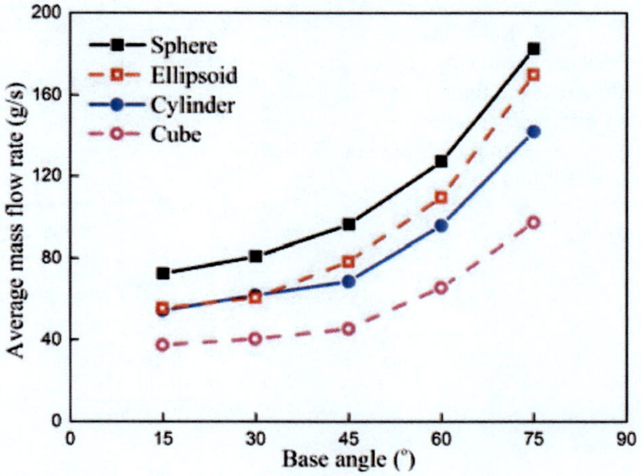

Fig. 2.17 Average mass flow rates for different shapes and different base angles

to cubes (e, f) within beds having $\theta = 30°$ (a, c, e) and 60° (b, d, f). The percentage of discharged mass increased almost linearly with time for different blockiness, and the discharge rate always decreased with increasing blockiness. Meanwhile, the flow of cube-like particles exhibited greater volatility than the flow of spheres and cylinders. As the blockiness increases, the contact pattern between the particles becomes

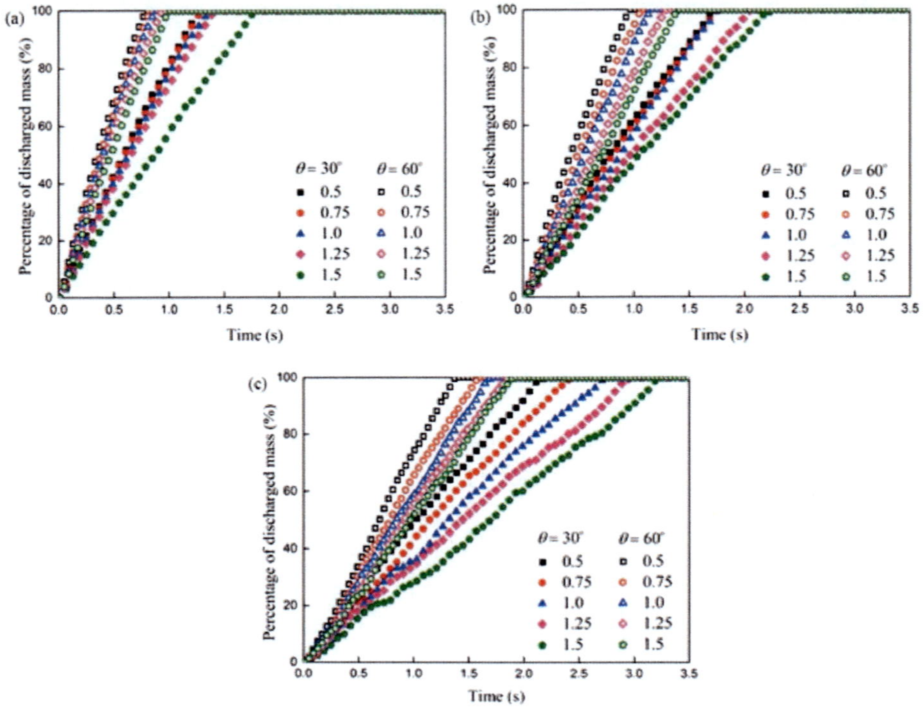

Fig. 2.18 Percentage of discharged mass as a function of time for different aspect ratios of the ellipsoids (**a**), cylinders (**b**), and cubes (**c**) within beds having $\theta = 30°$ and $60°$

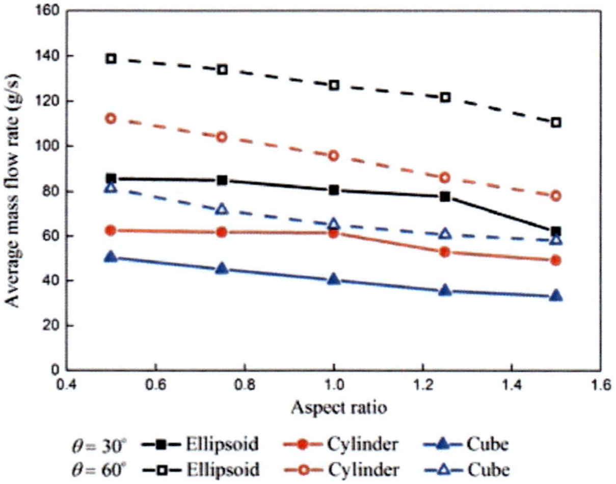

Fig. 2.19 Relationship between the average mass flow rate and the aspect ratio of differently shaped particles within beds having $\theta = 30°$ and $60°$

the face–face contact, which results in a more stable particle system and more non-continuous particle flow. Figure 2.21 shows the average mass flow rate of particles with different blockiness within beds having $\theta = 30°$ (a) and $60°$ (b). The influence of the blockiness on the discharge rate was greater when a base angle of 60° was used. However, this effect became insignificant when the blockiness parameter was greater than six.

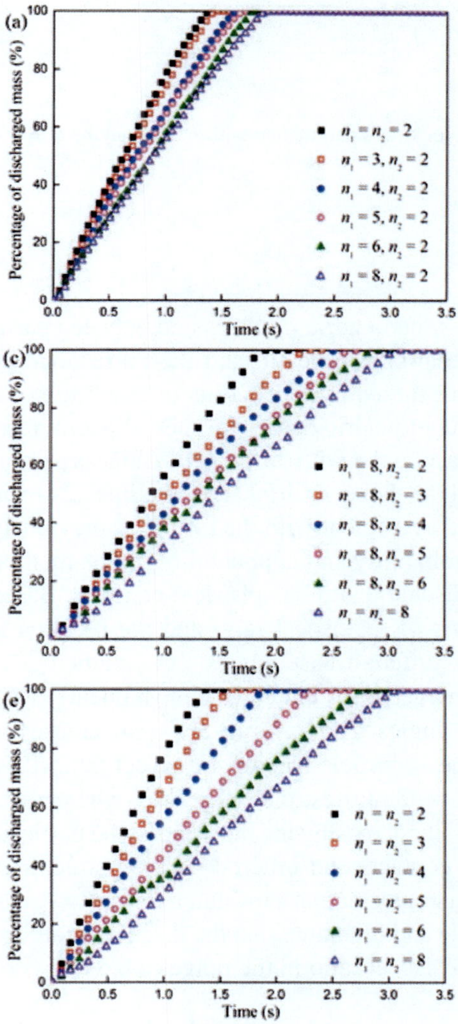

Fig. 2.20 Time variation of the percentage of discharged mass for different processes changing from spheres to cylinders (**a**, **b**), cylinders to cubes (**c**, **d**), and spheres to cubes (**e**, **f**) within beds having $\theta = 30°$ (**a**, **c**, **e**) and 60° (**b**, **d**, **f**)

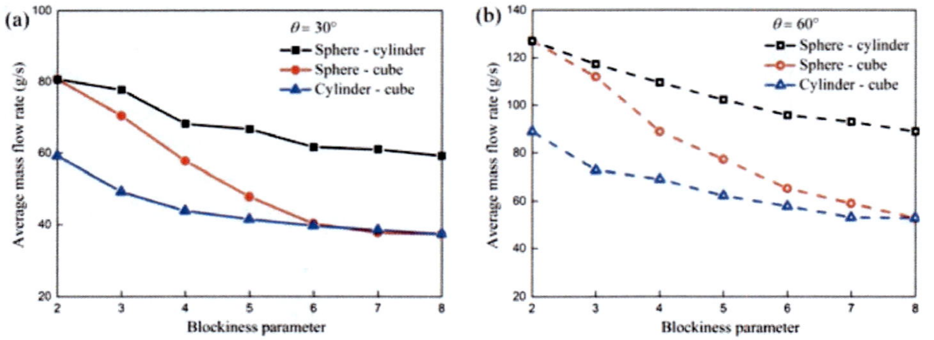

Fig. 2.21 Average mass flow rates for different blockiness parameters within beds having $\theta = 30°$ (**a**) and 60° (**b**)

2.4 Summary

In this chapter, contact detection between superquadric particles and between particles and triangular elements has been described in detail, and the corresponding nonlinear contact force model considering the equivalent radius of curvature at a local contact point has been applied for superquadric particles. The validation tests consisted of a flat wall impacted by a single cylinder with different blockiness parameters, the presence of arches of cubes, and the dynamic discharge of ellipsoids from a hopper. The simulated results show good agreement with the results of previously reported experiments, which verifies the reliability and applicability of this method and force model for the dynamic hopper discharge of non-spherical particles. Based on the aforementioned method, the effects of the aspect ratio and the blockiness on the discharge rate within beds having different base angles were studied. The results showed that the percentage of discharged mass increases almost linearly with time for different particle shapes and base angles. The flow rate always decreases as the base angle decreases or as the blockiness, particle friction, or aspect ratio (0.5–1.5) increases. Meanwhile, the spheres have the fastest discharge rate, whereas the discharge rate of the cubes is the lowest. With increasing the blockiness, the particles are more likely to form dense face–face contacts and ordered packing structures, which results in an increasingly non-continuous granular flow and a lower flow rate. In addition, the influences of the base angle and blockiness on the discharge rate are significant and cumulative; in contrast, the aspect ratio in the range of 0.5–1.5 is of secondary importance.

References

Barr AH (1981) Superquadrics and angle-preserving transformations. IEEE Comput Graph Appl 1:11–23

Chakraborty N, Peng J, Akella S, Mitchell JE (2008) Proximity queries between convex objects: an interior point approach for implicit surfaces. IEEE Trans Robot 24:211–220

Cleary PW (2004) Large scale industrial DEM modeling. Eng Comput 21:169–204

Cleary PW, Sawley ML (2002) DEM modelling of industrial granular flows: 3D case studies and the effect of particle shape on hopper discharge. Appl Math Model 26:89–111

Cundall PA, Strack ODL (1979) A discrete numerical mode for granular assemblies. Géotechnique 29:47–65

Dong K, Wang C, Yu A (2015) A novel method based on orientation discretization for discrete element modeling of non-spherical particles. Chem Eng Sci 126:500–516

Fraige FY, Langston PA, Chen GZ (2008) Distinct element modelling of cubic particle packing and flow. Powder Technol 186:224–240

Galindo-Torres SA, Pedroso DM (2010) Molecular dynamics simulations of complex-shaped particles using Voronoi-based spheropolyhedra. Phys Rev E 81:061303

Gan JQ, Zhou ZY, Yu AB (2017) Interparticle force analysis on the packing of fine ellipsoids. Powder Technol 320:610–624

Goldman R (2005) Curvature formulas for implicit curves and surfaces. Comput Aided Geom Des 22:632–658

Höhner D, Wirtz S, Scherer V (2013) Experimental and numerical investigation on the influence of particle shape and shape approximation on hopper discharge using the discrete element method. Powder Technol 235:614–627

Houlsby GT (2009) Potential particles: a method for modelling non-circular particles in DEM. Comput Geotech 36:953–959

Hu L, Hu GM, Fang ZQ, Zhang Y (2013) A new algorithm for contact detection between spherical particle and triangulated mesh boundary in discrete element method simulations. Int J Numer Meth Eng 94:787–804

Kildashti K, Dong K, Samali B (2018) A revisit of common normal method for discrete modelling of non-spherical particles. Powder Technol 326:1–6

Kodam M, Bharadwaj R, Curtis J, Hancock B, Wassgren C (2010) Cylindrical object contact detection for use in discrete element method simulations, part II—experimental validation. Chem Eng Sci 65:5863–5871

Kremmer M, Favier JF (2001) A method for representing boundaries in discrete element modelling—part I: geometry and contact detection. Int J Numer Meth Eng 51:1407–1421

Kruggel-Emden H, Rickelt S, Wirtz S, Scherer V (2008) A study on the validity of the multi-sphere discrete element method. Powder Technol 188:153–165

Liu SD, Zhou ZY, Zou RP, Pinson D, Yu AB (2014) Flow characteristics and discharge rate of ellipsoidal particles in a flat bottom hopper. Powder Technol 253:70–79

Lu G, Third JR, Müller CR (2012) Critical assessment of two approaches for evaluating contacts between super-quadric shaped particles in DEM simulations. Chem Eng Sci 78:226–235

Lu G, Third JR, Müller CR (2015) Discrete element models for non-spherical particle systems: from theoretical developments to applications. Chem Eng Sci 127:425–465

Podlozhnyuk A, Pirker S, Kloss C (2016) Efficient implementation of superquadric particles in discrete element method within an open-source framework. Comput Part Mech 4:101–118

Wellmann C, Lillie C, Wriggers P (2008) A contact detection algorithm for superellipsoids based on the common-normal concept. Eng Comput 25:432–442

Williams JR, Pentland AP (1992) Superquadrics and modal dynamics for discrete elements in interactive design. Eng Comput 9:115–127

Zhong W, Yu A, Liu X, Tong Z, Zhang H (2016) DEM/CFD-DEM modelling of non-spherical particulate systems: theoretical developments and applications. Powder Technol 302:108–152

Zhou JW, Liu Y, Du CL, Liu SY (2017) Effect of the particle shape and swirling intensity on the breakage of lump coal particle in pneumatic conveying. Powder Technol 317:438–448
Zhou W, Huang Y, Ng TT, Ma G (2018) A geometric potential-based contact detection algorithm for egg-shaped particles in discrete element modeling. Powder Technol 327:152–162
Zhu HP, Zhou ZY, Yang RY, Yu AB (2007) Discrete particle simulation of particulate systems: theoretical developments. Chem Eng Sci 62:3378–3396
Zhu HP, Zhou ZY, Yang RY, Yu AB (2008) Discrete particle simulation of particulate systems: a review of major applications and findings. Chem Eng Sci 63:5728–5770

Chapter 3
Multi-superquadric and Poly-superquadric DEM Models

The superquadric equation is a general method for describing spherical and irregular particles in a mathematical sense, and this equation can be used to construct particle morphologies with different surface sharpness and aspect ratio. However, the particles constructed by the superquadric equation are geometrically symmetric and strictly convex, which is different from the actual particle shape and limits the further engineering application of the superquadric model. In recent years, the non-spherical discrete element method based on the combined particle method has been widely developed and applied. Tetrahedral particles are formed by combining several spheres, and local cluster structures are randomly distributed in granular materials (Zhao et al. 2019). Considering the geometric symmetry of the ellipsoid element, one-eighth of the eight ellipsoids are spliced together to form a poly-ellipsoid with a smooth surface. Compared with ellipsoids, the numerical results obtained by this poly-ellipsoid model on axial compression are closer to the experimental results (Zhang et al. 2018). Moreover, one-eighth of the eight superquadric elements are spliced to form a continuous and smooth poly-superquadric particle, and two kinds of LM (Levenberg–Marquardt)-GJK (Gilbert–Johnson–Keerthi) are used to calculate the contact forces between particles (Zhao and Zhao 2019). Two circular planes and a cylindrical surface are combined to form a true cylindrical particle (Kodam et al. 2010). In the simulation of a single cylinder impacting a plane, the numerical results are basically consistent with the analytical results. Multiple superquadric elements are combined to construct tablet-shaped particles, and the numerical results obtained by the multiple superquadric model are in good agreement with the experimental results in terms of the accumulation fraction and mixing process in the rotating drum (Liu and Zhao 2020). A variety of convex elements such as spheres, cylinders, and polyhedrons are randomly combined to describe arbitrarily concave particles, and the GJK (Gilbert–Johnson–Keerthi) algorithm is used to calculate the contact forces between concave particles (Rakotonirina et al. 2019).

This chapter introduces two non-spherical discrete element methods, including the multi-superquadric DEM model and the poly-superquadric DEM model. Among

S. Wang and S. Ji, *Computational Mechanics of Arbitrarily Shaped Granular Materials*, Springer Tracts in Mechanical Engineering,
https://doi.org/10.1007/978-981-99-9927-9_3

them, the multi-superquadric model is a combination of multiple superquadric elements, and there is overlap between the basic elements. This model can be used to construct convex and concave particles. A poly-superquadric particle is composed of eight one-eighth superquadric elements, and the eight basic elements satisfy the smooth and continuous governing equation of the particle surface. This model can be used to construct geometrically asymmetric and smooth convex particles. Both discrete element methods transform the contact problem between particles into the numerical problem of the minimum distance between superquadric elements, and use the Newton iterative algorithm to calculate the contact points between particles. Meanwhile, the quaternion method is used to describe the spatial angle of the two models and the conversion relationship between the local coordinate system and the global coordinate system. The reliability of multi-superquadric model and poly-superquadric model is verified by DEM simulations of a single particle impacting a plane, particle accumulation and discharging process in silo, and the corresponding numerical results are compared with analytical results.

3.1 Discrete Element Method for Multi-superquadric Elements

In this section, the multi-superquadric model is introduced in detail. This model can be used to construct convex and concave shapes, and it is verified based on the good agreement between the numerical and analytical results. Moreover, the multi-superquadric approach is applied to the dynamic discharging process within a hopper.

3.1.1 Functional Representation of Multi-superquadric Elements

A multi-superquadric particle is constructed by assembling several superquadric elements (Liu and Zhao 2020), and the shape of each basic element is controlled by the superquadric equation. The function of a multi-superquadric particle can be expressed as follows:

$$\left(\left|\frac{x}{a_\alpha}\right|^{n_{2\alpha}} + \left|\frac{y}{b_\alpha}\right|^{n_{2\alpha}}\right)^{n_{1\alpha}/n_{2\alpha}} + \left|\frac{z}{c_\alpha}\right|^{n_{1\alpha}} - 1 = 0, \tag{3.1}$$

where a_α, b_α, c_α, $n_{1\alpha}$, and $n_{2\alpha}$ are shape parameters of the αth superquadric equation. If a multi-superquadric particle is composed of N_s superquadric elements, then $5N_s$ shape parameters are required to represent the particle shape. Multi-superquadric particles of different shapes are shown in Fig. 3.1.

Fig. 3.1 Multi-superquadric model composed of different numbers of superquadric elements

3.1.2 Calculation of Mass and Moment of Inertia

In the DEM simulation of multi-superquadric elements, the mass and moment of inertia are critical parameters. Considering the large overlap between particles, the background grid was used to calculate the mass and moment of inertia. This approach eliminates the effects of overlapping regions on the mass calculation by accumulating the mass of all effective grids; subsequently, the real information of multi-superquadric elements is reflected, as shown in Fig. 3.2. The calculation region of the spatial grid can be obtained by the positional relationship between the basic superquadric elements and the radius of the bounding sphere of each basic element. The size of the cubic grid is one-eighth of the minimum radius of the bounding sphere of all basic superquadric elements. The center of each cubic grid is used to determine if it is inside a multi-superquadric. The mass of multi-superquadric elements can be obtained by:

$$M_{\mathrm{s}} = \sum_{nz=1}^{N_z} \sum_{ny=1}^{N_y} \sum_{nx=1}^{N_x} \rho l_{nx} l_{ny} l_{nz}, \tag{3.2}$$

where M_{s} is the mass, and ρ is the particle density. l_{nx}, l_{ny}, and l_{nz} are the lengths of the cubic grid in the x-, y-, and z-directions, respectively. N_x, N_y, and N_z are the total numbers of effective grids in the x-, y-, and z-directions, respectively.

Subsequently, the centroid of the particle can be expressed as:

$$C_x = \frac{1}{M_{\mathrm{s}}} \sum_{nz=1}^{N_z} \sum_{ny=1}^{N_y} \sum_{nx=1}^{N_x} \rho l_{nx} l_{ny} l_{nz} x_{\mathrm{g}}, \tag{3.3}$$

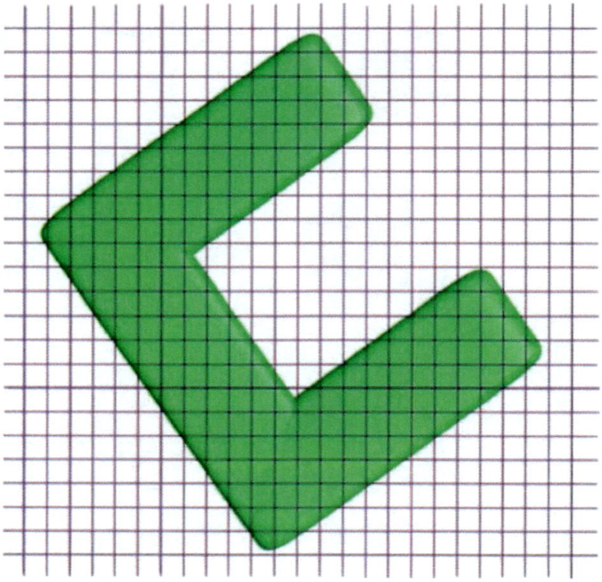

Fig. 3.2 Calculation method of mass and moment of inertia for multi-superquadric elements

$$C_y = \frac{1}{M_s} \sum_{nz=1}^{N_z} \sum_{ny=1}^{N_y} \sum_{nx=1}^{N_x} \rho l_{nx} l_{ny} l_{nz} y_g, \tag{3.4}$$

$$C_z = \frac{1}{M_s} \sum_{nz=1}^{N_z} \sum_{ny=1}^{N_y} \sum_{nx=1}^{N_x} \rho l_{nx} l_{ny} l_{nz} z_g, \tag{3.5}$$

where C_x, C_y, and C_z are the centroid coordinates of the particle in the x-, y-, and z-directions, respectively. x_g, y_g, and z_g are the center coordinates of the background grid in the x-, y-, and z-directions, respectively.

Moreover, the moment of inertia of the particle can be expressed as:

$$I_{xx} = \sum_{nz=1}^{N_z} \sum_{ny=1}^{N_y} \sum_{nx=1}^{N_x} \rho l_{nx} l_{ny} l_{nz} \left[(y_g - C_y)^2 + (z_g - C_z)^2 \right], \tag{3.6}$$

$$I_{yy} = \sum_{nz=1}^{N_z} \sum_{ny=1}^{N_y} \sum_{nx=1}^{N_x} \rho l_{nx} l_{ny} l_{nz} \left[(x_g - C_x)^2 + (z_g - C_z)^2 \right], \tag{3.7}$$

$$I_{zz} = \sum_{nz=1}^{N_z} \sum_{ny=1}^{N_y} \sum_{nx=1}^{N_x} \rho l_{nx} l_{ny} l_{nz} \left[(x_g - C_x)^2 + (y_g - C_y)^2 \right], \tag{3.8}$$

$$I_{xy} = I_{yx} = -\sum_{nz=1}^{N_z} \sum_{ny=1}^{N_y} \sum_{nx=1}^{N_x} \rho l_{nx} l_{ny} l_{nz} \left[(y_g - C_y)(z_g - C_z) \right], \tag{3.9}$$

$$I_{xz} = I_{zx} = -\sum_{nz=1}^{N_z}\sum_{ny=1}^{N_y}\sum_{nx=1}^{N_x} \rho l_{nx} l_{ny} l_{nz}\left[\left(x_g - C_x\right)\left(z_g - C_z\right)\right], \tag{3.10}$$

$$I_{yz} = I_{zy} = -\sum_{nz=1}^{N_z}\sum_{ny=1}^{N_y}\sum_{nx=1}^{N_x} \rho l_{nx} l_{ny} l_{nz}\left[\left(x_g - C_x\right)\left(y_g - C_y\right)\right]. \tag{3.11}$$

3.1.3 *Contact Detection Between Multi-superquadric Elements*

Considering the complexity of contact detection between multi-superquadric elements, the calculation method of the multi-sphere model is extended to the multi-superquadric approach. Therefore, a multi-superquadric element is divided into several superquadric elements; subsequently, the contact detection between the multi-superquadric elements is transformed into the contact detection between the superquadric elements. It is noteworthy that the contact pair is not established for two superquadric particles belonging to the same multi-superquadric element. Moreover, the positional relationship between the basic superquadric elements needs to be determined in advance for each multi-superquadric element. A potential contact pair between the basic element and an adjacent particle is not established if this basic element is completely inside another basic element.

In general, the bounding sphere and bounding box are employed to reduce the number of potential contact pairs between superquadric particles, thereby improving the computational efficiency of the algorithm (Wang et al. 2018). In addition, an accurate search algorithm between superquadric elements primarily includes the common normal method (Wellmann et al. 2008), geometric potential approach (Houlsby 2009), and midway point concept (Podlozhnyuk et al. 2016). Herein, the midway point algorithm is used to calculate the overlap between neighboring particles *i* and *j*. Therefore, the optimization problem between particles is transformed into nonlinear equations, which can be expressed as (Podlozhnyuk et al. 2016):

$$\begin{cases} \nabla F_i(\boldsymbol{X}) + \lambda^2 \nabla F_j(\boldsymbol{X}) = 0 \\ F_i(\boldsymbol{X}) - F_j(\boldsymbol{X}) = 0 \end{cases}, \tag{3.12}$$

where $\boldsymbol{X} = (x, y, z)^{\mathrm{T}}$ and λ is a multiplier. F_i and F_j are the superquadric equations of particles i and j, respectively. ∇F is the first derivative of the superquadric equation. Equation (3.12) can be solved using the Newton iterative method. If the midway point $\boldsymbol{X}_0$ satisfies $F_i(\boldsymbol{X}_0) < 0$ and $F_j(\boldsymbol{X}_0) < 0$, then particles i and j are in contact. The normal direction is obtained by $\boldsymbol{n} = \nabla F_i(\boldsymbol{X})/\|\nabla F_i(\boldsymbol{X})\|$, as shown in Fig. 3.3. Subsequently, the surface points of particles i and j can be expressed as $\boldsymbol{X}_i = \boldsymbol{X}_0 + \gamma \boldsymbol{n}$ and $\boldsymbol{X}_j = \boldsymbol{X}_0 + \beta \boldsymbol{n}$, respectively. The unknown parameters γ and β

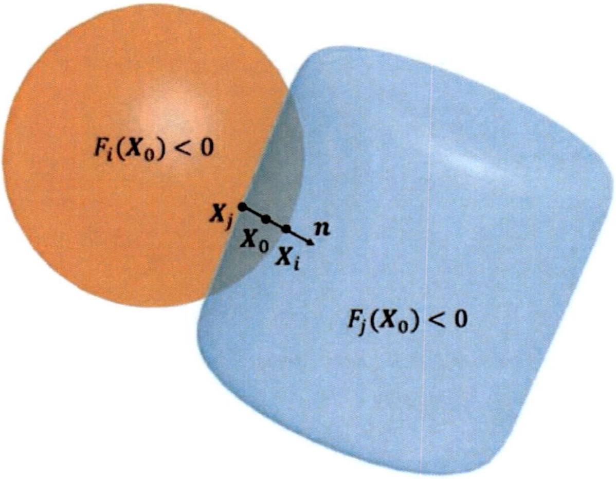

Fig. 3.3 Contact detection between superquadric particles

can be obtained using the Newton iterative method (Soltanbeigi et al. 2018): $\gamma^{(k+1)} = \gamma^{(k)} - F_i(\boldsymbol{X}_i^{(k)})/\left(\nabla F_i(\boldsymbol{X}_i^{(k)}) \cdot \boldsymbol{n}\right)$ and $\beta^{(k+1)} = \beta^{(k)} - F_j(\boldsymbol{X}_j^{(k)})/\left(\nabla F_j(\boldsymbol{X}_j^{(k)}) \cdot \boldsymbol{n}\right)$. Finally, the normal overlap can be obtained by $\boldsymbol{\delta}_{\mathrm{n}} = \boldsymbol{X}_i - \boldsymbol{X}_j$.

3.1.4 Position and Information Update of Elements

In non-spherical granular systems, the quaternion method is effective for describing the orientation of non-spherical particles and determining the transition relationship between the global and local coordinate systems. The transformation matrix $\boldsymbol{R}$ can be obtained by quaternions $\boldsymbol{Q}(q_0, q_1, q_2, q_3)$ and expressed as:

$$\boldsymbol{R} = \begin{bmatrix} q_0^2 + q_1^2 - q_2^2 - q_3^2 & 2(q_1q_2 + q_0q_3) & 2(q_1q_3 - q_0q_2) \\ 2(q_1q_2 - q_0q_3) & q_0^2 - q_1^2 + q_2^2 - q_3^2 & 2(q_2q_3 + q_0q_1) \\ 2(q_1q_3 + q_0q_2) & 2(q_2q_3 - q_0q_1) & q_0^2 - q_1^2 - q_2^2 + q_3^2 \end{bmatrix}. \tag{3.13}$$

Here, the transition relationship between the global coordinate system ($\boldsymbol{v}_{\mathrm{g}}$) and local coordinate system ($\boldsymbol{v}_{\mathrm{l}}$) can be obtained by:

$$\boldsymbol{v}_{\mathrm{l}} = \boldsymbol{R} \cdot \boldsymbol{v}_{\mathrm{g}} \tag{3.14}$$

In addition, the transformation matrix $\boldsymbol{R}$ exhibits the property of $\boldsymbol{R}^{\mathrm{T}} = \boldsymbol{R}^{-1}$. The total moment ($\boldsymbol{M}_{\mathrm{l},i}$) acting on the particle i in the local coordinate system can be obtained by $\boldsymbol{M}_{\mathrm{l},i} = \boldsymbol{R} \cdot \boldsymbol{M}_{\mathrm{g},i}$. Here, $\boldsymbol{M}_{\mathrm{g},i}$ is the total moment acting on the particle i in the global coordinate system. Therefore, the rotational velocity ($\boldsymbol{\omega}_{\mathrm{l},i}$) of particle i in the local coordinate system can be updated:

$$\begin{cases} \omega_{\mathrm{lx},i}^{(k+1)} = \omega_{\mathrm{lx},i}^{(k)} + \left[\boldsymbol{M}_{\mathrm{lx},i}^{(k)} + \left(I_{yy} - I_{zz}\right)\omega_{\mathrm{ly},i}^{(k)}\omega_{\mathrm{lz},i}^{(k)} \right] \cdot \mathrm{d}t / I_{xx} \\ \omega_{\mathrm{ly},i}^{(k+1)} = \omega_{\mathrm{ly},i}^{(k)} + \left[\boldsymbol{M}_{\mathrm{ly},i}^{(k)} + \left(I_{zz} - I_{xx}\right)\omega_{\mathrm{lz},i}^{(k)}\omega_{\mathrm{lx},i}^{(k)} \right] \cdot \mathrm{d}t / I_{yy} \\ \omega_{\mathrm{lz},i}^{(k+1)} = \omega_{\mathrm{lz},i}^{(k)} + \left[\boldsymbol{M}_{\mathrm{lz},i}^{(k)} + \left(I_{xx} - I_{yy}\right)\omega_{\mathrm{lx},i}^{(k)}\omega_{\mathrm{ly},i}^{(k)} \right] \cdot \mathrm{d}t / I_{zz} \end{cases}, \tag{3.15}$$

where the subscripts x, y, and z denote the components of the vector in the x-, y-, and z-directions, respectively. The updated quaternion can be expressed as:

$$\begin{bmatrix} q_{0,i} \\ q_{1,i} \\ q_{2,i} \\ q_{3,i} \end{bmatrix}^{(k+1)} = \begin{bmatrix} q_{0,i} \\ q_{1,i} \\ q_{2,i} \\ q_{3,i} \end{bmatrix}^{(k)} + 0.5\mathrm{d}t \cdot \begin{bmatrix} q_{0,i} & -q_{1,i} & -q_{2,i} & -q_{3,i} \\ q_{1,i} & q_{0,i} & -q_{3,i} & q_{2,i} \\ q_{2,i} & q_{3,i} & q_{0,i} & -q_{1,i} \\ q_{3,i} & -q_{2,i} & q_{1,i} & q_{0,i} \end{bmatrix}^{(k)} \begin{bmatrix} 0 \\ \omega_{lx,i} \\ \omega_{ly,i} \\ \omega_{lz,i} \end{bmatrix}^{(k+1)}. \tag{3.16}$$

Here, the quaternion update is relatively simple, and more detailed description of quaternion update can be found in Fritzer (2001). It is noteworthy that unlike the multi-sphere method, the multi-superquadric method involves two local coordinate systems (Wang and Ji 2022). The local coordinate systems of a multi-superquadric and superquadric element are separately established, as shown in Fig. 3.4. $\boldsymbol{R}_c$ is the transformation matrix between the global fixed and local coordinate systems of a multi-superquadric element. $\boldsymbol{R}_\alpha (\alpha = 1, 2, 3)$ is the transformation matrix between the local coordinate system of a multi-superquadric element and the local coordinate system of a superquadric element. $\boldsymbol{R}_{c\alpha} (\alpha = 1, 2, 3)$ is the transformation matrix between the global fixed and local coordinate systems of a superquadric element. $\boldsymbol{R}_\alpha$ is determined at the initial moment of particle generation and remains unchanged in the DEM simulations, while $\boldsymbol{R}_c$ is updated at each DEM time step and used to represent the orientation of the multi-superquadric element. $\boldsymbol{R}_{c\alpha}$ is determined by $\boldsymbol{R}_{c\alpha} = \boldsymbol{R}_\alpha \cdot \boldsymbol{R}_c$ and used for contact detection between superquadric elements.

3.2 Discrete Element Method for Poly-superquadric Elements

In this section, a poly-superquadric model that can construct asymmetrically shaped particles is introduced in detail. Moreover, eight bounding boxes are used for rough contact detection between basic elements, and the overlap between elements is calculated using the Newton iterative algorithm. Then, a hemisphere impacting a plane is simulated, and the numerical results are compared with the analytical results. Furthermore, the poly-superquadric model is used to simulate the dynamic accumulation and flow processes of the non-spherical particles.

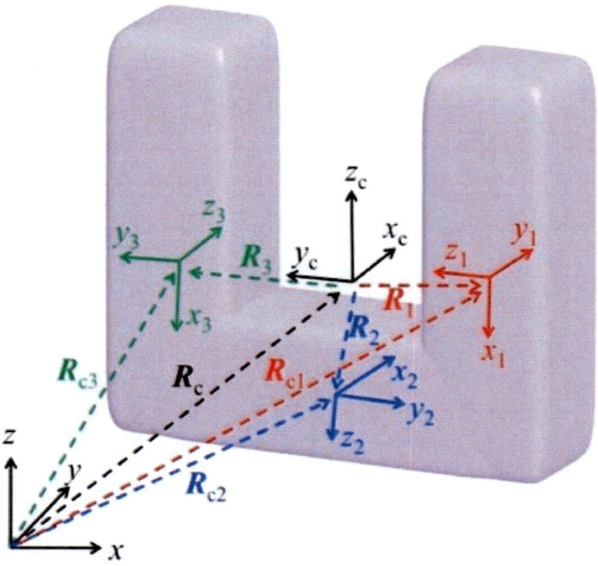

Fig. 3.4 Transition relationship between the global fixed coordinate system, the local coordinate system of the multi-superquadric element, and the local coordinate system of the superquadric element

3.2.1 Functional Representation of Poly-superquadric Elements

A poly-superquadric particle is composed of eight one-eighth superquadric elements, and Eq. (3.17) is used to determine each basic superquadric element. The function of a poly-superquadric element is expressed as (Zhao and Zhao 2019):

$$\left(\left|\frac{x}{a_k}\right|^{n_{2k}}+\left|\frac{y}{b_k}\right|^{n_{2k}}\right)^{n_{1k}/n_{2k}}+\left|\frac{z}{c_k}\right|^{n_{1k}}-1=0 \quad (k=1,2,3,\ldots,7,8), \tag{3.17}$$

where a_k, b_k, c_k, n_{1k}, and n_{2k} are the function parameters of the kth superquadric element. The value of k is equal to 8, and a poly-superquadric element is modeled by 40 function parameters. Meanwhile, the governing equation for the smooth and continuous particle surface can be expressed as:

$$a_1=a_4=a_5=a_8=r_x^+,\ (x>0), \tag{3.18}$$

$$a_2=a_3=a_6=a_7=r_x^-,\ (x<0), \tag{3.19}$$

$$b_1=b_2=b_5=b_6=r_y^+,\ (y>0), \tag{3.20}$$

$$b_3=b_4=b_7=b_8=r_y^-,\ (y<0), \tag{3.21}$$

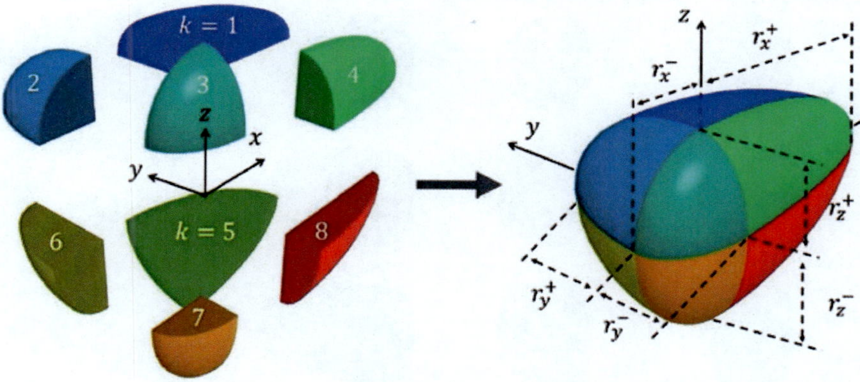

Fig. 3.5 A poly-superquadric element composed of eight one-eighth superquadric elements

$$c_1 = c_2 = c_3 = c_4 = r_z^+, (z > 0), \tag{3.22}$$

$$c_5 = c_6 = c_7 = c_8 = r_z^-, (z < 0), \tag{3.23}$$

$$n_{11} = n_{12} = n_{13} = n_{14} = n_{15} = n_{16} = n_{17} = n_{18} = n_1^*, \tag{3.24}$$

$$n_{21} = n_{22} = n_{23} = n_{24} = n_{25} = n_{26} = n_{27} = n_{28} = n_2^*, \tag{3.25}$$

where r_x^+, r_x^-, r_y^+, r_y^-, r_z^+, and r_z^- are the lengths of the poly-superquadric element in the positive and negative directions of the x, y, and z axes, respectively. n_1^* and n_2^* are the shape parameters of the poly-superquadric element. Therefore, only 8 parameters are needed to determine the shape of a poly-superquadric element. Figure 3.5 shows a poly-superquadric element composed of eight one-eighth superquadric elements. Moreover, poly-superquadric elements of different shapes are obtained by changing eight shape parameters, as shown in Fig. 3.6. Poly-ellipsoid elements are obtained if $n_1^* = n_2^* = 2$ (Zhang et al. 2018), and superquadric elements of different shapes are obtained if $r_x^+ = r_x^-$, $r_y^+ = r_y^-$, and $r_z^+ = r_z^-$.

3.2.2 Calculation of Mass and Moment of Inertia

The mass, mass center, and moment of inertia of a poly-superquadric element can be obtained by weighting the relevant parameters of eight one-eighth superquadric elements, and more detailed calculations can be found in Zhao and Zhao (2019). The background grid is used to simplify the calculation of the real information of the elements, as shown in Fig. 3.7. The size of each cube grid is one-eighth of the

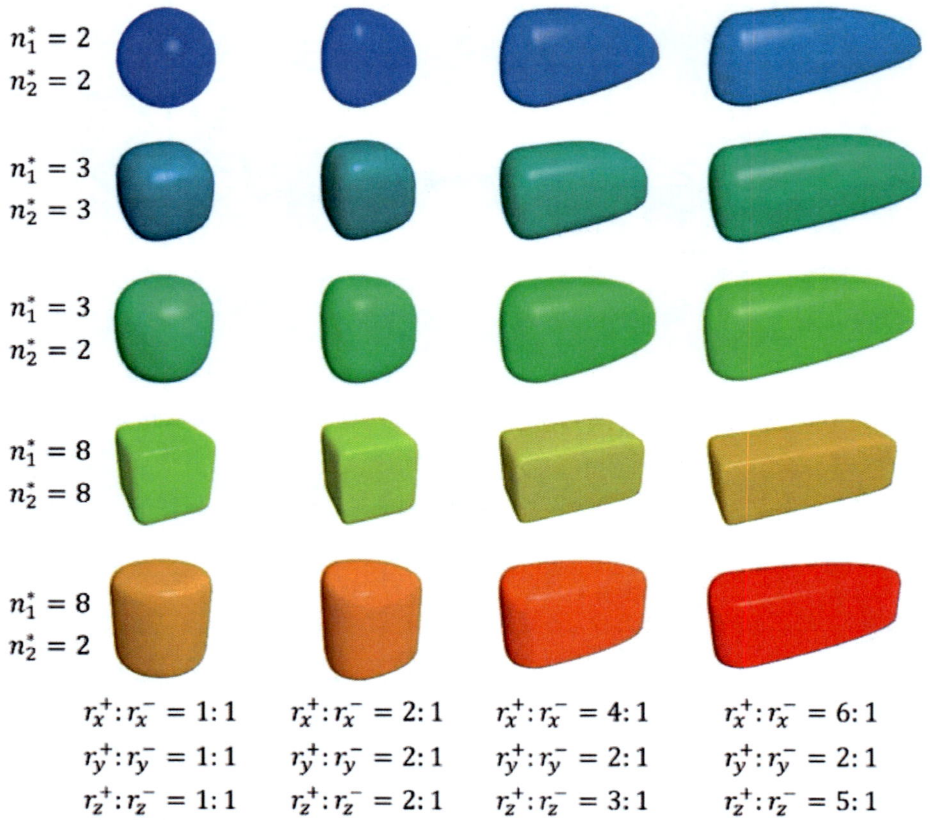

Fig. 3.6 Different shapes of particles represented by the poly-superquadric equation

minimum of the three semi-axis lengths of the eight superquadric elements. Subsequently, each effective grid is obtained through the positional relationship between the grid centroid and the element, and the mass of all effective grids is accumulated to represent the real information of the elements. Therefore, the mass of a poly-superquadric element can be expressed as

$$M_s = \sum_{nz=1}^{N_z} \sum_{ny=1}^{N_y} \sum_{nx=1}^{N_x} \rho l_{nx} l_{ny} l_{nz}, \tag{3.26}$$

where M_s and ρ are mass and particle density, respectively. l_n is the length of the grid, and N is the total number of effective grids. The subscripts x, y, and z indicate the components in the x-, y-, and z-directions, respectively. In addition, the calculation of the center of mass and moment of inertia of the poly-superquadric particle is the same as that of the multi-superquadric particle.

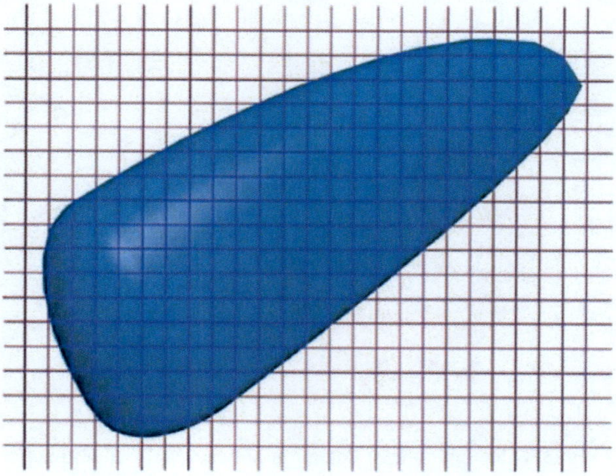

Fig. 3.7 Background grid method for calculating the mass and moment of inertia of poly-superquadric elements

3.2.3 Contact Detection Between Poly-superquadric Elements

A poly-superquadric element consists of eight one-eighth superquadric elements, and the contact algorithm of the poly-superquadric elements is transformed into the contact algorithm of the superquadric elements. Subsequently, the quadrant position of the contact point in the local coordinate system is used to determine whether the point is on the surface of the one-eighth superquadric element. Moreover, the search between elements is not performed for two basic superquadric elements belonging to the same poly-superquadric particle.

Generally, bounding spheres and orientated bounding boxes are used to reduce the number of contact pairs between elements and improve the computational efficiency of DEM simulations, as shown in Fig. 3.8. The search of the bounding sphere is performed as the first rough detection between poly-superquadric elements. If the distance between the centers of neighboring elements is less than the sum of the bounding radii, the two elements may be in contact. Subsequently, the oriented bounding box is used for the second rough detection between poly-superquadric elements. A poly-superquadric element has eight sub-bounding boxes, and a basic superquadric element is surrounded by one sub-bounding box. If the sum of the distances of the bounding boxes projected on the separating axis is less than the projection distance between the centroids of the two superquadric elements, the two elements are not in contact. More detailed information about bounding box calculations can be found in Eberly (2002).

Moreover, the accurate contact detection between superquadric elements has been well established, including the common normal approach (Wellmann et al. 2008), geometric potential concept (Houlsby 2009), and midway point method (Podlozhnyuk et al. 2016). Here, the overlap between two superquadric elements i and j is calculated by the midway point method. The nonlinear equations for calculating the midway point ($\boldsymbol{X}_0$) between the elements are expressed as (Soltanbeigi et al. 2018):

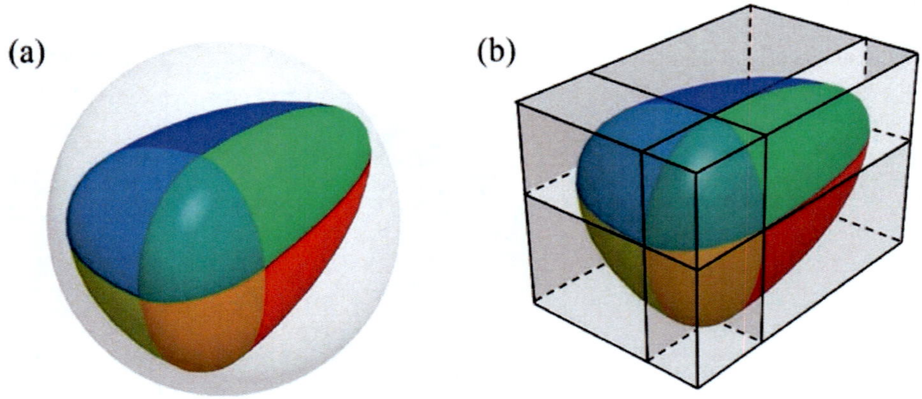

Fig. 3.8 Rough contact detection between poly-superquadric elements: **a** a bounding sphere and **b** an oriented bounding box

$$\begin{cases} \nabla F_i(\boldsymbol{X}) + \lambda^2 \nabla F_j(\boldsymbol{X}) = 0 \\ F_i(\boldsymbol{X}) - F_j(\boldsymbol{X}) = 0 \end{cases}, \tag{3.27}$$

where $\boldsymbol{X} = (x, y, z)^{\mathrm{T}}$ and λ is a multiplier. F_i and F_j are the functional equations of elements i and j, respectively. The Newton iteration algorithm can be used to solve nonlinear equations, expressed as (Podlozhnyuk et al. 2016):

$$\begin{pmatrix} \nabla^2 F_i(\boldsymbol{X}) + \lambda^2 \nabla F_j(\boldsymbol{X}) & 2\lambda \nabla F_j(\boldsymbol{X}) \\ \nabla F_i(\boldsymbol{X}) - \nabla F_j(\boldsymbol{X}) & 0 \end{pmatrix} \begin{pmatrix} \mathrm{d}\boldsymbol{X} \\ \mathrm{d}\lambda \end{pmatrix} = -\begin{pmatrix} \nabla F_i(\boldsymbol{X}) + \lambda^2 \nabla F_j(\boldsymbol{X}) \\ F_i(\boldsymbol{X}) - F_j(\boldsymbol{X}) \end{pmatrix}, \tag{3.28}$$

where $\boldsymbol{X}^{(k+1)} = \boldsymbol{X}^{(k)} + \mathrm{d}\boldsymbol{X}$ and $\lambda^{(k+1)} = \lambda^{(k)} + \mathrm{d}\lambda$. If the midway point $\boldsymbol{X}_0$ satisfies $F_i(\boldsymbol{X}_0) < 0$ and $F_j(\boldsymbol{X}_0) < 0$, element i may be in contact with element j, as shown in Fig. 3.9. Subsequently, the point $\boldsymbol{X}_0$ in the global coordinate system is transformed to the points $\boldsymbol{X}^1_{0,i}$ and $\boldsymbol{X}^1_{0,j}$ of elements i and j in the local coordinate system, respectively. If the quadrants of the points $\boldsymbol{X}^1_{0,i}$ and $\boldsymbol{X}^1_{0,j}$ are the same as the quadrants of the one-eighth superquadric elements i and j, the midway point $\boldsymbol{X}_0$ is the contact point between the elements i and j. The normal direction ($\boldsymbol{n}$) is obtained by $\boldsymbol{n} = \nabla F_i(\boldsymbol{X}_0)/\|\nabla F_i(\boldsymbol{X}_0)\|$ or $\boldsymbol{n} = -\nabla F_j(\boldsymbol{X}_0)/\left\|\nabla F_j(\boldsymbol{X}_0)\right\|$. Finally, the surface points $\boldsymbol{X}_i$ and $\boldsymbol{X}_j$ of elements i and j are expressed as $\mathbf{X}_i = \boldsymbol{X}_0 + \gamma\boldsymbol{n}$ and $\mathbf{X}_j = \boldsymbol{X}_0 + \beta\boldsymbol{n}$, respectively. The parameters γ and β are obtained by Newton's iterative algorithm: $\gamma^{(k+1)} = \gamma^{(k)} - F_i\left(\boldsymbol{X}_i^{(k)}\right)/\left(\nabla F_i(\boldsymbol{X}_i^{(k)}) \cdot \boldsymbol{n}\right)$ and $\beta^{(k+1)} = \beta^{(k)} - F_j\left(\boldsymbol{X}_j^{(k)}\right)/\left(\nabla F_j(\boldsymbol{X}_j^{(k)}) \cdot \boldsymbol{n}\right)$. Therefore, the normal overlap between elements can be obtained by $\boldsymbol{\delta}_{\mathrm{n}} = \mathbf{X}_i - \mathbf{X}_j$.

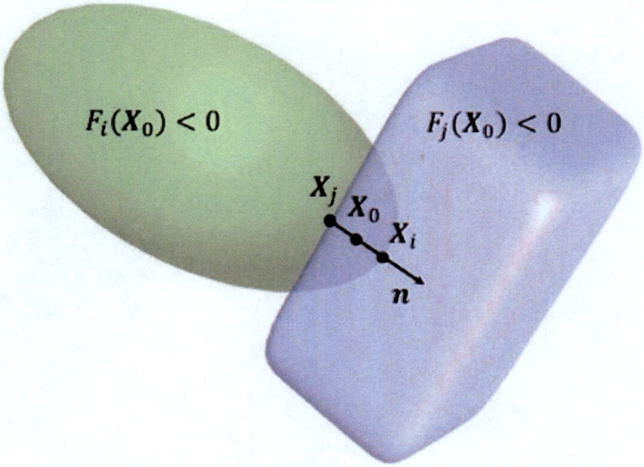

Fig. 3.9 Accurate contact detection between superquadric elements

3.2.4 Position and Information Update of Elements

For poly-superquadric elements, the quaternion method is used to describe the orientation of non-spherical elements. The transformation matrix $\boldsymbol{R}$ represented by the quaternion $\boldsymbol{Q}(q_0, q_1, q_2, q_3)$ can be expressed as:

$$\boldsymbol{R} = \begin{bmatrix} q_0^2 + q_1^2 - q_2^2 - q_3^2 & 2(q_1q_2 + q_0q_3) & 2(q_1q_3 - q_0q_2) \\ 2(q_1q_2 - q_0q_3) & q_0^2 - q_1^2 + q_2^2 - q_3^2 & 2(q_2q_3 + q_0q_1) \\ 2(q_1q_3 + q_0q_2) & 2(q_2q_3 - q_0q_1) & q_0^2 - q_1^2 - q_2^2 + q_3^2 \end{bmatrix}. \tag{3.29}$$

It is worth noting that the coordinate origin and centroid of the poly-superquadric element are different. At the initial moment, the positional relationship between the origin coordinate (O_0) and mass center (O_c) should be determined, as shown in Fig. 3.10. $\boldsymbol{R}_0$ is the transformation matrix of a poly-superquadric element, which is used to represent the orientation of the poly-superquadric element. The vector $\boldsymbol{v}_c^l$ directs from point O_c to point O_0 and remains unchanged in the DEM simulation. Furthermore, the centroid position ($\boldsymbol{X}_c$) of the poly-superquadric element in the global coordinate system is calculated at each DEM step. The origin coordinate ($\boldsymbol{X}_0$) is obtained by $\boldsymbol{X}_0 = \boldsymbol{X}_c + \boldsymbol{R}_0^T \cdot \boldsymbol{v}_c^l$ and used to update the position of the basic superquadric elements.

3.3 Numerical Validation and Applications

In this section, the validity of the multi-superquadric model and the poly-superquadric model is examined. The simulation result of a single particle impacting a flat wall is compared with the theoretical results, and then the influences of particle shape on the flow characteristics of the granular materials are investigated.

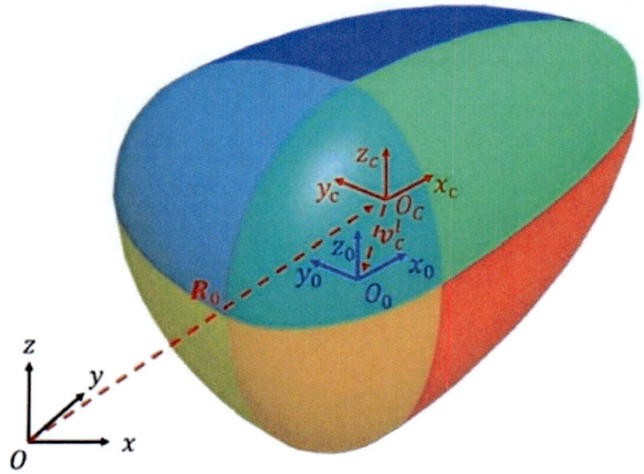

Fig. 3.10 Relationship between the mass center and origin coordinate of the poly-superquadric element

3.3.1 Theoretical Verification of a Single Particle Impacting a Plane

To validate the multi-superquadric algorithm, a single spherocylinder impacting a flat wall was simulated using the DEM. A spherocylinder is composed of a cylinder and two spheres, as shown in Fig. 3.11a. A cylinder-like particle has parameters of $a = b = c = 6\,\text{mm}$, $n_1 = 10$, and $n_2 = 2$. A sphere has parameters of $a = b = c = 6\,\text{mm}$ and $n_1 = n_2 = 2$. Therefore, a spherocylinder has parameters of $r_0 = c_0 = 6\,\text{mm}$. Moreover, no gravity or friction exists during the impact. The rebound angular velocity (w_y^+) and translational velocity (v_z^+) at different impact angles can be theoretically expressed as follows (Kodam et al. 2012):

$$w_y^+ = mv_z^-(1+\varepsilon)c_0\sin(\theta_0)/\left[I_{yy} + mc_0^2\sin^2(\theta_0)\right], \tag{3.30}$$

$$v_z^+ = w_y^+ c_0 \sin(\theta_0) - \varepsilon v_z^-, \tag{3.31}$$

where m is the mass of the spherocylidner; I_{yy} is the moment of inertia about the y-axis; ε is the coefficient of restitution that is defined as 0.85; θ_0 is the angle between any diametrical plane of the spherocylinder and wall; c_0 is the distance between the centroids of the cylinder and sphere; v_z^- is the initial translational velocity that is defined as − 1 m/s. The main calculation parameters are listed in Table 3.1.

The dimensionless rebound velocity (v_z^+/v_z^-) and rebound angular velocity ($r_0 w_y^+/v_z^-$) are obtained by numerical simulations and compared with the analytical results (Kodam et al. 2012), as shown in Fig. 3.11b–c. Although the spherocylinder model constructed by multi-superquadric elements has a slight difference from the real spherocylinder model, the DEM results agree well with the analytical results, which indicates that the multi-superquadric DEM model can be effectively used to simulate the dynamic behavior of complex-shaped particles.

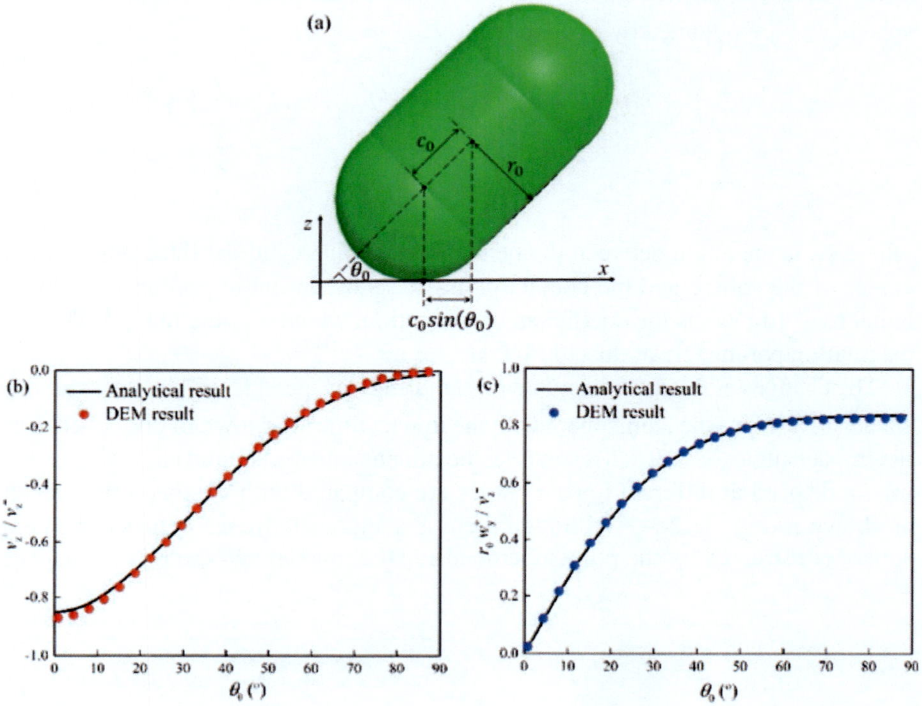

Fig. 3.11 Comparison between analytical results (Kodam et al. 2012) and numerical results for spherocylinder-wall impact: **a** schematics of a spherocylinder-wall contact, **b** dimensionless rebound translational speed, v_z^+/v_z^-, and **c** dimensionless rebound angular speed, $r_0 w_y^+/v_z^-$

Table 3.1 Major computational parameters of spherocylinder impact

Definitions	Value	Definitions	Value
Density (kg/m^3)	2500	Normal damping coefficient	0.04
Young's modulus (Pa)	1×10^9	Tangential damping coefficient	0.0
Poisson's ratio	0.3	Sliding friction coefficient	0.0

To verify the poly-superquadric DEM model, a hemisphere impacting a plane is simulated. An approximate hemisphere is represented by a poly-superquadric element, as shown in Fig. 3.12a. A poly-superquadric element has the parameters of $r_x^+ = r_y^+ = r_y^- = r_z^+ = r_z^- = r_0 = 6\,\text{mm}$, $r_x^- = 0.01\,\text{mm}$, and $n_1^* = n_2^* = 2$. Moreover, the distance (c_0) between the mass center (O_c) and the center (O_0) of the sphere is expressed as $c_0 = 3r_0/8$, and the density of the element is 2500 kg/m^3. The mass (m) and the y-axis moment $\left(I_{yy}\right)$ of inertia of the actual hemisphere are 1.131×10^{-3} kg and 1.056×10^{-8} (kg m^2), respectively, and the mass and the y-axis moment of inertia of the approximate hemisphere are 1.131×10^{-3} kg and 1.056×10^{-8} (kg m^2), respectively. Besides, gravity and particle–wall friction are not considered in the impact process. The rebound angular speed $\left(w_y^+\right)$ and translational speed $\left(v_z^+\right)$ are obtained by (Kodam et al. 2010):

$$w_y^+ = -mv_z^-(1+\varepsilon)c_0\cos(\varphi_0)/\left(I_{yy} + mc_0^2\cos^2(\varphi_0)\right), \tag{3.32}$$

$$v_z^+ = -w_y^+ c_0\cos(\varphi_0) - \varepsilon v_z^-, \tag{3.33}$$

where φ_0 is the angle between the hemispherical plane and the line connecting the center of the sphere and the contact point. v_z^- is the initial impact speed, which is equal to $-$ 1 m/s. ε is the coefficient of restitution, which is equal to 0.85. The main calculation parameters are listed in Table 3.1.

The changes of the dimensionless rebound angular speed $\left(r_0 w_y^+/v_z^-\right)$ and rebound speed $\left(v_z^+/v_z^-\right)$ of the hemisphere with the impact time are shown in Fig. 3.12b. Here, the impact angle $\varphi_0 = 30°$. Meanwhile, the dimensionless rebound angular speed and rebound speed at different impact angles are compared with the theoretical results, as shown in Fig. 3.12c–d. Although there is a slight difference between the hemisphere constructed by the poly-superquadric DEM model and the real hemisphere,

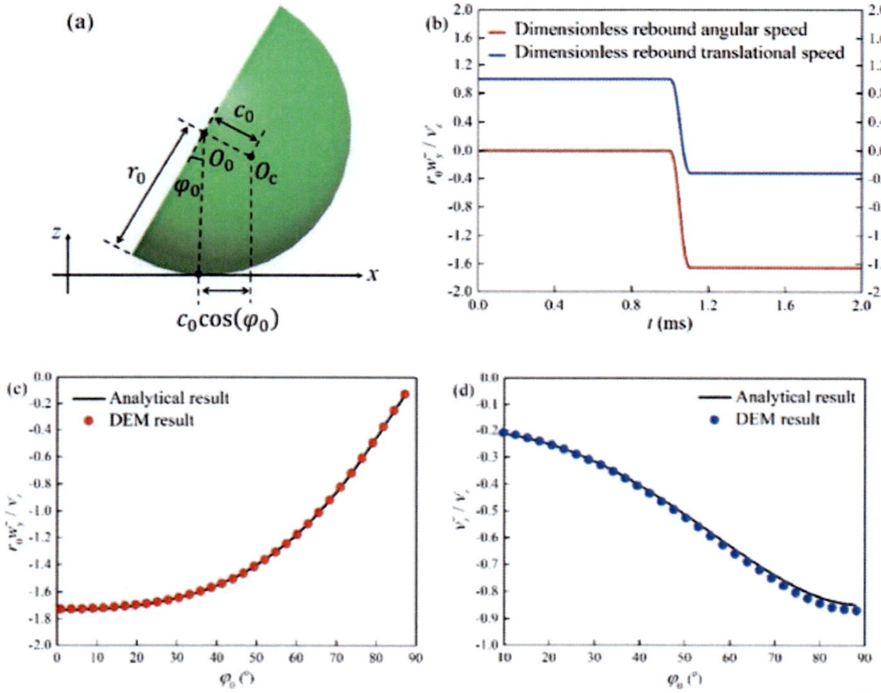

Fig. 3.12 Comparison between analytical results (Kodam et al. 2010) and DEM results for the hemisphere-wall impact: **a** sketch of the hemisphere-wall impact, **b** time variation of the dimensionless rebound angular speed and translational speed, **c** dimensionless rebound angular speed at different impact angles, $r_0 w_y^+/v_z^-$, and **d** dimensionless rebound translational speed at different impact angles, v_z^+/v_z^-

the DEM results are in good agreement with the analytical results (Kodam et al. 2010), which indicates that the aforementioned poly-superquadric DEM model is suitable for simulating the motion behavior of non-spherical particles.

3.3.2 DEM Simulation of Flow Processes of Multi-superquadric Elements

The particle shape significantly affects the macro- and micro-characteristics of granular materials. Therefore, multi-superquadric models were used to simulate the formation of the granular bed. Figure 3.13 shows the differently shaped particles constructed by superquadric and multi-superquadric elements. The ratio of the particle length (L), width (W), and height (H) is 1:1:1, and the ratio of the particle length (L) to the diameter (D) is 3:1. The volume of this multi-superquadric element can be obtained by the spatial grid method and can be denoted as V_{m0}. Meanwhile, the volume of a sphere with a diameter of 5 mm can be denoted as V_0. Generally, V_{m0} is not equal to V_0, and the volume ratio can be expressed as $\sigma_0 = (V_0/V_{m0})^{1/3}$. Then, the initial half-axis lengths (a_0, b_0, and c_0) of all basic superquadric elements need to be updated: $a_1 = a_0\sigma_0$, $b_1 = b_0\sigma_0$, and $c_1 = c_0\sigma_0$. The new positions of the centroids of the basic elements are re-determined based on the initial positional relationship between the basic elements. Finally, a new multi-superquadric element is reconstructed and the corresponding volume can be denoted as V_{m1}. The volume ratio is expressed as $\sigma_1 = (V_0/V_{m1})^{1/3}$ and satisfies $|\sigma_1 - 1| \leq 10^{-8}$. Therefore, the particles of different shapes have the same mass or volume. The diameter (D_0) of the volume-equivalent sphere is 5 mm, and the total number of particles is 4000.

The particle shape significantly affects the flow characteristics of the granular materials. Figure 3.14 shows a hopper model used in the DEM simulations. The upper hopper has a length (L_0) of 75 mm and a width (W_0) of 75 mm. The orifice has a length (L_1) of 30 mm and a width (W_1) of 30 mm. The lower container has a length (L_2) of 225 mm, a width (W_2) of 225 mm, and a height (H_2) of 75 mm. The hopper angle (θ_0) is 30°. Particles of different shapes have random positions and orientations and were dropped into the upper hopper under gravity. Subsequently, the granular bed was categorized into three parts and represented by red, green, and blue. Finally, the orifice was opened until all particles remained motionless.

Figure 3.15 shows the snapshots of discharging processes of differently shaped particles at different moments. When the orifice was opened, a V-shaped flow pattern appeared gradually. The vertical velocity of the particles located in the center of the hopper was higher than that of the particles near the wall, and this flow pattern became significant for concave particles. This was primarily because the concave particles close to the wall formed an interlock, resulting in a stable packing structure. Finally, the V-shaped pattern disappeared as the number of particles in the hopper decreased. It is noteworthy that the flow of spheres and spherocylinders was continuous, while the flow of concave particles was intermittent. The interlocking between the concave

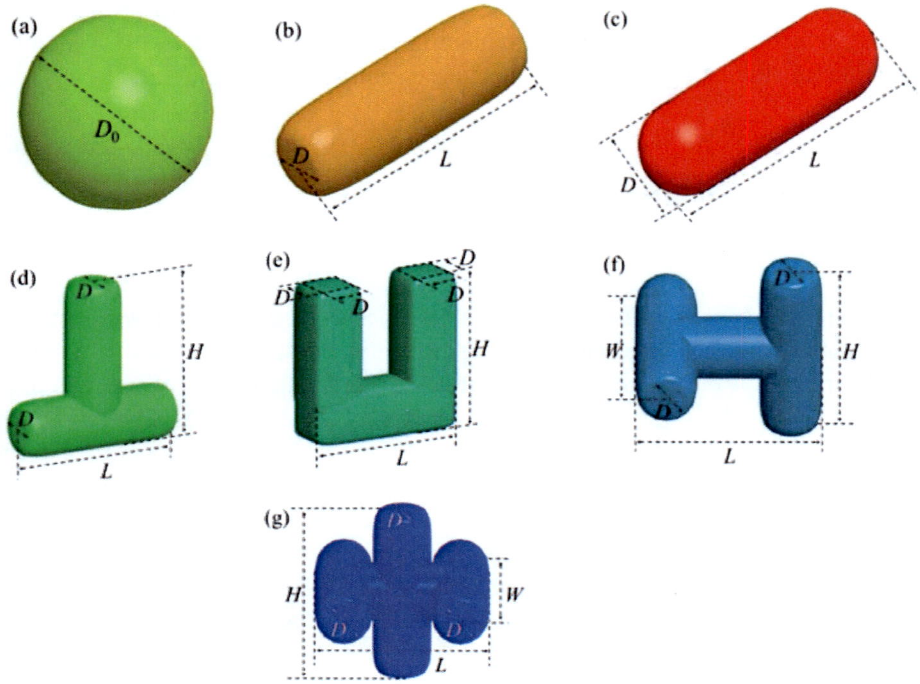

Fig. 3.13 Differently shaped particles constructed by different models: **a–b** superquadric elements and **c–g** multi-superquadric elements

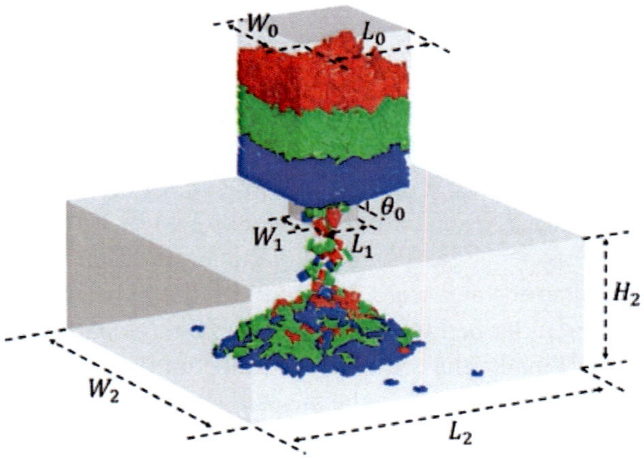

Fig. 3.14 Sketch of the hopper model used in the DEM simulations

particles yielded a local arch structure and hindered the granular flow. In addition, a layered pattern of granular beds was observed in the lower container. The blue particles in the lower part of the hopper were primarily located at the bottom and upper parts of the granular bed, while the red particles in the upper part of the hopper were primarily located in the middle part of the granular bed. However, this layered pattern was vague for spheres and spherocylinders, and those particles could slide

Fig. 3.15 Snapshots of discharging process of the differently shaped particles at different moments: **a** superquadric elements and **b–d** multi-superquadric elements

and rotate easily. Therefore, a large number of blue particles at the bottom and upper parts of the granular bed moved toward the wall of the container.

Furthermore, the effects of particle shape on the mass flow rate and angle of repose of the granular bed are shown in Fig. 3.16a, b, respectively. The spherical particles exhibit the fastest flow rate and smallest angle of repose. Both the cylinders and spherocylinders have a faster flow rate than the concave particles, but their angles of repose are smaller than those of the concave particles. It is noteworthy that the spherocylinders have a smoother surface and can slide or rotate easier than the cylindrical particles. Consequently, the spherocylinders flow faster than the cylindrical particles, while their angle of repose is smaller than that of cylindrical particles. Moreover, the shape complexity of the concave particles significantly affects the flow rate and angle of repose of the granular bed. In general, the discharge rate of the concave particles decreases and the angle of repose increases as the complexity of the particle shape increases.

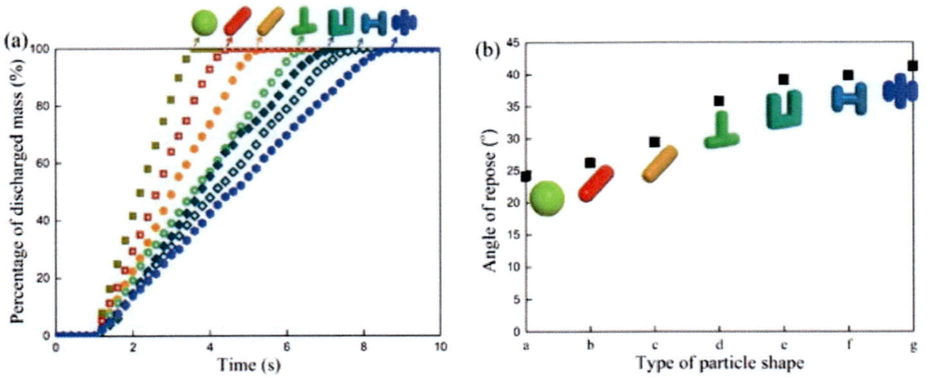

Fig. 3.16 Effect of particle shapes on the mass flow rate and angle of repose: **a** mass flow rate; **b** angle of repose

3.3.3 DEM Simulation of Flow Processes of Poly-superquadric Elements

The particle shape significantly affects the flow characteristics of the particles. Firstly, the formation of spherical and non-spherical granular beds is simulated by the superquadric and poly-superquadric models. A sphere constructed by the superquadric model has parameters of $a : b : c = 1 : 1 : 1$ and $n_1 = n_2 = 2$. A pebble-like particle constructed by the poly-superquadric model has parameters of $r_x^+ : r_x^- = 0.7, r_y^+ : r_y^- = 0.4, r_z^+ : r_z^- = 0.5$, and $n_1^* = n_2^* = 2$. A cube-like particle constructed by the superquadric model has parameters of $a : b : c = 1 : 1 : 1$ and $n_1 = n_2 = 6$. The total number of particles is 6000, and differently shaped particles have the same volume. The diameter of the volume-equivalent sphere is 5 mm.

The container has parameters of $L_0 = W_0 = 75$ mm, and the height is 180 mm. Initially, the superquadric and poly-superquadric elements have arbitrary positions and orientations and are dropped into the container, as shown in Fig. 3.17. It is worth noting that the bounding spheres of particles of different shapes are used to determine the initial contact between the elements, thereby generating a granular system with no overlap between the elements. Particles of different shapes have different radii of the bounding spheres, which results in different generation rates of the particles. The color of the particle represents the height of the particle from the bottom. Red means the height of the particles is higher, and blue means the height of the particles is lower. Finally, all particles are motionless and form a stable system.

Furthermore, the flow processes of non-spherical particles within a cuboid hopper are simulated using superquadric and poly-superquadric DEM models. The hopper has parameters of $L_0 = W_0 = 75$ mm, and the orifice has parameters of $L_1 = W_1 = 25$ mm, as shown in Fig. 3.18. For particles of different shapes, the hopper angle varies from 15 to 75° to investigate the superimposed influence of the hopper structure. The formation of a stable particle system is shown in Fig. 3.17. Then,

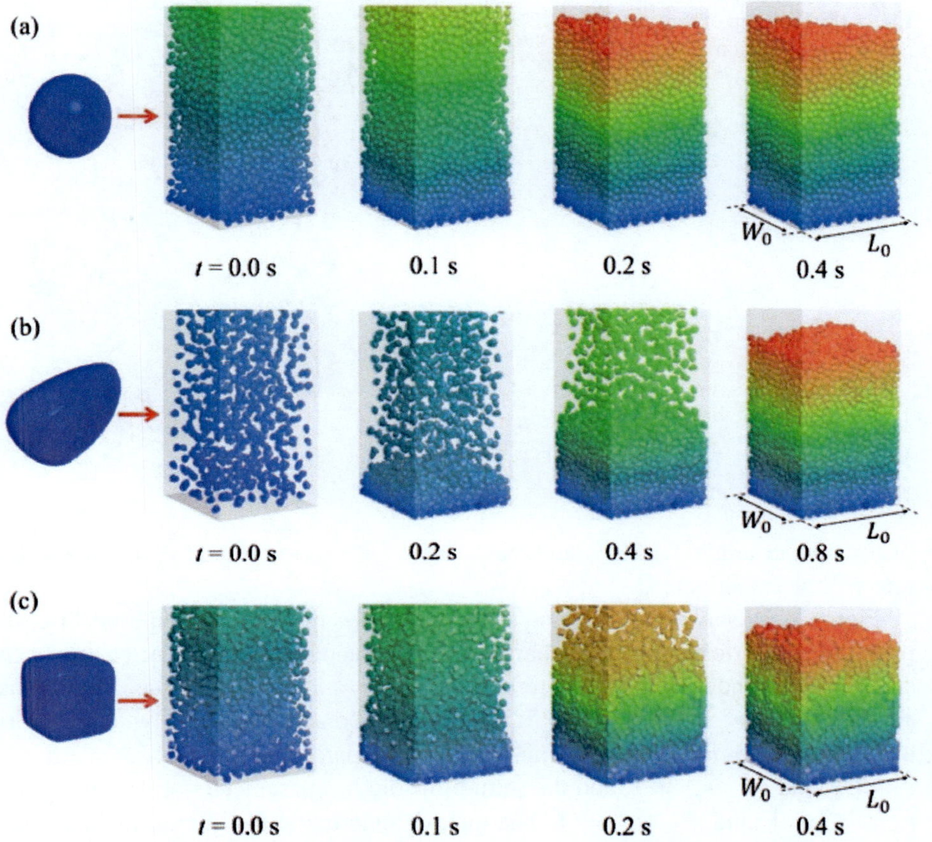

Fig. 3.17 A stable system composed of particles of different shapes: **a**, **c** superquadric particles and **b** poly-superquadric particles

all the particles are divided into four parts according to height and represented by different colors. Eventually, the orifice is opened when all particles are motionless.

Figure 3.19 shows snapshots of the discharge processes of particles of different shapes. Here, the base angle is 30°. After the orifice is opened, a V-shaped flow pattern is observed. Meanwhile, the particle shape has a significant influence on the motion behaviors of granular materials. The spherical particles basically have a uniform speed, and the V-shaped pattern is not clearly observed. Moreover, the speed of the cubes near the wall is significantly lower than that of the cubes located at the center of the container. As a result, this flow pattern strengthens for cubes. For the pebble-like particles constructed by poly-superquadric models, the speed of the particles located on the upper layer of the system is uniform, while the velocity distribution of the particles located on the lower layer of the granular bed is not uniform. Moreover, the percentage of the discharged mass of differently shaped particles varies with time as shown in Fig. 3.20. It is obvious that the particle morphology is a key factor affecting the discharge rate of the granular material. In the following sections, the influence

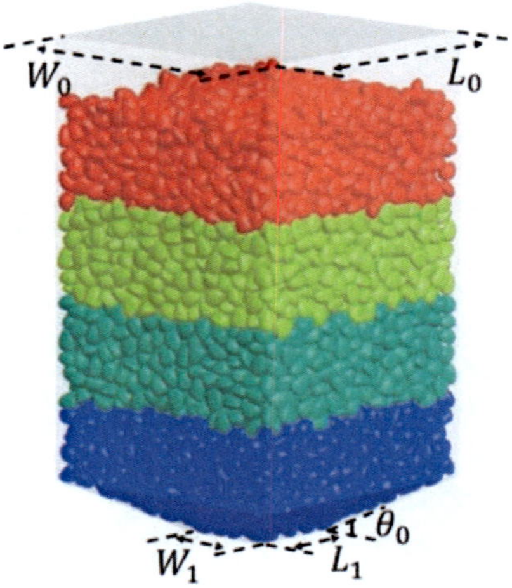

Fig. 3.18 Diagram of the hopper configuration simulated by DEM

of the hopper angle, surface blockiness, and aspect ratio on the average mass flow rate will be investigated.

Figure 3.21 shows particles of different shapes modeled by superquadric and poly-superquadric elements. For particles constructed by superquadric elements, the sphere and cylinder have parameters of $a : b : c = 1 : 1 : 1$, and the ellipsoid has parameters of $a : b : c = 3 : 4 : 5$. For particles constructed by poly-superquadric elements, the hemisphere and semi-cylinder have parameters of $r_x^+ : r_x^- = 3.5$, $r_y^+ : r_y^- = 1$, and $r_z^+ : r_z^- = 1$, and the semi-ellipsoid has parameters of $r_x^+ : r_x^- = 1.5$, $r_y^+ : r_y^- = 1$, and $r_z^+ : r_z^- = 1$. The quarter sphere and the quarter cylinder have the parameters of $r_x^+ : r_x^- = 3.5$, $r_y^+ : r_y^- = 3.5$, and $r_z^+ : r_z^- = 1$, and the quarter ellipsoid has parameters of $r_x^+ : r_x^- = 1.5$, $r_y^+ : r_y^- = 2$, and $r_z^+ : r_z^- = 1$. The eighth sphere and eighth cylinder have the parameters of $r_x^+ : r_x^- = 3.5$, $r_y^+ : r_y^- = 3.5$, and $r_z^+ : r_z^- = 3.5$, and the eighth ellipsoid has parameters of $r_x^+ : r_x^- = 1.5$, $r_y^+ : r_y^- = 2$, and $r_z^+ : r_z^- = 2.5$. The radius of the volume-equivalent sphere of differently shaped particles is 5 mm.

Figure 3.22 shows the effect of aspect ratio on the average mass flow rate of particles. The spheres have the fastest flow rate, and the flow rate of the ellipsoids is faster than that of the cylinders. Meanwhile, the flow rate of particles represented by superquadric elements is faster than that of approximately half, one-quarter, and one-eighth particles represented by poly-superquadric elements. Moreover, the approximately one-quarter particles have the lowest flow rate. This is mainly because differently shaped particles have the same volume, which results in the one-quarter particles having the longest length in the z-axis direction. The relative movement between elongated particles is restricted by local clusters and interlocking structures, which reduces the flowability of the non-spherical granular materials.

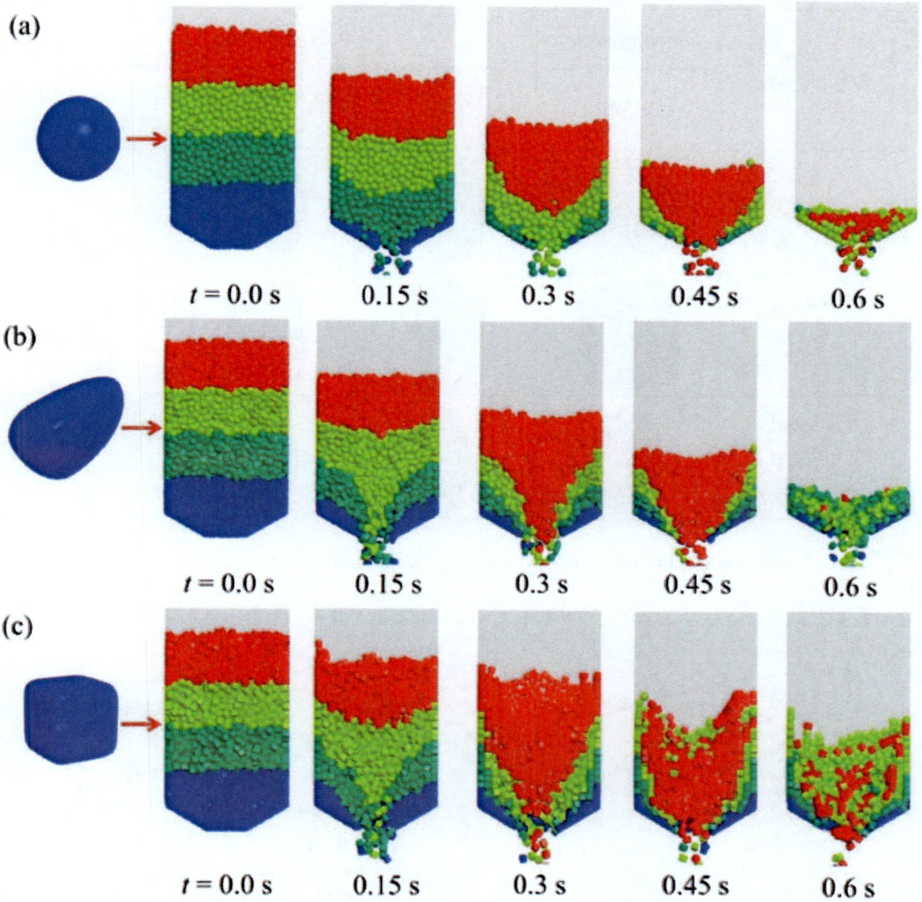

Fig. 3.19 Snapshots of the discharge process of particles of different shapes: **a**, **c** superquadric elements and **b** poly-superquadric elements

Fig. 3.20 Time variation of the percentage of discharged mass for particles of different shapes

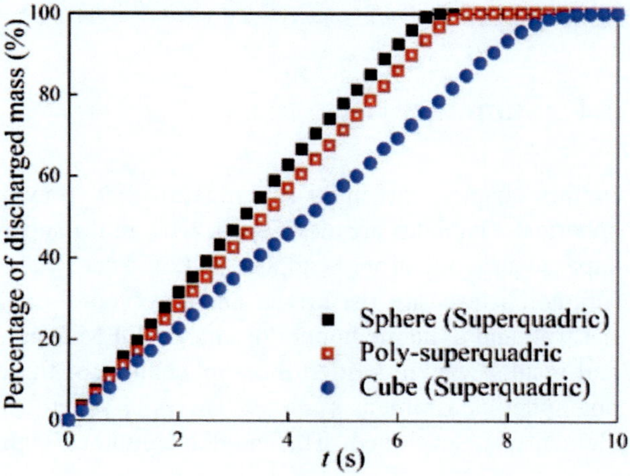

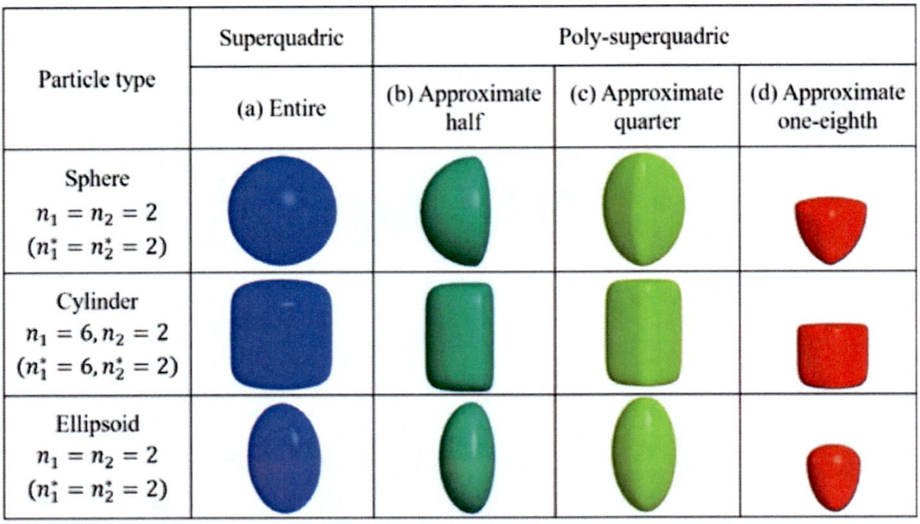

Fig. 3.21 Superquadric and poly-superquadric particles with different aspect ratios

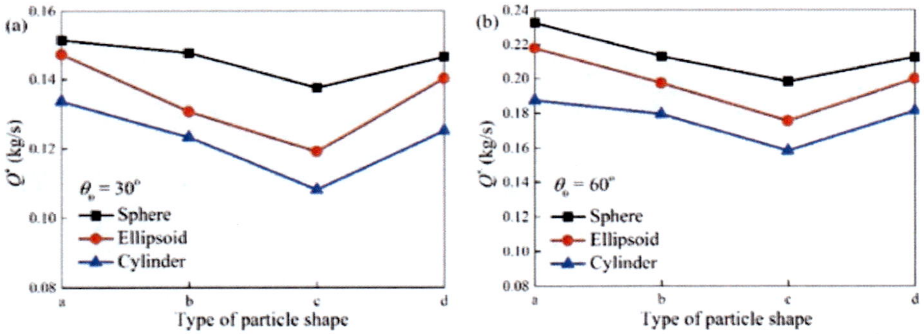

Fig. 3.22 Average mass flow rates for differently shaped particles within beds having $\theta = 30°$ (**a**) and $60°$ (**b**)

3.4 Summary

In this chapter, the multi-superquadric and poly-superquadric models for non-spherical elements are developed. The multi-superquadric model, based on the superquadric equation, is applied to construct concave and convex particle shapes. Validation tests are performed and involved a single spherocylinder impacting a flat wall and dynamic hopper discharge. DEM results were compared with theoretical results, which verified the applicability of the multi-superquadric models for non-spherical granular systems. The poly-superquadric method for non-spherical elements is developed. This model combines eight superquadric equations and eight shape parameters are used to describe the convex and geometrically asymmetric particle shapes. The verification tests consist of a plane impacted by a hemisphere, the packing and discharging process within a cuboid hopper. The DEM results are compared with the theoretical results, which verifies the applicability of poly-superquadric models for the non-spherical granular systems.

References

Eberly D (2002) Dynamic collision detection using oriented bounding boxes, geometric tools

Fritzer HP (2001) Molecular symmetry with quaternions. Spectrochim Acta Part A 57:1919–1930

Houlsby GT (2009) Potential particles: a method for modelling non-circular particles in DEM. Comput Geotech 36:953–959

Kodam M, Bharadwaj R, Curtis J, Hancock B, Wassgren C (2010) Cylindrical object contact detection for use in discrete element method simulations, part II—experimental validation. Chem Eng Sci 65:5863–5871

Kodam M, Curtis J, Hancock B, Wassgren C (2012) Discrete element method modeling of bi-convex pharmaceutical tablets: contact detection algorithms and validation. Chem Eng Sci 69:587–601

Liu Z, Zhao Y (2020) Multi-super-ellipsoid model for non-spherical particles in DEM simulation. Powder Technol 361:190–202

Podlozhnyuk A, Pirker S, Kloss C (2016) Efficient implementation of superquadric particles in discrete element method within an open-source framework. Comput Part Mech 4:101–118

Rakotonirina AD, Delenne J-Y, Radjai F, Wachs A (2019) Grains3D, a flexible DEM approach for particles of arbitrary convex shape—part III: extension to non-convex particles modelled as glued convex particles. Comput Part Mech 6:55–84

Soltanbeigi B, Podlozhnyuk A, Papanicolopulos SA, Kloss C, Pirker S, Ooi JY (2018) DEM study of mechanical characteristics of multi-spherical and superquadric particles at micro and macro scales. Powder Technol 329:288–303

Wang S, Ji S (2022) Poly-superquadric model for DEM simulations of asymmetrically shaped particles. Comput Part Mech 9:299–313

Wang S, Fan Y, Ji S (2018) Interaction between super-quadric particles and triangular elements and its application to hopper discharge. Powder Technol 339:534–549

Wellmann C, Lillie C, Wriggers P (2008) A contact detection algorithm for superellipsoids based on the common-normal concept. Eng Comput 25:432–442

Zhang B, Regueiro R, Druckrey A, Alshibli K (2018) Construction of poly-ellipsoidal grain shapes from SMT imaging on sand, and the development of a new DEM contact detection algorithm. Eng Comput 35:733–771

Zhao S, Zhao J (2019) A poly-superellipsoid-based approach on particle morphology for DEM modeling of granular media. Int J Numer Anal Meth Geomech 43:2147–2169

Zhao B, An X, Zhao H, Shen L, Sun X, Zhou Z (2019) DEM simulation of the local ordering of tetrahedral granular matter. Soft Matter 15:2260–2268

Chapter 4
Smoothed Polyhedral DEM Model Based on Minkowski Sum Algorithm

In recent years, an increasing number of numerical studies have focused on the actual morphology of particles (Nie et al. 2020), and DEMs have been developed for arbitrarily shaped particles (Jha et al. 2021; Gui et al. 2016; Kafashan et al. 2019). Ellipsoidal and super-ellipsoidal models are mathematical formulations for constructing non-spherical particles (Zhao et al. 2018), and ellipsoidal, cylindrical, and cubic particles with various aspect ratios and sharp surfaces are obtained by varying the function parameters (Kildashti et al. 2020). The contact problem between particles is solved by transforming it into a nonlinear optimization equation with constraints, and the two DEM models have a high computational efficiency (Zhao et al. 2019). The polyhedral model is a general method for describing non-spherical particles based on geometric topology, and the normal overlap between particles is calculated using the common plane method, GJK, EPA, and contact volume (Govender et al. 2014, 2018; Feng and Tan 2019). The spherocylinders, spheropolygons, spheropolyhedrons, dilated polyhedrons, and smooth polyhedrons are mathematical methods for modeling irregular-shaped particles with smooth surfaces based on the Minkowski sum concept. The contact point between particles is obtained by geometric detections, solution of envelope functions, and multi-sphere methods (Nye et al. 2014; Liu and Ji 2020; Alonso-Marroquín et al. 2009). Besides, non-spherical particles constructed by polyhedral and smooth polyhedral models have significantly different mechanical response and flow characteristics (Höhner et al. 2014).

This chapter introduces a unified contact algorithm based on the Minkowski sum concept to simulate mixed smooth granular materials consisting of multiple dilated DEM models. Arbitrarily shaped particles with smooth surfaces are constructed using the dilated super-ellipsoidal model, dilated spherical harmonic model, and dilated polyhedral model. Fibonacci and automatic mesh simplification methods are employed to obtain the minimum number of surface meshes for differently shaped particles. Subsequently, single or several contact points of the particles in static or dynamic structures are determined using the virtual sphere method. The mechanical responses of two or multiple particles are tested numerically using the proposed

S. Wang and S. Ji, *Computational Mechanics of Arbitrarily Shaped Granular Materials*, Springer Tracts in Mechanical Engineering,
https://doi.org/10.1007/978-981-99-9927-9_4

Minkowski sum contact algorithm, including the elastic and inelastic collisions between arbitrarily shaped particles, static packing, and dynamic flow of complex granular systems. The DEM results agree well with the theoretical results, validating the ability of the proposed Minkowski sum contact algorithm to accurately reflect the dynamic properties of arbitrarily shaped particles with smooth surfaces.

4.1 Dilated DEM Models of Arbitrarily Shaped Particles

In this section, arbitrarily shaped particles with convex and concave surfaces are constructed using super-ellipsoidal, spherical harmonic, and polyhedral models. Subsequently, the surface meshes of the super-ellipsoidal and spherical harmonic particles are obtained using the Fibonacci method. By applying the Minkowski sum algorithm, a dilated radius is added to the surface meshes of each particle to construct differently shaped particles with smooth surfaces. Finally, an automatic mesh simplification method is introduced to optimize the number of meshes on the particle surface.

4.1.1 Polyhedral Model Based on Geometric Topology

The polyhedral model is based on geometric topological representation and consists of a series of planes, edges, and vertices. It has been reported that polynomial summation based on the potential function is only applicable for describing convex polyhedrons, and its functional form is expressed as follows (Liu and Ji 2020; Houlsby 2009):

$$F(x, y, z) = \sum_{i=1}^{M} \langle a_i x + b_i x + c_i x - d_i \rangle \tag{4.1}$$

where a_i, b_i, and c_i are the components of the unit-normal vector of plane i in the x, y, and z directions, respectively. d_i is the perpendicular distance between the coordinate origin and plane i. $\langle x \rangle$ is the Macaulay bracket. If $x \geq 0$, $\langle x \rangle = x$; if $x < 0$, $\langle x \rangle = 0$. However, to date, there is no envelope function for describing concave polyhedrons. In the present study, arbitrarily shaped particles were constructed, and information about the particle surfaces consisting of several triangular elements was obtained using the finite element softwares ANSYS and ABAQUS, as shown in Fig. 4.1.

Fig. 4.1 Convex and concave particles described by polyhedral models

4.1.2 Dilated DEM Models Based on Fibonacci and Minkowski Sum Algorithms

Considering the complexity of inter-particle contact detection in a complex granular system, three dilated DEM models were developed based on the Fibonacci and Minkowski sum algorithms. The Fibonacci series is a mathematical golden mean series and can be used for the generation of nodes or meshes on the surface of a particle/structure (Gei 2010; Liu et al. 2020). The surfaces of the particles constructed using the super-ellipsoidal and spherical harmonic models were meshed using the Fibonacci algorithm. The Fibonacci sequence is expressed as follows:

$$(\theta, \phi)_i = \left(\frac{2\pi F_1}{F_2} i, \arccos\left(1 - \frac{2i}{F_2}\right) \right) \tag{4.2}$$

where F_1 and F_2 are the adjacent Fibonacci numbers that satisfy $F_2 > F_1$. F_2 denotes the total number of nodes. i denotes the index of a node that satisfies $0 \leq i < F_2$. It has been reported that F_2/F_1 converges to a golden ratio of 1.618 as F_1 and F_2 increase (Feng 2021d). Thus, the optimized Fibonacci series is expressed as follows (Feng 2021d):

$$(\theta, \phi)_i = \left(\frac{2\pi}{\lambda} i, 1 - \frac{2(i + \varepsilon)}{M - 1 + 2\varepsilon} \right) \tag{4.3}$$

where ε is an empirically determined parameter for improving the point distribution, which satisfies $\varepsilon = 0.36$. M is the number of nodes.

Subsequently, the Minkowski sum algorithm was used to model the smooth particles with dilated radii. The functional form of the Minkowski sum is expressed as follows (Galindo-Torres et al. 2012):

$$\boldsymbol{A} \oplus \boldsymbol{B} = \{\boldsymbol{x} + \boldsymbol{y} | \boldsymbol{x} \in \boldsymbol{A}, \boldsymbol{y} \in \boldsymbol{B}\} \tag{4.4}$$

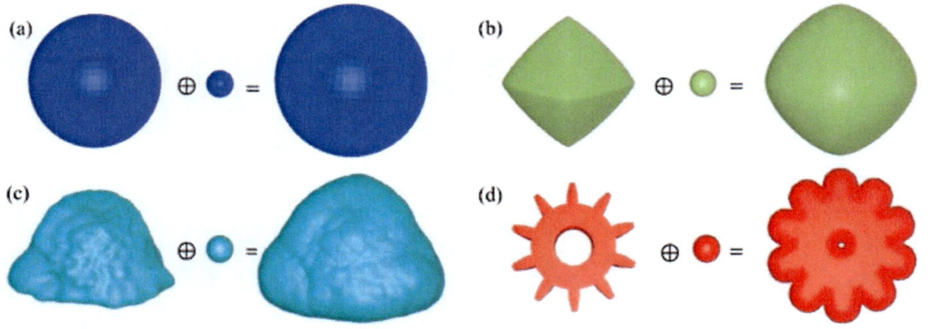

Fig. 4.2 Different dilated DEM models based on the Minkowski sum algorithm: **a** dilated sphere; **b** dilated super-ellipsoid; **c** dilated spherical harmonic particle; **d** dilated polyhedron

where $\boldsymbol{A}$ and $\boldsymbol{B}$ denote the polyhedron and dilated sphere, respectively. The outer surface of polyhedron $\boldsymbol{A}$ is swept by a dilated sphere $\boldsymbol{B}$, and a new dilated polyhedron is the Minkowski sum of a polyhedron $\boldsymbol{A}$ and a dilated sphere $\boldsymbol{B}$. Moreover, different dilated DEM models have been developed, that is, the Minkowski sum of a basic particle constructed by a spherical or non-spherical DEM model and a dilated sphere, as shown in Fig. 4.2. Note that an arbitrarily shaped particle constructed using the dilated DEM model is composed of several dilated triangular elements, cylinders, and spheres, whereas a particle constructed using the traditional DEM model is composed of several triangular elements, edges, and vertices. Thus, the particles constructed by the dilated DEM model have smoother surfaces than those constructed by the traditional DEM model.

4.1.3 *Automatic Mesh Simplification Method for Dilated DEM Models*

The runtime consumption of DEM simulations depends on the number of particles and the number of surface meshes per particle, and more surface meshes increase runtime consumption. Therefore, an automatic mesh simplification method was developed to reduce the number of surface meshes of the particles and improve the computational efficiency of DEM simulations (Feng 2021d). In this method, two adjacent surface nodes were merged and the edges connecting these two nodes were removed. Meanwhile, the triangular elements on the particle surface were automatically obtained according to the reduction in the nodes. The merging of nodes and deletion of edges were not random, and the cost of merging two nodes was calculated as follows:

$$F(A, B) = \|\boldsymbol{X}_A - \boldsymbol{X}_B\| \cdot \max_{\boldsymbol{f}_{\mathrm{p}} \in T_A} \left\{ \min_{\boldsymbol{n}_{\mathrm{p}} \in T_{AB}} \left[\left(1 - \boldsymbol{f}_{\mathrm{p}} \cdot \boldsymbol{n}_{\mathrm{p}}\right)/2 \right] \right\} \tag{4.5}$$

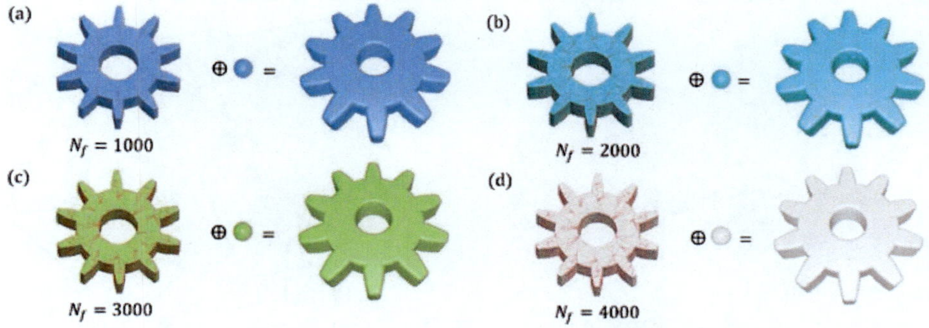

Fig. 4.3 Gear-shaped particles with different numbers of surface meshes obtained by the automatic mesh simplification method: **a** $N_{\mathrm{f}} = 1000$; **b** $N_{\mathrm{f}} = 2000$; **c** $N_{\mathrm{f}} = 3000$; **d** $N_{\mathrm{f}} = 4000$

where $\boldsymbol{X}_A$ and $\boldsymbol{X}_B$ are the position vectors of points A and B, respectively. T_A is the set of normal vectors of the triangular elements containing point A. T_{AB} is the set of normal vectors of the triangular elements containing edge AB. The merging cost $F(A, B)$ is calculated for two nodes in each edge, and nodes A and B are merged when the value of $F(A, B)$ is the minimum. This indicates that only two nodes are merged in each iteration step. The mesh simplification of the particle is completed when the number of surface meshes of the particle is reduced to the predefined number N_{f} of surface meshes. Meanwhile, the merging process considers the following two aspects: (i) the deleted edge cannot cross other edges and (ii) the merged point cannot be a sharp vertex of the particle. Figure 4.3 shows the gear-shaped particles with different numbers of surface meshes obtained using the automatic mesh simplification method. Here, N_{f} denotes the number of surface meshes of the particles. The merged points and simplified triangular elements are essentially located on the particle surface with less curvature, whereas particle surfaces with greater curvature retain their basic shape characteristics.

4.2 A Novel Minkowski Sum Contact Algorithm Between Particles

In this section, the Minkowski sum contact algorithm for arbitrary morphological particles is presented. Before the accurate detection of particles, the bounding sphere and oriented bounding box were employed to reduce the potential number of contact pairs and improve the computational efficiency of DEM simulations, as shown in Fig. 4.4. The bounding sphere was used as the first rough contact detection between particles and is expressed as follows:

$$S_{ij} = \left|\boldsymbol{X}_i - \boldsymbol{X}_j\right| - r_i - r_j \begin{cases} \leq 0, \text{ Contact} \\ > 0, \text{ Separation} \end{cases} \tag{4.6}$$

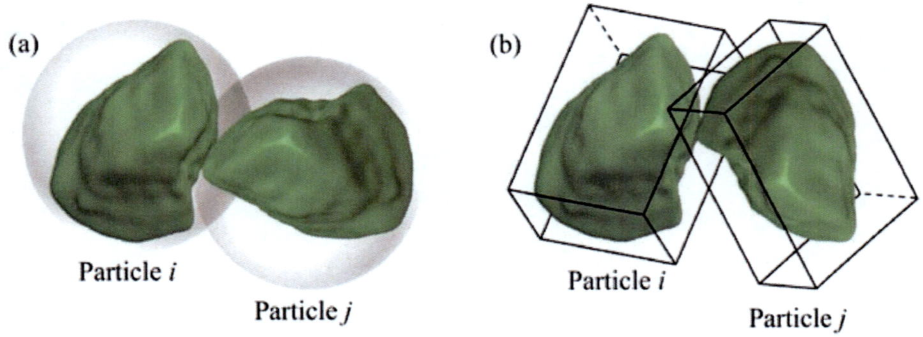

Fig. 4.4 Rough contact detection between particles: **a** the bounding sphere; **b** the oriented bounding box

where $\boldsymbol{X}_i$ and $\boldsymbol{X}_j$ are the mass centers of particles i and j, respectively; r_i and r_j are the bounding radii of particles i and j, respectively.

Subsequently, the oriented bounding box was used to detect the contact state between the particles. The length, width, and height of the bounding box were obtained through the inertia matrix and the longest lengths in the three principal axis directions. The contact detection between the bounding boxes is based on the separation axis concept, and is expressed as follows:

$$C_{ij} = \left|(\boldsymbol{X}_i - \boldsymbol{X}_j)\overline{\boldsymbol{l}}\right| - \sum_{k=1}^{3}\left|a_k\overline{\boldsymbol{r}}_{ik}\overline{\boldsymbol{l}}\right| - \sum_{k=1}^{3}\left|b_k\overline{\boldsymbol{r}}_{jk}\overline{\boldsymbol{l}}\right| \begin{cases} \leq 0, \text{ Contact} \\ > 0, \text{ Separation} \end{cases} \tag{4.7}$$

where $\overline{\boldsymbol{l}}$ is the unit vector of the separation axis; a_k and b_k are the kth semi-axis lengths of particles i and j, respectively; $\overline{\boldsymbol{r}}_{ik}$ and $\overline{\boldsymbol{r}}_{jk}$ are the unit vectors of the kth semi-axis length of particles i and j, respectively.

An arbitrarily shaped particle with a dilated radius is composed of several dilated triangular elements, and each dilated triangular element is essentially a Minkowski sum of a basic triangular element and dilated sphere. Thus, the contact state between arbitrarily shaped particles is determined by calculating the closest distance between dilated triangular elements. Because the contact detection between two dilated triangular elements can be individually resolved without the information of surrounding elements, this solution method is well suitable for GPU parallelism. Similar to the rough contact detections between particles, the rough contact detections between dilated triangular elements are shown in Fig. 4.5. A dilated triangular element is composed of a dilated triangular face, three cylinders, and three spheres. There are six contact forms between the two dilated triangular elements: face-sphere, cylinder-sphere, sphere-sphere, face-cylinder, cylinder-cylinder, and face-face contacts. The overlap calculations for the six contact modes between the dilated triangular elements are described in detail below.

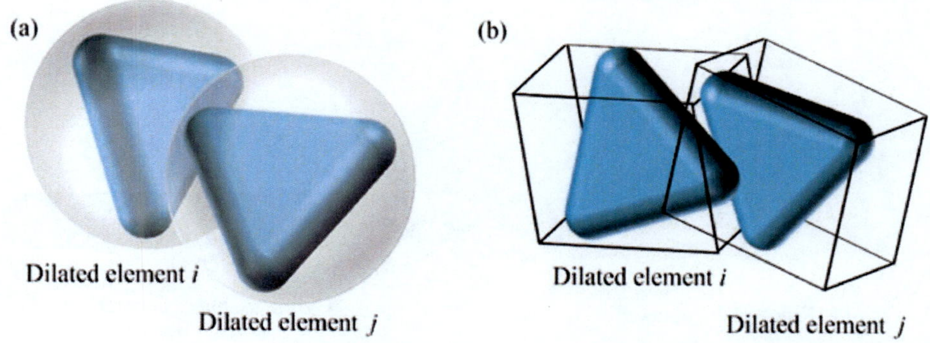

Fig. 4.5 Rough contact detection between dilated triangular elements: **a** the bounding sphere; **b** the orientation bounding box

4.2.1 Contact Detection Between a Dilated Triangular Element and a Sphere

There are three forms of contact between a dilated triangular element and a sphere: face-sphere, cylinder-sphere, and sphere-sphere contacts. Dilated triangular element B has three vertices, denoted as B_1, B_2 and B_3, as shown in Fig. 4.6a. A_1 is the mass center of the sphere, and R_A is the radius of the dilated element A. Q_1 is the mass center of the virtual sphere formed by the projection point of sphere A in dilated element B, and R_B is the radius of dilated element B. $\boldsymbol{n}_B$ is the normal unit vector of dilated element B. First, it is necessary to determine whether sphere A_1 is in contact with the plane where dilated element B is located. If $\left|\left(\boldsymbol{X}_A - \boldsymbol{X}_{B_1}\right) \cdot \boldsymbol{n}_B\right| - R_A - R_B < 0$, sphere A_1 is in contact with plane.

Second, the contact modes between sphere A_1 and dilated element B must be determined, such as the face, cylinder, and sphere contacts. The position of projection point Q_1 was obtained using $\boldsymbol{X}_{Q_1} = \left[\left(\boldsymbol{X}_{B_1} - \boldsymbol{X}_{A_1}\right) \cdot \boldsymbol{n}_B\right] \cdot \boldsymbol{n}_B + \boldsymbol{X}_{A_1}$. In addition, the position of point Q_1 can be expressed by the unknown parameters s and t for points $\boldsymbol{X}_{B_1}$, $\boldsymbol{X}_{B_2}$, and $\boldsymbol{X}_{B_3}$, as follows:

$$\boldsymbol{X}_{Q_1} = \boldsymbol{X}_{B_1} + s \cdot \left(\boldsymbol{X}_{B_2} - \boldsymbol{X}_{B_1}\right) + t \cdot \left(\boldsymbol{X}_{B_3} - \boldsymbol{X}_{B_1}\right) \tag{4.8}$$

Equation (4.8) is multiplied by $\left(\boldsymbol{X}_{B_2} - \boldsymbol{X}_{B_1}\right)$ and $\left(\boldsymbol{X}_{B_3} - \boldsymbol{X}_{B_1}\right)$, obtained as follows:

$$\begin{cases} (\boldsymbol{X}_{Q_1} - \boldsymbol{X}_{B_1})(\boldsymbol{X}_{B_2} - \boldsymbol{X}_{B_1}) = [s \cdot (\boldsymbol{X}_{B_2} - \boldsymbol{X}_{B_1}) + t \cdot (\boldsymbol{X}_{B_3} - \boldsymbol{X}_{B_1})](\boldsymbol{X}_{B_2} - \boldsymbol{X}_{B_1}) \\ (\boldsymbol{X}_{Q_1} - \boldsymbol{X}_{B_1})(\boldsymbol{X}_{B_3} - \boldsymbol{X}_{B_1}) = [s \cdot (\boldsymbol{X}_{B_2} - \boldsymbol{X}_{B_1}) + t \cdot (\boldsymbol{X}_{B_3} - \boldsymbol{X}_{B_1})](\boldsymbol{X}_{B_3} - \boldsymbol{X}_{B_1}) \end{cases} \tag{4.9}$$

Here, vector $\boldsymbol{b}_{12}$ denotes $\left(\boldsymbol{X}_{B_2} - \boldsymbol{X}_{B_1}\right)$, and vector $\boldsymbol{b}_{13}$ denotes $\left(\boldsymbol{X}_{B_3} - \boldsymbol{X}_{B_1}\right)$. The unknown parameters s and t can be obtained as follows:

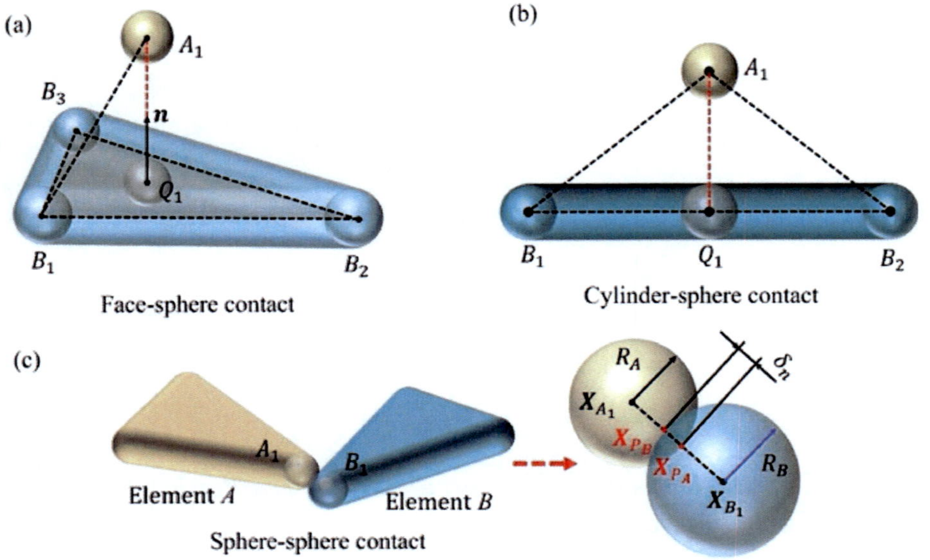

Fig. 4.6 Three contact modes between a dilated triangular element and a sphere: **a** face-sphere contact; **b** cylinder-sphere contact; **c** sphere-sphere contact

$$\begin{cases} s = \dfrac{(\boldsymbol{b}_{13}\cdot\boldsymbol{b}_{13})\left[\left(\boldsymbol{X}_{Q_1}-\boldsymbol{X}_{B_1}\right)\cdot\boldsymbol{b}_{12}\right]-(\boldsymbol{b}_{13}\cdot\boldsymbol{b}_{12})\left[\left(\boldsymbol{X}_{Q_1}-\boldsymbol{X}_{B_1}\right)\cdot\boldsymbol{b}_{13}\right]}{(\boldsymbol{b}_{12}\cdot\boldsymbol{b}_{12})(\boldsymbol{b}_{13}\cdot\boldsymbol{b}_{13})-(\boldsymbol{b}_{12}\cdot\boldsymbol{b}_{13})(\boldsymbol{b}_{13}\cdot\boldsymbol{b}_{12})} \\ t = \dfrac{(\boldsymbol{b}_{12}\cdot\boldsymbol{b}_{12})\left[\left(\boldsymbol{X}_{Q_1}-\boldsymbol{X}_{B_1}\right)\cdot\boldsymbol{b}_{13}\right]-(\boldsymbol{b}_{12}\cdot\boldsymbol{b}_{13})\left[\left(\boldsymbol{X}_{Q_1}-\boldsymbol{X}_{B_1}\right)\cdot\boldsymbol{b}_{12}\right]}{(\boldsymbol{b}_{12}\cdot\boldsymbol{b}_{12})(\boldsymbol{b}_{13}\cdot\boldsymbol{b}_{13})-(\boldsymbol{b}_{12}\cdot\boldsymbol{b}_{13})(\boldsymbol{b}_{13}\cdot\boldsymbol{b}_{12})} \end{cases} \tag{4.10}$$

If there is face contact between sphere A_1 and dilated element B, the parameters s and t must satisfy the conditions as follows:

$$s > 0 \quad \cap \quad t > 0 \quad \cap \quad s + t < 1 \tag{4.11}$$

The overlap between sphere A_1 and dilated element B was obtained as $\boldsymbol{\delta}_n = \left(R_A + R_B - \left|\boldsymbol{X}_{A_1} - \boldsymbol{X}_{Q_1}\right|\right) \cdot \boldsymbol{n}_B$. Otherwise, the cylinder and sphere contacts must be further detected. Taking cylinder B_1B_2 as an example, point Q_1 is the mass center of the virtual sphere formed by the projection point of sphere A_1, as shown in Fig. 4.6b. The position relationship between point Q_1 and cylinder B_1B_2 is expressed as follows:

$$F(\boldsymbol{X}_{A1}) = \left[\left(\boldsymbol{X}_{A1} - \boldsymbol{X}_{B_1}\right) \cdot \left(\boldsymbol{X}_{B_2} - \boldsymbol{X}_{B_1}\right)\right] \cdot \left[\left(\boldsymbol{X}_{B_2} - \boldsymbol{X}_{B_1}\right)\left(\boldsymbol{X}_{A1} - \boldsymbol{X}_{B_2}\right)\right] \tag{4.12}$$

If $F(\boldsymbol{X}_{A1}) \leq 0$, point Q_1 is located within cylinder B_1B_2. Then, the distance between the sphere A_1 and the point Q_1 is obtained as follows:

$$d_{A_1Q_1} = \left|\boldsymbol{X}_{A_1} - \boldsymbol{X}_{Q_1}\right|$$

$$= \left| (\boldsymbol{X}_{A_1} - \boldsymbol{X}_{B_1}) - \frac{(\boldsymbol{X}_{A_1} - \boldsymbol{X}_{B_1}) \cdot (\boldsymbol{X}_{B_2} - \boldsymbol{X}_{B_1})}{|\boldsymbol{X}_{B_2} - \boldsymbol{X}_{B_1}|} \cdot \frac{(\boldsymbol{X}_{B_2} - \boldsymbol{X}_{B_1})}{|\boldsymbol{X}_{B_2} - \boldsymbol{X}_{B_1}|} \right| \tag{4.13}$$

If $d_{A_1Q_1} \leq R_A + R_B$, sphere A_1 and dilated element B are in cylinder-sphere contact. The overlap between sphere A_1 and cylinder B_1B_2 was obtained as $\boldsymbol{\delta}_n = (R_A + R_B - |\boldsymbol{X}_{A_1} - \boldsymbol{X}_{Q_1}|) \cdot (\boldsymbol{X}_{A_1} - \boldsymbol{X}_{Q_1})/|\boldsymbol{X}_{A_1} - \boldsymbol{X}_{Q_1}|$. If sphere A_1 is not in contact with the three cylinders of dilated element B, the sphere-sphere contact needs to be detected.

Taking sphere B_1 as an example, and $\boldsymbol{X}_{A_1}$ and $\boldsymbol{X}_{B_1}$ are the positions of the mass centers of spheres A_1 and B_1, respectively, as shown in Fig. 4.6c. $\boldsymbol{X}_{P_A}$ and $\boldsymbol{X}_{P_B}$ are the positions of the contact points between two spheres. If $R_A + R_B - |\boldsymbol{X}_{A_1} - \boldsymbol{X}_{B_1}| > 0$, the two spheres, A_1 and B_1 are in contact. The overlap between the two spheres was obtained by $\boldsymbol{\delta}_n = (R_A + R_B - |\boldsymbol{X}_{A_1} - \boldsymbol{X}_{B_1}|) \cdot (\boldsymbol{X}_{A_1} - \boldsymbol{X}_{B_1})/|\boldsymbol{X}_{A_1} - \boldsymbol{X}_{B_1}|$, and the contact points between the two spheres were obtained as follows:

$$\begin{cases} \boldsymbol{X}_{P_A} = \boldsymbol{X}_{A_1} + \frac{\boldsymbol{X}_{B_1} - \boldsymbol{X}_{A_1}}{\|\boldsymbol{X}_{B_1} - \boldsymbol{X}_{A_1}\|} \cdot R_A \\ \boldsymbol{X}_{P_B} = \boldsymbol{X}_{B_1} + \frac{\boldsymbol{X}_{A_1} - \boldsymbol{X}_{B_1}}{\|\boldsymbol{X}_{A_1} - \boldsymbol{X}_{B_1}\|} \cdot R_B \end{cases} \tag{4.14}$$

4.2.2 Contact Detection Between a Dilated Triangular Element and a Cylinder

There are two contact modes between a dilated triangular element and a cylinder, including cylinder-cylinder contact and face-cylinder contact. Taking the example of cylinder A_1A_2 of dilated element A and cylinder B_1B_2 of dilated element B, there are two types of contact between the two cylinders: cross contact and parallel contact, as shown in Fig. 4.7. For cross contact between the two cylinders, two virtual spheres were generated, and their mass centers (Q_A and Q_B) were located on lines A_1A_2 and B_1B_2, respectively. Meanwhile, line Q_AQ_B was perpendicular to lines A_1A_2 and B_1B_2, respectively. The positions of points Q_A and Q_B are expressed as follows:

$$\begin{cases} \boldsymbol{X}_{Q_A} = \boldsymbol{X}_{A_1} + s_c(\boldsymbol{X}_{A_2} - \boldsymbol{X}_{A_1}) \\ \boldsymbol{X}_{Q_B} = \boldsymbol{X}_{B_1} + t_c(\boldsymbol{X}_{B_2} - \boldsymbol{X}_{B_1}) \end{cases} \tag{4.15}$$

where s_c and t_c are the unknown parameters. Equation (4.15) is multiplied by $(\boldsymbol{X}_{A_2} - \boldsymbol{X}_{A_1})$ and $(\boldsymbol{X}_{B_2} - \boldsymbol{X}_{B_1})$, obtained as follows:

$$\begin{cases} [\boldsymbol{X}_{A_1} + s_c(\boldsymbol{X}_{A_2} - \boldsymbol{X}_{A_1}) - \boldsymbol{X}_{B_1} - t_c(\boldsymbol{X}_{B_2} - \boldsymbol{X}_{B_1})] \cdot (\boldsymbol{X}_{A_2} - \boldsymbol{X}_{A_1}) = 0 \\ [\boldsymbol{X}_{A_1} + s_c(\boldsymbol{X}_{A_2} - \boldsymbol{X}_{A_1}) - \boldsymbol{X}_{B_1} - t_c(\boldsymbol{X}_{B_2} - \boldsymbol{X}_{B_1})] \cdot (\boldsymbol{X}_{B_2} - \boldsymbol{X}_{B_1}) = 0 \end{cases} \tag{4.16}$$

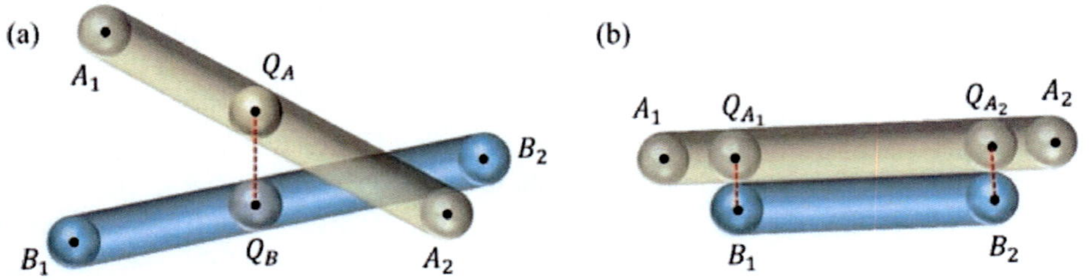

Fig. 4.7 Two types of contact between two cylinders: **a** the cross contact of cylinders; **b** the parallel contact of cylinders

Here, $\boldsymbol{d} = (\boldsymbol{X}_{A_2} - \boldsymbol{X}_{A_1})$ and $\boldsymbol{e} = (\boldsymbol{X}_{B_2} - \boldsymbol{X}_{B_1})$. Thus, the unknown parameters s_c and t_c are obtained as follows:

$$\begin{cases} s_c = \dfrac{[(\boldsymbol{X}_{A_1} - \boldsymbol{X}_{B_1})\cdot\boldsymbol{d}](\boldsymbol{e}\cdot\boldsymbol{e}) - [(\boldsymbol{X}_{A_1} - \boldsymbol{X}_{B_1})\cdot\boldsymbol{e}](\boldsymbol{d}\cdot\boldsymbol{e})}{(\boldsymbol{d}\cdot\boldsymbol{d})(\boldsymbol{e}\cdot\boldsymbol{e}) - (\boldsymbol{d}\cdot\boldsymbol{e})^2} \\ t_c = \dfrac{[(\boldsymbol{X}_{A_1} - \boldsymbol{X}_{B_1})\cdot\boldsymbol{d}](\boldsymbol{d}\cdot\boldsymbol{e}) - [(\boldsymbol{X}_{A_1} - \boldsymbol{X}_{B_1})\cdot\boldsymbol{e}](\boldsymbol{d}\cdot\boldsymbol{d})}{(\boldsymbol{d}\cdot\boldsymbol{d})(\boldsymbol{e}\cdot\boldsymbol{e}) - (\boldsymbol{d}\cdot\boldsymbol{e})^2} \end{cases} \tag{4.17}$$

Two cylinders, A_1A_2 and B_1B_2 are in contact if $0 \leq s_c \leq 1$, $0 \leq t_c \leq 1$, and $|\boldsymbol{X}_{Q_A} - \boldsymbol{X}_{Q_B}| - R_A - R_B \leq 0$. Here, R_A and R_B are the dilated radii of dilated triangular elements A and B, respectively. If the two cylinders A_1A_2 and B_1B_2 are parallel to each other, the cylinder-cylinder contact is replaced by two sphere-sphere contacts, as shown in Fig. 4.7b. Points Q_{A_1} and Q_{A_2} are the mass centers of the virtual spheres formed by the projection of points B_1 and B_2 in cylinder A_1A_2, respectively. The contact point between the two spheres can be obtained using Eq. (4.14).

When one of the cylinders of dilated triangular element A is parallel to dilated triangular element B, dilated triangular element B and cylinder A_1A_2 are in face-cylinder contact, and the number of contact points is two. There are three contact types depending on the location of the contact points between dilated element B and cylinder A_1A_2. In the first contact type, the projection points of both endpoints of cylinder A_1A_2 on the plane are inside the triangular face, as shown in Fig. 4.8a. This face-cylinder contact can be replaced by two face-sphere contacts. In the second contact type, the projection point of one endpoint of the cylinder is inside the triangular face, whereas the projection point of the other endpoint is outside the triangular face, as shown in Fig. 4.8b. This face-cylinder contact can be replaced by face-sphere and cylinder-sphere contacts. In the third contact type, the projection points of both endpoints of the cylinder are located outside the triangular face, as shown in Fig. 4.8c. This face-cylinder contact can be replaced by two cylinder-cylinder contacts.

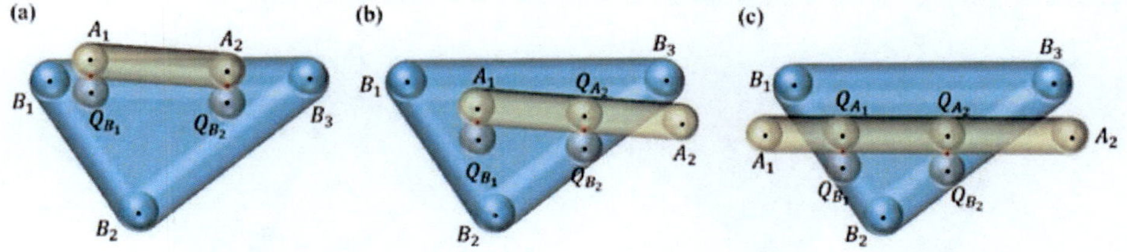

Fig. 4.8 Three types of contact between a dilated triangular face and a cylinder: **a** both contact points are inside the face; **b** one contact point is inside the face, and the other contact point is on the cylinder; **c** both contact points are on the cylinder

4.2.3 Contact Detection Between Two Dilated Triangular Faces

When the two dilated triangular faces are parallel to each other, the two dilated triangular elements are in face-face contact, and the number of contact points may range from three to six. There are four types of contacts depending on the number of contact points between the two dilated faces, as shown in Fig. 4.9. Considering that each dilated triangular element has different dimensions, any of the contacts between the two dilated faces may be face-sphere contact, cylinder-sphere contact, sphere-sphere contact, or cylinder-cylinder contact. Therefore, multiple contact modes between two dilated faces may be a random combination of the above four contacts. The calculations of the overlaps for the four contacts are described in detail in Sects. 3.1 and 3.2.

In DEM simulations, the theoretical models for the maximum overlap between spherical particles have been successfully established, which are related to the yield strength of the material and to the loading and unloading processes (Wojtkowski et al. 2010). However, the theoretical models for the maximum overlap between non-spherical particles are still not well developed (Sinnott and Cleary 2009). When the dilated radius and contact stiffness of the particles are relatively small, the overlap between particles may be greater than the sum of the dilated radii of the two particles. Thus, each particle has a maximum dilated radius $R^{\max}$ and an actual dilated radius R, as shown in Fig. 4.10a. Here, the maximum dilated radius is predefined and is greater than the actual dilated radius. The maximum dilated layer is obtained by sweeping the maximum dilated radius along the particle surface, while the dilated layer is obtained by sweeping the dilated radius along the particle surface. Note that the center position of the virtual sphere of the particle calculated from the maximum dilated radius is the same as that of the virtual sphere of the particle calculated from the actual dilated radius. The maximum dilated radius and the maximum dilated layer are used to calculate the center positions of the virtual spheres of two neighboring particles i and j, and the detailed calculations have been introduced in Sects. 3.1 and 3.2.

Taking a virtual sphere A_1 of a dilated triangular element of particle i as an example, this sphere A_1 is in contact with a virtual sphere B_1 of a dilated triangular

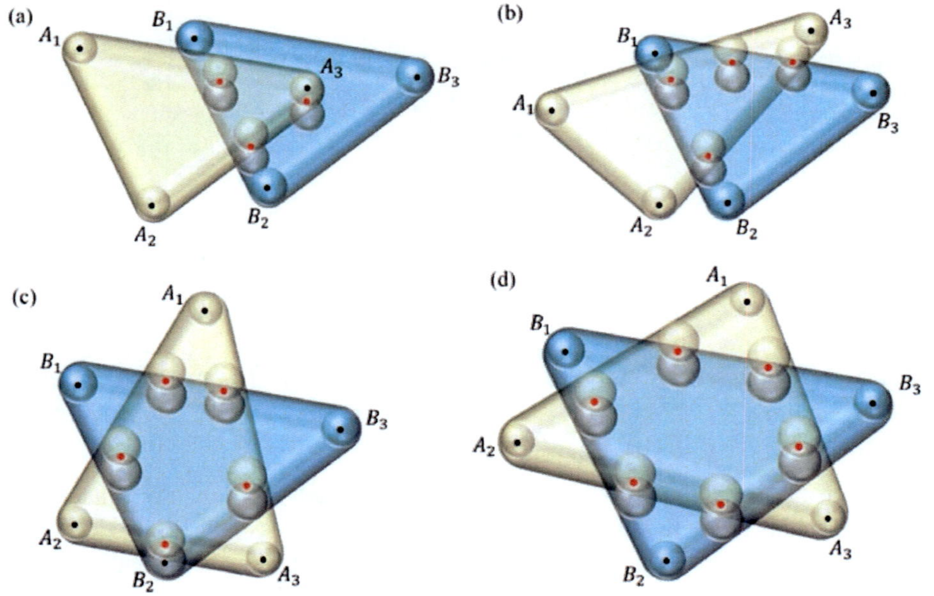

Fig. 4.9 Four types of contact between two dilated triangular faces: **a** three contact points; **b** four contact points; **c** five contact points; **d** six contact points

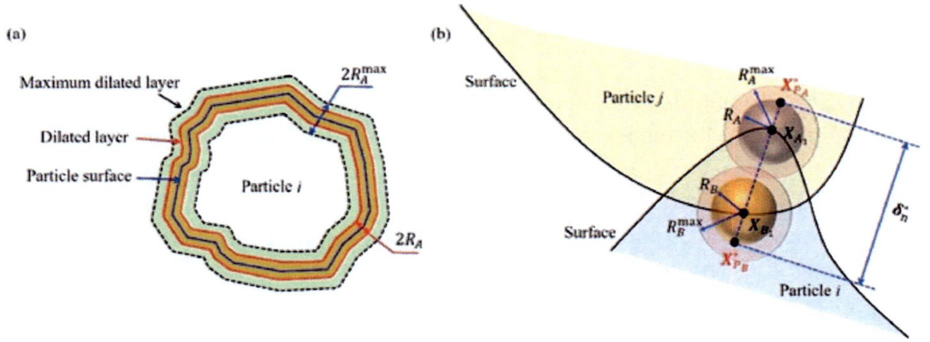

Fig. 4.10 **a** Schematic diagram of the maximum dilated layer and the actual dilated layer of particle i; **b** schematic diagram of the overlap between particles i and j when they penetrate each other

element of particle j. Here, the radii of the two virtual spheres are the maximum dilated radii $R_A^{\max}$ and $R_B^{\max}$, respectively. The center positions of the two spheres are denoted as $\boldsymbol{X}_{A1}$ and $\boldsymbol{X}_{B1}$, respectively. Subsequently, an odd–even rule algorithm is used to determine whether this sphere A_1 is located within the surface of particle j. If the number of intersection points between the ray passing through the center position $\boldsymbol{X}_{A1}$ of this sphere and the surface of particle j is even, this sphere is located outside the surface of particle j. If $R_A + R_B - \left|\boldsymbol{X}_{A_1} - \boldsymbol{X}_{B_1}\right| > 0$, the two spheres, A_1 and B_1 are in contact. The overlap between the two spheres was obtained by $\boldsymbol{\delta}_n = \left(R_A + R_B - \left|\boldsymbol{X}_{A_1} - \boldsymbol{X}_{B_1}\right|\right) \cdot \left(\boldsymbol{X}_{A_1} - \boldsymbol{X}_{B_1}\right) / \left|\boldsymbol{X}_{A_1} - \boldsymbol{X}_{B_1}\right|$, and the contact point between the two spheres was expressed as Eq. (4.14). If the number of intersection points is odd, this sphere is located inside the surface of particle j, as shown in Fig. 4.10b. The overlap between the two spheres was obtained by $\boldsymbol{\delta}_n^* = \left(R_A + R_B + \left|\boldsymbol{X}_{A_1} - \boldsymbol{X}_{B_1}\right|\right) \cdot \left(\boldsymbol{X}_{A_1} - \boldsymbol{X}_{B_1}\right) / \left|\boldsymbol{X}_{A_1} - \boldsymbol{X}_{B_1}\right|$, and the contact point between the two spheres was expressed as follows:

$$\begin{cases} \boldsymbol{X}_{P_A}^* = \boldsymbol{X}_{A_1} - \frac{\boldsymbol{X}_{B_1} - \boldsymbol{X}_{A_1}}{\|\boldsymbol{X}_{B_1} - \boldsymbol{X}_{A_1}\|} \cdot R_A \\ \boldsymbol{X}_{P_B}^* = \boldsymbol{X}_{B_1} - \frac{\boldsymbol{X}_{A_1} - \boldsymbol{X}_{B_1}}{\|\boldsymbol{X}_{A_1} - \boldsymbol{X}_{B_1}\|} \cdot R_B \end{cases} \tag{4.18}$$

Although the maximum dilated radius is used to effectively solve the problem of large overlaps between particles due to particle penetration, the actual dilated radius still cannot be given a value that tends to zero. When the actual dilated radius tends to zero, a dilated polyhedron with smooth surfaces approximates a polyhedral particle with sharp edges and vertices, and the contact points and overlaps between particles are not accurately calculated using the virtual sphere method. Feng (2021a, b, c) proposed an energy conservation algorithm that can be used to accurately and stably calculate the contact forces between arbitrarily shaped polyhedrons. Therefore, the efficient and stable numerical simulations of arbitrarily shaped granular materials with smooth surfaces using the Minkowski sum contact algorithm are the main emphasis in this paper.

4.2.4 Contact Force Model Between Particles

In the DEM simulation, the total contact force ($\boldsymbol{F}$) of a particle contains both the normal and tangential forces (Zhu et al. 2007). The normal force ($\boldsymbol{F}_\mathrm{n}$) contains the normal elastic force ($\boldsymbol{F}_\mathrm{n}^\mathrm{e}$) and normal damping force ($\boldsymbol{F}_\mathrm{n}^\mathrm{d}$), which can be expressed as follows:

$$\boldsymbol{F}_\mathrm{n}^\mathrm{e} = 4/3 E^* \sqrt{R^*} \boldsymbol{\delta}_\mathrm{n}^{3/2} / N_\mathrm{p} \tag{4.19}$$

$$\boldsymbol{F}_\mathrm{n}^\mathrm{d} = C_\mathrm{n} \sqrt{8 m^* E^* \sqrt{R^* \boldsymbol{\delta}_\mathrm{n}}} \cdot \boldsymbol{v}_{\mathrm{n},ij} / N_\mathrm{p} \tag{4.20}$$

where $E^* = \frac{E}{2(1-\nu^2)}$, $R^* = \frac{R_i \cdot R_j}{R_i + R_j}$, and $m^* = \frac{m_i \cdot m_j}{m_i + m_j}$. E is Young's modulus, ν is Poisson's ratio, m is the particle mass, C_n is the normal damping coefficient, $\boldsymbol{v}_{\mathrm{n},ij}$ is the normal relative velocity, R is the radius of the particle's volume-equivalent sphere, N_p is the shared number of contact points.

The tangential force ($\boldsymbol{F}_\mathrm{t}$) consists of the tangential elastic force ($\boldsymbol{F}_\mathrm{t}^\mathrm{e}$) and tangential damping force ($\boldsymbol{F}_\mathrm{t}^\mathrm{d}$), which can be expressed as follows:

$$\boldsymbol{F}_\mathrm{t}^\mathrm{e} = \mu_\mathrm{s} \left|\boldsymbol{F}_\mathrm{n}^\mathrm{e}\right| \left(1 - \left(1 - \min\left(\boldsymbol{\delta}_\mathrm{t}, \boldsymbol{\delta}_{\mathrm{t,max}}\right) / \boldsymbol{\delta}_{\mathrm{t,max}}\right)^{3/2}\right) \cdot \boldsymbol{\delta}_\mathrm{t} / |\boldsymbol{\delta}_\mathrm{t}| \tag{4.21}$$

$$\boldsymbol{F}_{\mathrm{t}}^{\mathrm{d}} = C_{\mathrm{t}}\left(6\mu_{\mathrm{s}} m^{*}\left|\boldsymbol{F}_{\mathrm{n}}^{\mathrm{e}}\right|\sqrt{1-\min\left(\boldsymbol{\delta}_{\mathrm{t}}, \boldsymbol{\delta}_{\mathrm{t,max}}\right)/\boldsymbol{\delta}_{\mathrm{t,max}}}\right)^{1/2} \cdot \boldsymbol{v}_{\mathrm{t},ij} \tag{4.22}$$

where μ_{s} is the sliding friction coefficient, C_{t} the tangential damping coefficient, $\boldsymbol{v}_{\mathrm{t},ij}$ the relative tangential velocity. $\boldsymbol{\delta}_{\mathrm{t}}$ is the tangential overlap and $\boldsymbol{\delta}_{\mathrm{t,max}} = \mu_{\mathrm{s}}(2-\nu)/2(1-\nu) \cdot \boldsymbol{\delta}_{\mathrm{n}}$.

When there is a relative rotation between the two particles, the rolling friction moment ($\boldsymbol{M}_{\mathrm{r}}$) is obtained as follows:

$$\boldsymbol{M}_{\mathrm{r}} = \mu_{\mathrm{r}} R^{*}|\boldsymbol{F}_{\mathrm{n}}|\hat{\boldsymbol{\omega}}_{ij} \tag{4.23}$$

where μ_{r} denotes the rolling friction coefficient. $\hat{\boldsymbol{\omega}}_{ij}$ is the relative rotational speed obtained by $\hat{\boldsymbol{\omega}}_{ij} = \boldsymbol{\omega}_{ij}/\left|\boldsymbol{\omega}_{ij}\right|$.

The DEM time step for arbitrarily shaped particles is calculated based on the time step of the spheres, obtained as follows (Kremmer and Favier 2001):

$$t_{\max} = \frac{\pi r_{\min}\sqrt{\rho/G}}{0.163\nu + 0.8766} \tag{4.24}$$

where ρ is the particle density, G is the shear modulus, $r_{\min}$ is the minimum dilated radius of the arbitrary morphological particles in a complex granular system. In this study, $\mathrm{d}t \leq 0.01 t_{\max}$ was applied to the current time step.

4.3 Numerical Validation and Applications

In this section, numerical examples of single and multiple particles were performed to validate the conservation and accuracy of the proposed Minkowski sum contact algorithm. Moreover, mixed granular flows containing multiple dilated DEM models were simulated to verify the robustness of the algorithm.

4.3.1 Elastic Collision Between Two Particles

The elastic collisions between two particles were used as the first fundamental example to validate the conservation of the Minkowski sum contact algorithm. Figure 4.11 shows arbitrarily shaped particles with smooth surfaces constructed by a dilated sphere, dilated super-ellipsoid, dilated spherical harmonic function, and dilated polyhedron. Particles of different shapes had the same volume and material parameters. The radii of the volume-equivalent sphere and dilated radius were 20 mm and 2 mm, respectively. The Young's modulus and Poisson's ratio of the particles were

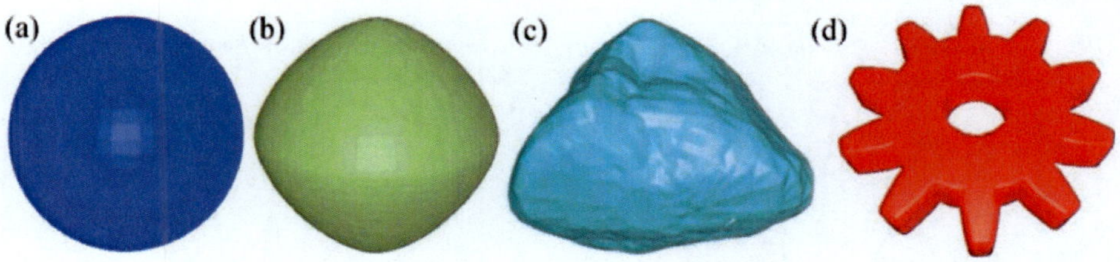

Fig. 4.11 Arbitrarily shaped particles with smooth surfaces constructed by different dilated DEM models: **a** dilated sphere; **b** dilated super-ellipsoid; **c** dilated spherical harmonic function; **d** dilated polyhedron

5 GPa and 0.3, respectively, and their density was 2500 kg/m^3. Two identical particles with arbitrary shapes collided elastically with a horizontal velocity of 0.5 m/s. Damping, friction, and gravity were not considered during collision.

Figure 4.12 shows the variation of the translational, rotational, and total kinetic energies of two particles with a normalized amount of time during elastic collision. Two illustrations represent the spatial orientation and velocity of the two particles before and after the collision. The color indicates the velocity magnitude of the particles. At the initial moment, the two particles have a random spatial angle. During the collision of spherical particles, the variation of the translational kinetic energy with time is the same as the variation of the total kinetic energy, which agrees with the theoretical results. After single or multiple collisions of irregular particles, the translational kinetic energy of the granular system was reduced, and part of the translational kinetic energy was converted to rotational kinetic energy. The total kinetic energy of two arbitrarily shaped particles with smooth surfaces modeled by various dilated DEM models remains in conservation.

To validate the effect of the number (N_f) of surface meshes of arbitrarily shaped particles on the conservation of the total kinetic energy of two particles during elastic collisions, the number of surface meshes of gear-shaped particles was reduced by the automatic mesh simplification method, as shown in Fig. 4.13a. The number of surface meshes of the gear-shaped particles was reduced from 5000 to 1000, and the variation in the total kinetic energy of the particles with a normalized amount of time during the first elastic collision of the two gear-shaped particles was compared, as shown in Fig. 4.13b. The dilated radius was 2 mm. The results show that the relationship between kinetic energy and time was slightly different for gear-shaped particles with different numbers of surface meshes. Different numbers of meshes affected the surface curvature and location of the contact points of the gear-shaped particles, which further changed the translational and rotational motion of the particles. Moreover, particles with different mesh numbers maintained total kinetic energy conservation after collision. This indicates that the proposed Minkowski sum contact algorithm is applicable for simulating the dynamic response of arbitrarily shaped granular materials.

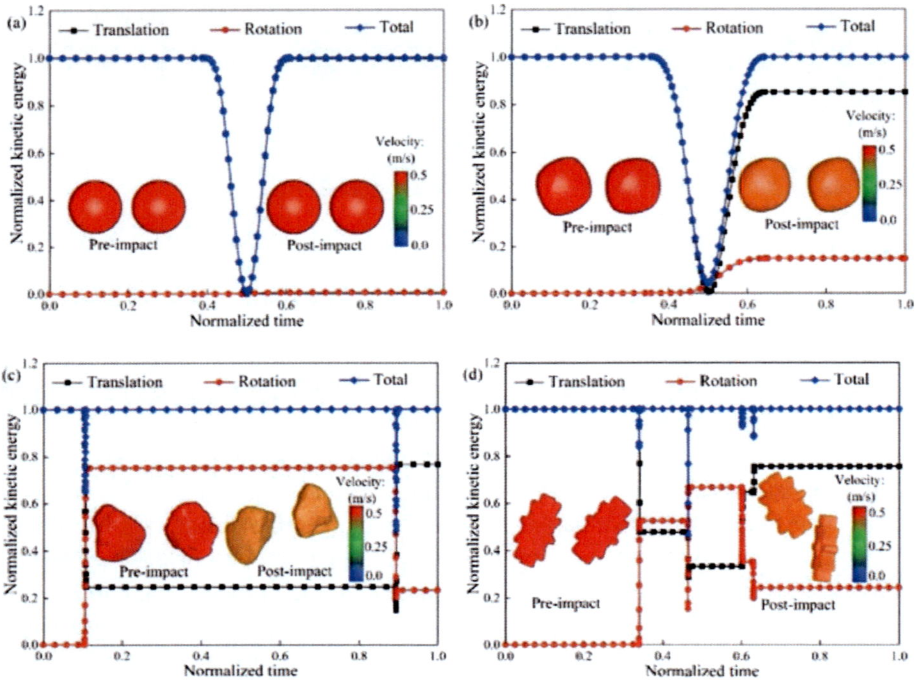

Fig. 4.12 Variation in the translational, rotational, and total kinetic energies of particles with time during the collision between two particles, as modeled by different dilated DEM models: **a** dilated sphere; **b** dilated super-ellipsoid; **c** dilated spherical harmonic function; **d** dilated polyhedron

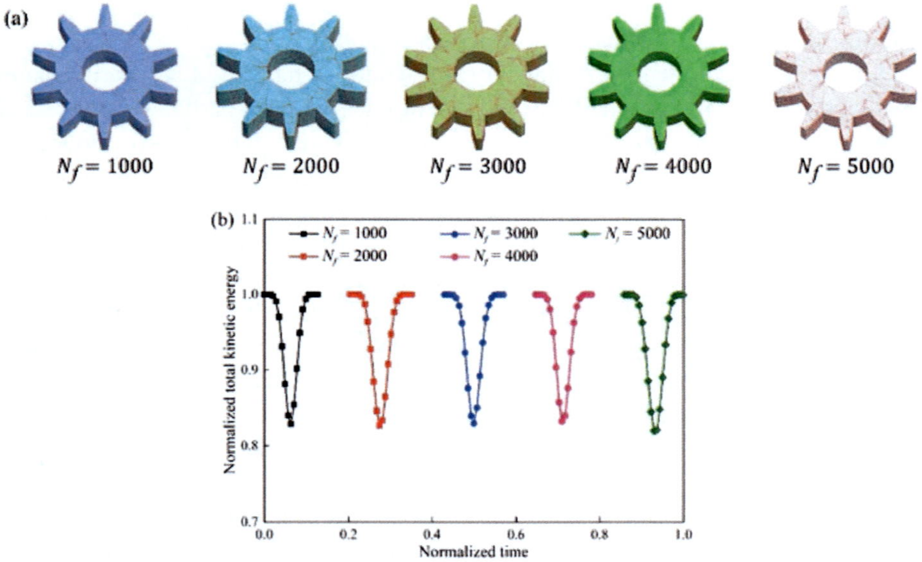

Fig. 4.13 **a** Four types of gear-shaped particles with different numbers of surface meshes. **b** Effects of the number of surface meshes of the particles on the variation of the normalized total kinetic energy with normalized time

To validate the effect of the dilated radius (R_d) on the conservation of the total kinetic energy of arbitrarily shaped particles during elastic collisions, gear-shaped particles with different dilated radii were constructed using a smooth polyhedral model, as shown in Fig. 4.14a. Here, $N_f = 1000$. Figure 4.14b shows the variation in the normalized total kinetic energy with normalized time for gear-shaped particles with different dilated radii during the first elastic collision. The results show that the relationship between kinetic energy and time was slightly different for particles with different dilated radii. The dilated radius changed the smoothness of the particle surface, which further changed the position of the contact point when particles collide. The time step of the DEM simulation depended on the dilated radius of the particles, and a smaller dilated radius corresponded to a smaller DEM time step. Thus, a larger dilated radius was chosen to improve the calculation efficiency of the numerical simulation with guaranteed computational accuracy. Arbitrarily shaped particles with different dilated radii maintained the total kinetic energy conservation after collision, which indicates that the current contact algorithm is suitable for simulating granular materials with different surface smoothnesses.

In addition, a dilated polyhedral model was applied to describe a tablet-shaped particle, and the numerical results of a single particle impacting a plane were compared with the theoretical results to verify the computational accuracy of the proposed algorithm. The rim diameter, total thickness, and band thickness of this tablet-shaped particle were 10.42 mm, 6.75 mm, and 4.25 mm, respectively, as shown in Fig. 4.15a. Young's modulus and Poisson's ratio were 1 GPa and 0.3, respectively. The particle density and dilated radius were 958.9 kg/m^3 and 1 mm, respectively. Damping, friction, and gravity were not considered during the particle collisions. The

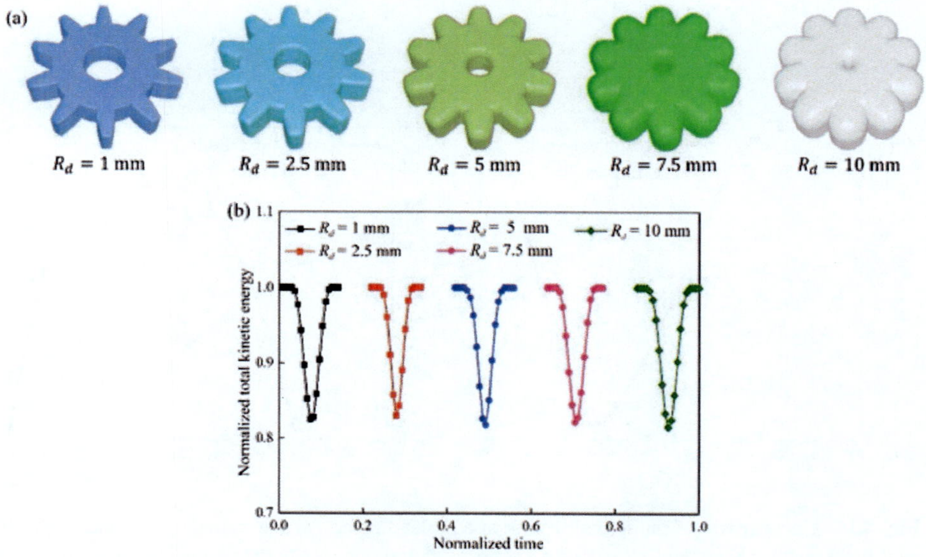

Fig. 4.14 **a** Four types of gear-shaped particles with different dilated radii. **b** Effects of the dilated radius of the particles on the variation in the total kinetic energy with time

analytical solutions for the rebound angular velocity (w_z^+) and translational velocity (v_y^+) of a tablet-shaped particle impacting a plane are expressed as follows (Kureck et al. 2019):

$$w_z^+ = mv_y^-(1+\gamma)r_x/\left(I_{zz} + mr_x^2\right) \tag{4.25}$$

$$v_y^+ = w_z^+ r_x - \gamma v_y^- \tag{4.26}$$

where m is the mass of the particle; v_y^- is the translational velocity of the particle in the y direction at the initial moment, which is -1.0 m/s; γ is the restitution coefficient, which is 1.0; I_{zz} is the moment of inertia of the particle in the z direction. r_x is the distance between the mass center of the particle and the contact point in the x direction, expressed as follows:

$$r_x = \begin{cases} 0 & \text{if}\,\theta \le \alpha; \\ R_1 \cos\left(\frac{\pi}{2} - \theta + \alpha\right) & \text{if}\,\alpha < \theta \le \frac{\pi}{2} - \beta; \\ \left(R_2 - \frac{L_2}{2}\right)\sin\left(\frac{\pi}{2} - \theta\right) & \text{if}\,\frac{\pi}{2} - \beta < \theta \le \frac{\pi}{2}; \end{cases} \tag{4.27}$$

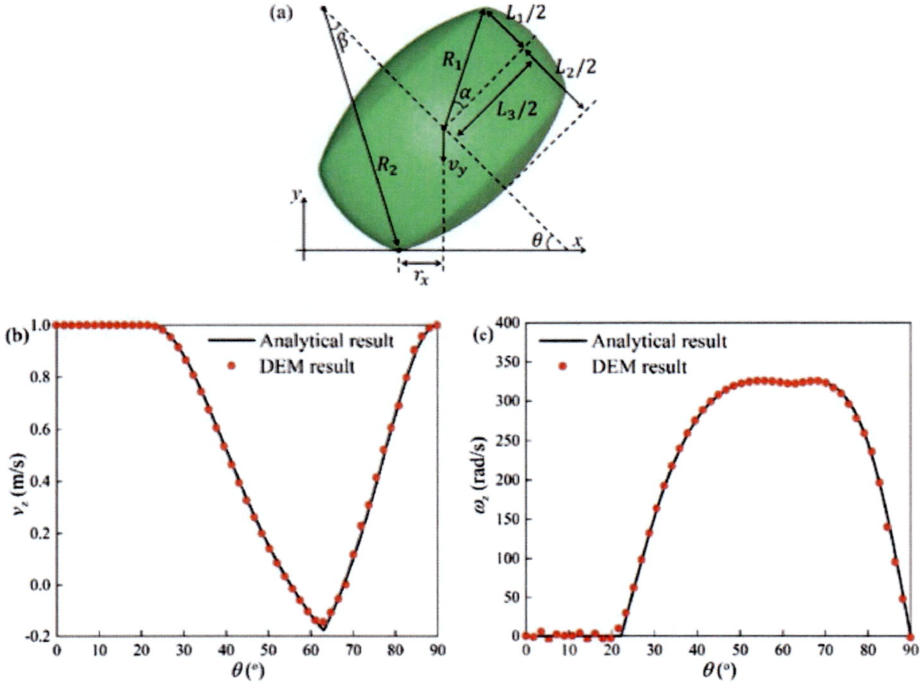

Fig. 4.15 Comparison of numerical results of a tablet-shaped particle impacting a plane with the analytical results (Kureck et al. 2019): **a** a schematic diagram of a tablet-shaped particle impacting a plane; **b** the variation of the rebound translational velocity of the particles with the impact angle; **c** the variation of the rebound angular velocity of the particles with the impact angle

where R_1 and R_2 are the radii of curvature obtained by $R_1 = 0.5\sqrt{L_1^2 + L_3^2}$ and $R_2 = 0.25\left[L_3^2/(L_2 - L_1) + (L_2 - L_1)\right]$, respectively. θ is the impact angle. α and β are the basic angles of a tablet-shaped particle and are expressed as $\alpha = \arcsin(0.5L_1/R_1)$ and $\beta = \arcsin(0.5L_3/R_2)$, respectively. Here, α and β were about 22° and 63°, respectively.

The numerical results of the rebound angular speed and translational speed of a tablet-shaped particle impacting a plane were compared with the analytical results, as shown in Fig. 4.15b, c. When $0° \le \theta \le 22°$, the collision force between the tablet-shaped particle and the plane passed through the mass center of the particle, which caused the particle to have only translational velocity, but not rotational velocity. However, the surface of a tablet-shaped particle was essentially composed of several triangular elements, which allowed the particle to have a slight rotational speed after the collision. Thus, the numerical results for the rebound angular velocity were slightly different from the analytical results. When $22° < \theta \le 90°$, the numerical results obtained by the proposed Minkowski sum contact algorithm were in agreement with the analytical results, which indicates that the current algorithm captured the motion behaviors of arbitrarily shaped particles.

4.3.2 Inelastic Collision Between Two Particles

Inelastic collisions between two particles were used in the second test to verify the stability of the proposed Minkowski sum contact algorithm. A plane with a dilated radius was constructed using the dilated polyhedral model, and the contact forces between a particle and a plane were calculated using the proposed contact algorithm between the particles. Four particles were constructed using dilated spheres, dilated super-ellipsoids, dilated spherical harmonic particles, and dilated polyhedrons. The diameter of the volume-equivalent sphere was 20 mm and the dilated radius was 1 mm. The Young's modulus and Poisson's ratio of the particles were 1 GPa and 0.3, respectively, and the density was 2500 kg/m^3. The coefficients of damping, sliding friction, and rolling friction between a particle and a plane were 0.3, 0.3, and 0.001, respectively. The initial distance between the particle and the plane was 120 mm, and the particle had a random spatial angle. The particle had no initial translational velocity and had angular velocities of 1 rad/s around the x, y, and z axes. Figure 4.16 shows the falling and stabilization processes of differently shaped particles with smooth surfaces under gravity. The results show that particles of different shapes undergo different dynamic processes. The particle collides with the plane several times, and finally remains stationary.

When the dilated radii of the particles were 1, 5, 10, and 20 mm, the spatial positions of the surface nodes of particles were linearly scaled so that these particles had the same volume. Subsequently, the variation in the total kinetic energy of the differently shaped particles with time was counted, as shown in Fig. 4.17. The results show that the relationships between the total energy of particles with different dilated

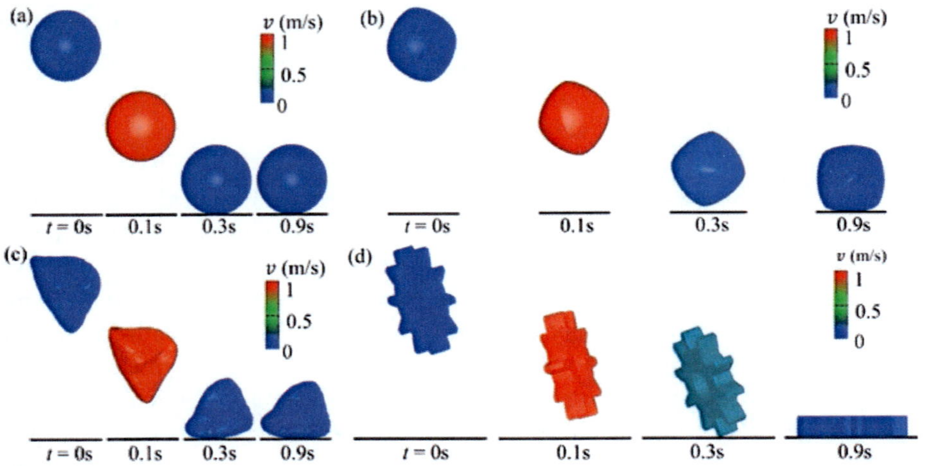

Fig. 4.16 The falling and stabilization processes of particles modeled by various dilated DEM models under gravity: **a** dilated sphere; **b** dilated super-ellipsoid; **c** dilated spherical harmonic function; **d** dilated polyhedron

radii and time were identical during the falling process. When the particle collided with the plane, the different dilated radii changed the position of the contact point between the non-spherical particle and plane. Eventually, the total kinetic energy of the differently shaped particles was equal to zero. Therefore, the proposed contact algorithm exhibits good numerical reliability for simulating the collision processes of arbitrarily shaped particles with different smooth surfaces. In future works, it is necessary to further investigate the effects of the impact velocity and damping coefficient on the mechanical response of particles during the rebound process.

4.3.3 Packing Process of Multiple Particles

The packing of granular materials was applied as a third example to validate the computational stability of the proposed Minkowski sum contact algorithm. Cylindrical particles with dilated radii of 0.1 mm and 5 mm were constructed using dilated super-ellipsoidal functions, and the diameter and height of the particles were 60 mm and 10 mm, respectively. Rectangular particles with dilated radii of 0.1 mm and 5 mm were constructed using dilated polyhedrons, and the length, width, and height of the particles were 80, 80, and 10 mm, respectively. The cylindrical and rectangular particles with dilated radii of 0.1 and 5 mm were stacked vertically, as shown in Fig. 4.18a. Meanwhile, mixed vertical packing of cylindrical and rectangular particles with a dilated radius of 0.1 mm was simulated using the discrete element method. The particle color indicates the distance of the particles from the bottom. Red indicates that the particles were far from the bottom, whereas blue indicates that the particles were close to the bottom.

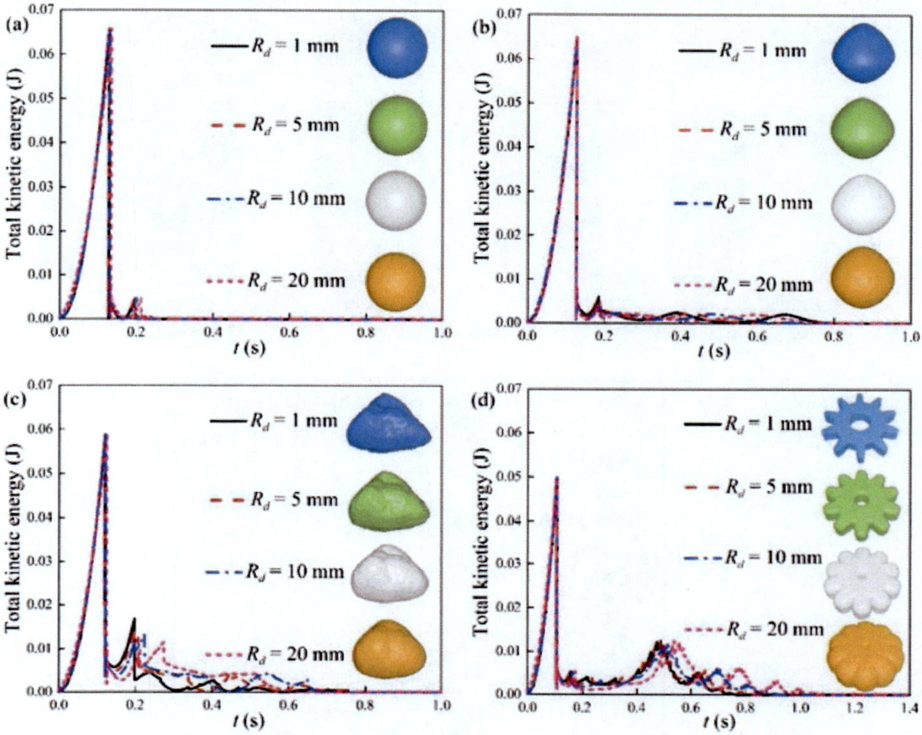

Fig. 4.17 The time history of the total kinetic energy of arbitrarily shaped particles with different dilated radii constructed by different dilated DEM models: **a** dilated sphere; **b** dilated super-ellipsoid; **c** dilated spherical harmonic function; **d** dilated polyhedron

Figure 4.18b shows a comparison of the numerical and analytical results of the normal contact forces between the particles. In the numerical simulation, the contact force between the particles was the sum of the contact forces contributed by multiple contact points between the particles. The analytical solution of the contact force between particles was calculated by accumulating the gravitational forces of all the particles located on the target particle. The results show that the numerical results of the contact forces between cylindrical and rectangular particles with different dilated radii are in good agreement with the analytical results. This indicates that the proposed Minkowski sum contact algorithm can accurately calculate the contact forces between particles of arbitrary shapes modeled by various dilated DEM models.

Moreover, particles of different shapes were constructed using dilated spheres, dilated super-ellipsoidal equations, dilated spherical harmonic functions, and dilated polyhedrons, and these particles had the same dilated radius and volume. The particles constructed by dilated spheres, dilated super-ellipsoidal equations and dilated spherical harmonic functions all had 200 nodes and 396 triangular elements. For particles constructed by dilated polyhedrons, the tetrahedron had 4 nodes and 4 triangular elements; the hexahedron had 8 nodes and 12 triangular elements; the

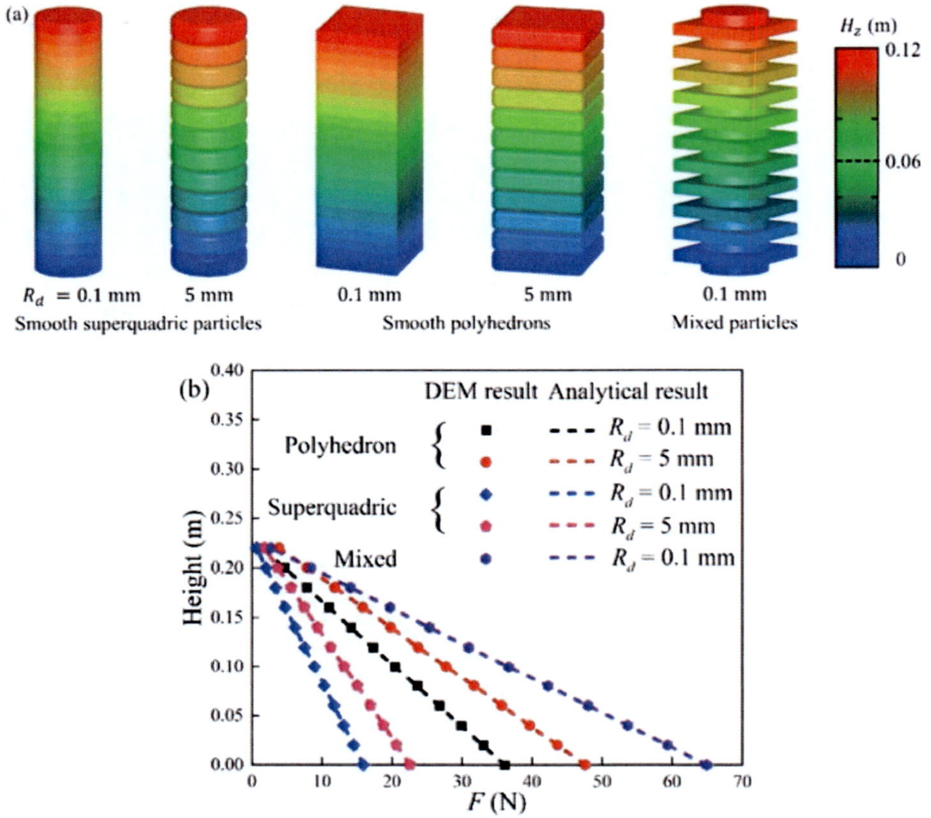

Fig. 4.18 **a** Illustration of the vertical packing of cylindrical and rectangular particles constructed by dilated super-ellipsoidal equations and dilated polyhedrons. **b** Comparison of numerical results with analytical results for the contact force between cylindrical and rectangular particles with different dilated radii in the vertical packing of particles

concave tetrahedron had 5 nodes and 6 triangular elements; the concave hexahedron without internal cavities had 12 nodes and 20 triangular elements; the concave hexahedron with internal cavities had 16 nodes and 32 triangular elements; and the ring-shaped particle had 192 nodes and 368 triangular elements. The diameter of the volume-equivalent sphere and dilated radius of the particles were 40 mm and 5 mm, respectively. The Young's modulus and Poisson's ratio of the particles were 1×10^7 Pa and 0.3, respectively, and the density was 2500 kg/m^3. The coefficients of damping, sliding friction, and rolling friction between a particle and a plane were 0.3, 0.3, and 0.001, respectively. The total number of particles was 2000. The packing process of gear-shaped particles constructed using dilated polyhedrons and the mixed packing process of granular materials modeled using various dilated DEM models were simulated separately, as shown in Fig. 4.19. The length, width, and height of the rectangular container were 0.6, 0.6, and 2 m, respectively. Initially, all particles had random positions and spatial orientations, and there was no overlap between the particles. The color indicates the particle shape. The differently shaped particles fell into the container, and there was no relative motion between the particles at the final moment.

The variation in the kinetic, potential, and total energies of the arbitrarily shaped particles over time during the packing process is shown in Fig. 4.20. As the time increased, the kinetic energy of the granular material increased and subsequently decreased, whereas both the potential energy and total energy decreased. Eventually, the kinetic energy of the granular materials approached zero, and the total energy was equal to the potential energy. This indicates that the proposed Minkowski sum contact algorithm had good numerical stability for simulating mixed granular systems consisting of particles of different shapes modeled by various dilated DEM models.

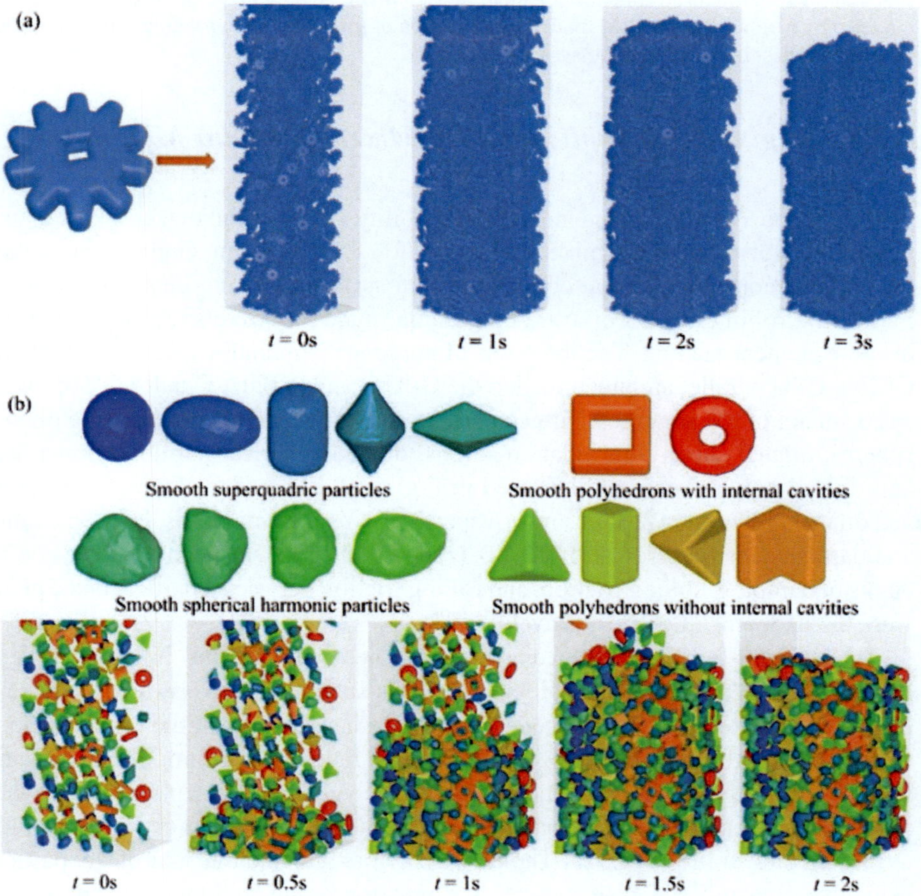

Fig. 4.19 **a** The packing process of gear-shaped particles constructed by dilated polyhedrons; **b** mixed packing process of particles of different shapes modeled by the dilated sphere, dilated super-ellipsoid, dilated spherical harmonic function, and dilated polyhedron

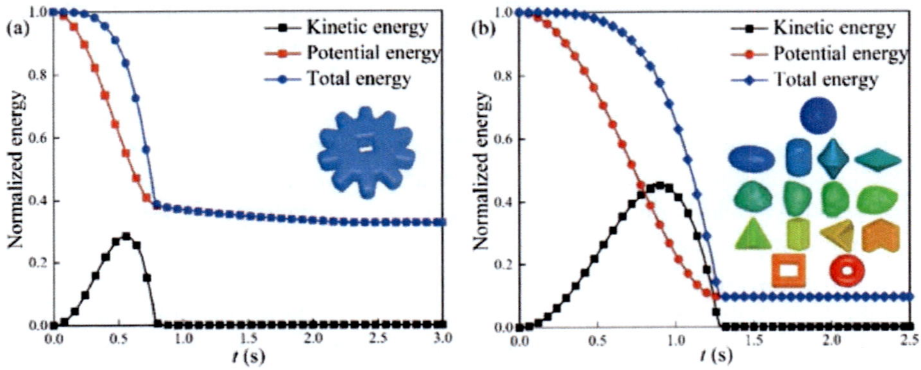

Fig. 4.20 The normalized kinetic, potential, and total energies of particles of different shapes modeled by various dilated DEM models as a function of time during the packing process: **a** gear-shaped particles; **b** mixed granular systems

4.3.4 Engineering Applications of Mixed Granular Materials

To verify the robustness and numerical stability of the Minkowski sum contact algorithms for simulating mixed granular flows in different engineering fields, the stabilization processes on complex terrain surfaces, mixing behaviors inside a mixer, and rolling motion of a tire on granular materials were simulated using the discrete element method. The three sets of numerical simulations were based on the CUDA-GPU parallel architecture, and the GPU is an NVIDIA Quadro GV100 with 5120 stream processor cores. Fifteen particle shapes were constructed using dilated spheres, dilated super-ellipsoidal equations, dilated spherical harmonic functions, and dilated polyhedrons, as illustrated in Fig. 4.19b. The equivalent sphere radius and dilated radius were 20 and 5 mm, respectively. Young's modulus, Poisson's ratio, and particle density were 1×10^7 Pa, 0.3, and 2500 kg/m^3, respectively. The coefficients of damping, sliding friction, and rolling friction between a particle and a plane were 0.3, 0.3, and 0.001, respectively.

The complex terrain surface composed of plateaus, mountains, hills, and valleys is widely applied to predict geological hazards such as collapses, landslides, and mudflows. The total number of nodes and triangular elements of the terrain surface was 1597 and 3178, respectively, and the dilated radius was 5 mm. The length and width of the terrain were 1.8 m and 1.2 m, respectively. Initially, 1000 particles with random shapes, positions, and spatial orientations were generated over the complex terrain, as shown in Fig. 4.21a. The particle colors indicate the sum of the particle velocities, and red and blue indicate the maximum and minimum speeds of the particles, respectively. Differently shaped particles fell under gravity and were in contact with the terrain surface, as shown in Fig. 4.21b, c. The total simulation time was 3 s, while the actual consumption time was about 27.4 h. The friction and viscosity between the particles and terrain surface caused the energy of the granular

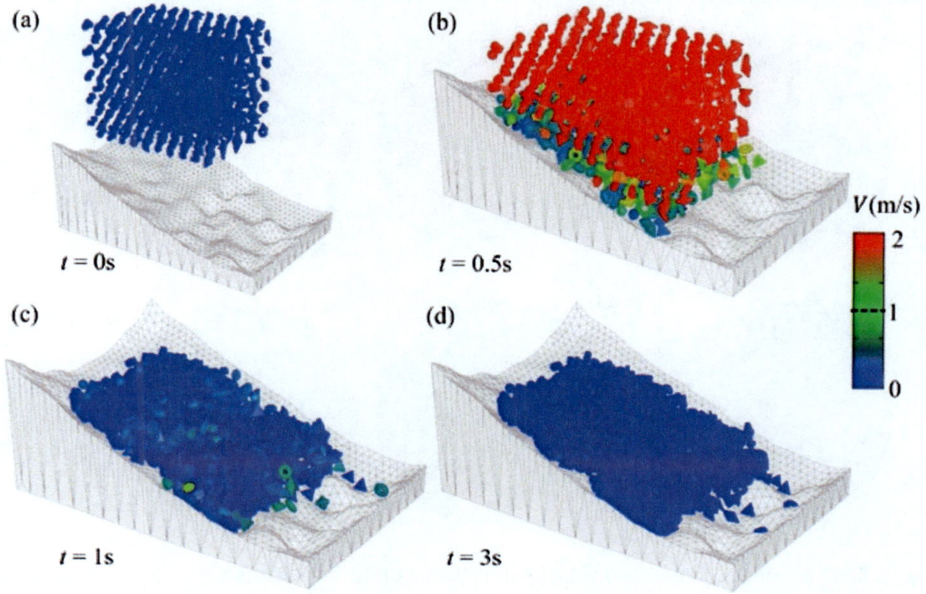

Fig. 4.21 Falling and stabilization processes of mixed granular materials on a complex terrain surface at different moments: **a** $t = 0$ s; **b** $t = 0.5$ s; **c** $t = 1$ s; **d** $t = 3$ s

materials to rapidly dissipate. Meanwhile, the particles located on the hill moved toward the valley, and all particles remained motionless, as shown in Fig. 4.21d.

The mixing process of irregular granular materials in a mixer is critical in industrial production and equipment design, including pharmaceuticals, plastics, product packaging, and food processing. The mixer contained a thin cylindrical shell, two rectangular-bladed impellers, and a cylindrical bottom. The ends and bottom of the blade impeller were almost in contact with the wall and bottom of the mixer, respectively; thus, the intermediate clearance was negligible. The radius and height of the cylindrical container were 0.5 m and 1.0 m, respectively. The angle between the impeller and bottom was 45°, and the rotation speed of the impeller was 30 rpm. A total of 1500 particles of different shapes were generated in the mixer, forming a mixed system under gravity, as shown in Fig. 4.22a, d. The differently shaped particles did not initially mix. As the blade impeller started to rotate, differently shaped particles were gradually mixed and randomly distributed in the mixer, as shown in Fig. 4.22b–c and e–f. The total simulation time was 3 s, while the actual consumption time was about 34.6 h. In addition, particles close to the blade impeller had a large translational velocity owing to the force of the blade impeller, which is similar to the numerical results of previous studies (Sinnott and Cleary 2015; Cleary and Sinnott 2008).

As essential equipment for transportation, resource exploration, agricultural production, and mineral collection, off-road vehicles tires encounter complex terrain consisting of arbitrarily shaped granular materials. Therefore, an in-depth study of the

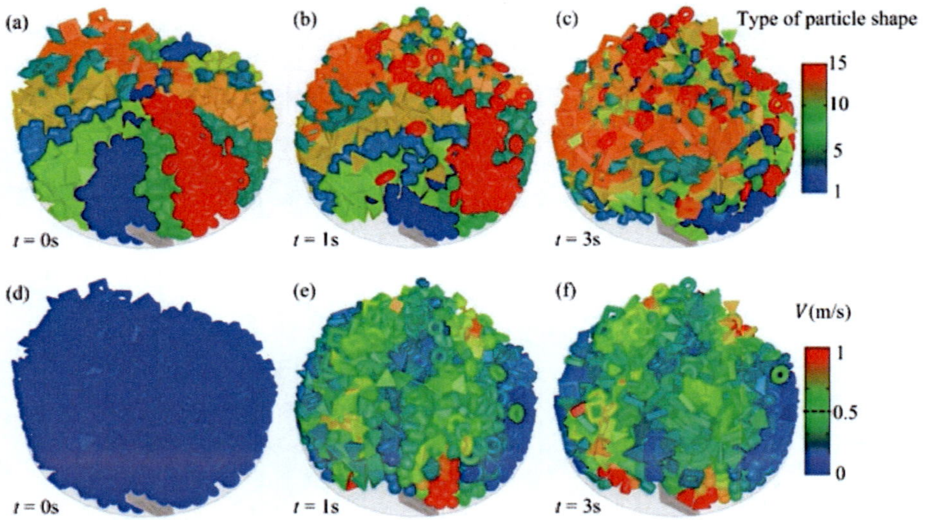

Fig. 4.22 Mixing processes and velocity distribution of the mixed granular materials in the mixer at different moments: **a–c** mixing processes; **d–e** velocity distribution

interaction between tires and granular terrain is vital for the design and safety evaluation of off-road vehicles with good traction. Initially, 2100 particles with random shapes, positions, and spatial orientations were generated above the bottom and formed a stable system of mixed particles under gravity. Subsequently, a tire with a tread pattern was generated above the mixed granular bed, as shown in Fig. 4.23a. The total number of nodes and triangular elements of the tire were 2769 and 5538, respectively, and the dilated radius was 5 mm. The tire was rotated at 30 rpm. The particles had a large translational velocity in the contact area with the tire, and the kinetic energy of the particles was quickly dissipated through friction and damping between the particles, as shown in Fig. 4.23b–c. The total simulation time was 2 s, while the actual consumption time was about 31.5 h. Arbitrarily shaped particles have strong interlocking in a mixed granular system, and it is difficult for particles to slide and rotate relative to each other. Thus, there were insignificant rolling traces at the center of the mixed granular bed, which was slightly different from the numerical results obtained from the spherical granular bed in previous references (Michael et al. 2015; Yang et al. 2020; Zeng et al. 2020).

4.4 Summary

In this chapter, a novel Minkowski sum contact algorithm is proposed to simulate the dynamic properties of mixed granular systems containing various dilated DEM models in static or dynamic structures. In this algorithm, particles of different shapes are modeled using dilated super-ellipsoids, dilated spherical harmonic functions,

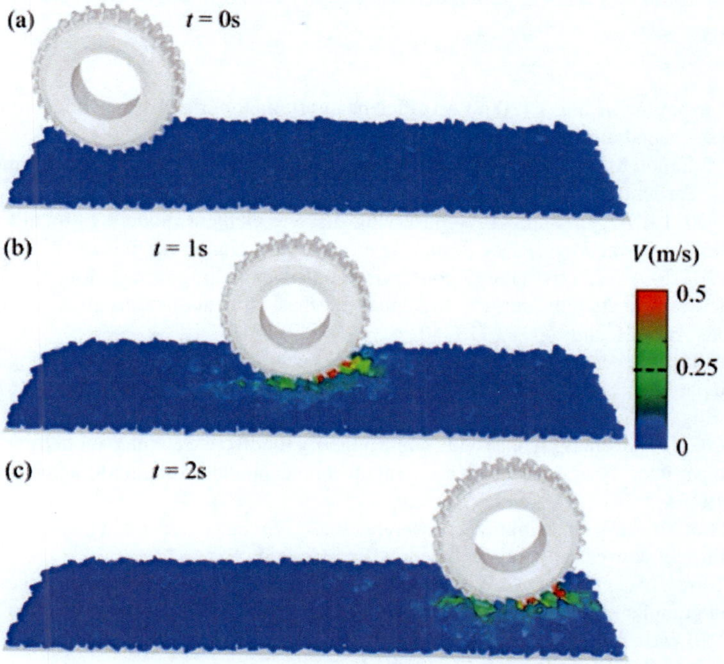

Fig. 4.23 The rolling process of a tire on the mixed granular bed at different moments: **a** $t = 0$ s; **b** $t = 1$ s; **c** $t = 2$ s

and dilated polyhedrons. Subsequently, single or multiple contact points between differently shaped particles and between particles and complex structures are determined using the Minkowski sum contact algorithm. Automatic mesh simplification and GPU parallel methods are developed to improve the computational efficiency of DEM simulations. The application of this algorithm enables the dynamic characterization of mixed granular systems that contain multiple dilated DEM models in static or dynamic structures. Finally, four sets of numerical samples are obtained to verify the conservation, accuracy, and robustness of the proposed Minkowski sum contact algorithm, including elastic collisions between particles, inelastic collisions between particles, packing of multiple particles, and mixed granular flows. The numerical results agree well with the analytical results, indicating that the proposed algorithm is applicable and robust for simulating arbitrarily shaped particles and mixed granular systems constructed using different dilated DEM models. This algorithm is broadly applicable to mixed granular flows in various engineering fields.

References

Alonso-Marroquín F, Wang Y (2009) An efficient algorithm for granular dynamics simulations with complex-shaped objects. Granular Matter 11:317–329

Cleary PW, Sinnott MD (2008) Assessing mixing characteristics of particle-mixing and granulation devices. Particuology 6:419–444

Feng YT (2021a) A generic energy-conserving discrete element modeling strategy for concave particles represented by surface triangular meshes. Int J Numer Meth Eng 122:2581–2597

Feng YT (2021b) An effective energy-conserving contact modelling strategy for spherical harmonic particles represented by surface triangular meshes with automatic simplification. Comput Methods Appl Mech Eng 379:113750

Feng YT (2021c) An energy-conserving contact theory for discrete element modelling of arbitrarily shaped particles: contact volume based model and computational issues. Comput Methods Appl Mech Eng 373:113493

Feng YT (2021d) An energy-conserving contact theory for discrete element modelling of arbitrarily shaped particles: basic framework and general contact model. Comput Methods Appl Mech Eng 373:113454

Feng YT, Tan Y (2019) On Minkowski difference-based contact detection in discrete/discontinuous modelling of convex polygons/polyhedral. Eng Comput 37:54–72

Galindo-Torres SA, Pedroso DM, Williams DJ, Li L (2012) Breaking processes in three-dimensional bonded granular materials with general shapes. Comput Phys Commun 183:266–277

Gei M (2010) Wave propagation in quasiperiodic structures: stop/pass band distribution and prestress effects. Int J Solids Struct 47:3067–3075

Govender N, Wilke DN, Kok S, Els R (2014) Development of a convex polyhedral discrete element simulation framework for NVIDIA Kepler based GPUs. J Comput Appl Math 270:386–400

Govender N, Wilke DN, Pizette P, Abriak N-E (2018) A study of shape non-uniformity and poly-dispersity in hopper discharge of spherical and polyhedral particle systems using the Blaze-DEM GPU code. Appl Math Comput 319:318–336

Gui N, Yang X, Tu J, Jiang S (2016) An extension of hard-particle model for three-dimensional non-spherical particles: mathematical formulation and validation. Appl Math Model 40:2485–2499

Höhner D, Wirtz S, Scherer V (2014) A study on the influence of particle shape and shape approximation on particle mechanics in a rotating drum using the discrete element method. Powder Technol 253:256–265

Houlsby GT (2009) Potential particles: a method for modelling non-circular particles in DEM. Comput Geotech 36:953–959

Jha PK, Desai PS, Bhattacharya D, Lipton R (2021) Peridynamics-based discrete element method (PeriDEM) model of granular systems involving breakage of arbitrarily shaped particles. J Mech Phys Solids 151:104376

Kafashan J, Wiącek J, Abd Rahman N, Gan J (2019) Two-dimensional particle shapes modelling for DEM simulations in engineering: a review. Granular Matter 21:1–19

Kildashti K, Dong K, Samali B (2020) An accurate geometric contact force model for super-quadric particles. Comput Methods Appl Mech Eng 360:112774

Kremmer M, Favier JF (2001) A method for representing boundaries in discrete element modelling—part II: kinematics. Int J Numer Meth Eng 51:1423–1436

Kureck H, Govender N, Siegmann E, Boehling P, Radeke C, Khinast JG (2019) Industrial scale simulations of tablet coating using GPU based DEM: a validation study. Chem Eng Sci 202:462–480

Liu L, Ji S (2020) A new contact detection method for arbitrary dilated polyhedra with potential function in discrete element method. Int J Numer Meth Eng 121:5742–5765

Liu X, Zhou A, Shen S-L, Li J, Sheng D (2020) A micro-mechanical model for unsaturated soils based on DEM. Comput Methods Appl Mech Eng 368:113183

Michael M, Vogel F, Peters B (2015) DEM–FEM coupling simulations of the interactions between a tire tread and granular terrain. Comput Methods Appl Mech Eng 289:227–248

Nie Z, Fang C, Gong J, Yin ZY (2020) Exploring the effect of particle shape caused by erosion on the shear behaviour of granular materials via the DEM. Int J Solids Struct 202:1–11

Nye B, Kulchitsky AV, Johnson JB (2014) Intersecting dilated convex polyhedra method for modeling complex particles in discrete element method. Int J Numer Anal Meth Geomech 38:978–990

Sinnott MD, Cleary PW (2009) Vibration-induced arching in a deep granular bed. Granular Matter 11:345–364

Sinnott MD, Cleary PW (2015) The effect of particle shape on mixing in a high shear mixer. Comput Part Mech 3:477–504

Wojtkowski M, Pecen J, Horabik J, Molenda M (2010) Rapeseed impact against a flat surface: physical testing and DEM simulation with two contact models. Powder Technol 198:61–68

Yang P, Zang M, Zeng H, Guo X (2020) The interactions between an off-road tire and granular terrain: GPU-based DEM-FEM simulation and experimental validation. Int J Mech Sci 179:105634

Zeng H, Xu W, Zang M, Yang P (2020) Calibration of DEM-FEM model parameters for traction performance analysis of an off-road tire on gravel terrain. Powder Technol 362:350–361

Zhao S, Evans TM, Zhou X (2018) Shear-induced anisotropy of granular materials with rolling resistance and particle shape effects. Int J Solids Struct 150:268–281

Zhao Y, Xu L, Umbanhowar PB, Lueptow RM (2019) Discrete element simulation of cylindrical particles using super-ellipsoids. Particuology 46:55–66

Zhu HP, Zhou ZY, Yang RY, Yu AB (2007) Discrete particle simulation of particulate systems: theoretical developments. Chem Eng Sci 62:3378–3396

Chapter 5
Arbitrarily Shaped DEM Model Based on Level Set Method

The level set method, first introduced by Osher and Sethian (1988), is a common and efficient method for calculating the motion of an interface and tracking its evolution (Osher and Sethian 1988; Caselles et al. 1993; Osher and Fedkiw 2001; Sethian 2001). Because these interfaces easily form sharp corners, internal cracks, and merge together in a robust way (Sukumar et al. 2001; Wang et al. 2007; Tran et al. 2011), the level set method has a wide range of applications, including solid modeling, crack characterization, image processing, and segmentation (Hettich and Ramm 2006; Hettich et al. 2008; Legrain et al. 2011). For the complex morphology of non-spherical granular materials, XRCT is used to obtain the image data of arbitrarily shaped particles, and the geometrical morphology of particles is mathematically characterized by the level set method according to the gradient of the X-ray attenuation (Vlahini et al. 2014; Macedo et al. 2018). A smooth distance function between two particles is established by the level set method, and the translation and rotation of the particles and the next collision between particles are accurately predicted (Stafford and Jackson 2010; Vlahinic et al. 2017). Recently, the level set method combined with the discrete element method can be used to reasonably simulate the motion behaviors of non-spherical granular materials (Tahmasebi 2018; Duriez and Bonelli 2021; Duriez and Galusinski 2021). The shape parameters of non-spherical particles such as aspect ratio, circularity, and main geometric direction are extracted by the level set method and then input into the DEM simulations (Jerves et al. 2016, 2017; Harmon et al. 2020). The combination of level set methods and discrete element methods is used not only to capture mechanical behaviors as macroscopic scales (such as stress–strain and volume-strain results of the triaxial test), but also to reproduce shear bands in a similar way to experiments, as well as local and particle-scale quantities (such as local deviatoric stress and particle rotation) (Kawamoto et al. 2016, 2018; Lim et al. 2016). Meanwhile, the combination of these two methods has the ability to reflect the contact force distribution within granular systems composed

S. Wang and S. Ji, *Computational Mechanics of Arbitrarily Shaped Granular Materials*, Springer Tracts in Mechanical Engineering,
https://doi.org/10.1007/978-981-99-9927-9_5

of arbitrarily shaped particles and provides quantitative estimates of the evolution of force chains and fabric orientations (Li et al. 2019; Bhattacharya et al. 2021; Harmon et al. 2021).

This chapter introduces a unified method for creating level set functions of arbitrarily shaped particles constructed by different non-spherical DEM models, and the contact points between arbitrarily shaped particles in a mixed granular flow are determined by the level set method. Moreover, the complex structure is discretized into a series of triangular elements, and a contact algorithm between level set functions and triangular elements is introduced and used for simulating mixed granular flows in moving or static complex structures. Three sets of numerical tests are performed by applying the present DEM model, including single-particle collision with a wall, granular accumulation, and mixed granular flow.

5.1 Modeling Methodology of Arbitrarily Shaped Elements

In this section, three non-spherical DEM models are introduced, including the superquadric equation, spherical harmonic function, and polyhedral method. Superquadric equations and spherical harmonic functions are used to construct particles with smooth or concave-convex surface characteristics. Polyhedral elements are used to construct particles with sharp edges and/or flat planes. When the polyhedron model has a sufficient number of planes, edges, and vertices, it can theoretically be used to construct particles of any shape.

5.1.1 *Superquadric Equation*

The superquadric equation is an extension of the quadric equation, which is used to construct 80% of the shapes of solids in nature (Zhong et al. 2016). The superquadric equation is expressed as Bar (1981):

$$\left(\left|\frac{x}{a}\right|^{n_2}+\left|\frac{y}{b}\right|^{n_2}\right)^{n_1/n_2}+\left|\frac{z}{c}\right|^{n_1}-1=0 \tag{5.1}$$

where a, b, and c are the half-axis lengths of the superquadric particle along the x, y, and z directions, respectively. n_1 and n_2 are the shape parameters that determine the sharpness of the particle surface. The parametric form of the superquadric equation is expressed as:

$$\begin{cases} x = a(\sin\theta)^{2/n_1}(\cos\varphi)^{2/n_2} \\ y = b(\sin\theta)^{2/n_1}(\sin\varphi)^{2/n_2} \\ z = c(\cos\theta)^{2/n_1} \end{cases} \tag{5.2}$$

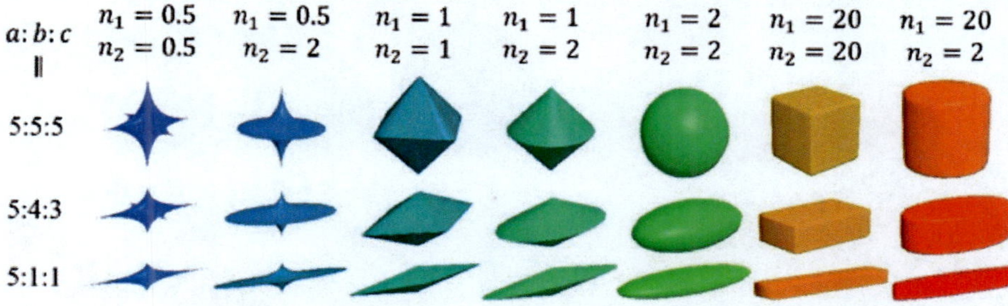

Fig. 5.1 Differently shaped particles constructed by superquadric equations

where $0 \le \theta \le \pi$ and $-\pi \le \varphi \le \pi$. A sphere or an ellipsoid is obtained if $n_1 = n_2 = 2$; cylinders with different aspect ratios are obtained if $n_1 > 2$ and $n_2 = 2$; cube-like particles with different aspect ratios are obtained if $n_1 = n_2 > 2$. Figure 5.1 shows differently shaped particles determined by the superquadric equation.

5.1.2 Spherical Harmonic Function

The spherical harmonic function is an effective way to represent arbitrarily shaped particles mathematically and is used to model concave and convex particles. The spherical harmonic function is expressed as Wang et al. (2021):

$$r(\theta, \varphi) = \sum_{n=0}^{N} \sum_{m=-n}^{n} A_{nm} Y_n^m(\theta, \varphi) \tag{5.3}$$

where r is the distance between the particle surface and the origin of the coordinates. θ is the angle between the line connecting the origin and the surface and the positive direction of the z-axis. φ is the angle between the projection of the line connecting the origin and the surface on the x–y plane and the positive direction of the x-axis, as shown in Fig. 5.2a. A_{nm} is the coefficient of the spherical harmonic function obtained by $A_{nm} = R_{nm} + C_{nm}$. R_{nm} and C_{nm} are the real and imaginary components of A_{nm}, respectively. $Y_n^m(\theta, \varphi)$ is a series of spherical harmonic functions, expressed as:

$$Y_n^m(\theta, \varphi) = \sqrt{\frac{(2n+1)(n-m)!}{4\pi(n+m)!}} P_n^m(\cos(\theta)) e^{im\varphi} \tag{5.4}$$

where m and n are the degree and order of P_n^m. P_n^m is the associated Legendre function, defined as:

$$P_n^m(x) = \begin{cases} (-1)^m \frac{1}{2^n n!} \left(1 - x^2\right)^{\frac{m}{2}} \frac{d^{m+n}}{dx^{m+n}} \left(x^2 - 1\right)^n, & (m \ge 0) \\ (-1)^{-m} \frac{(n+m)!}{(n-m)!} P_n^{-m}(x), & (m < 0) \end{cases} \tag{5.5}$$

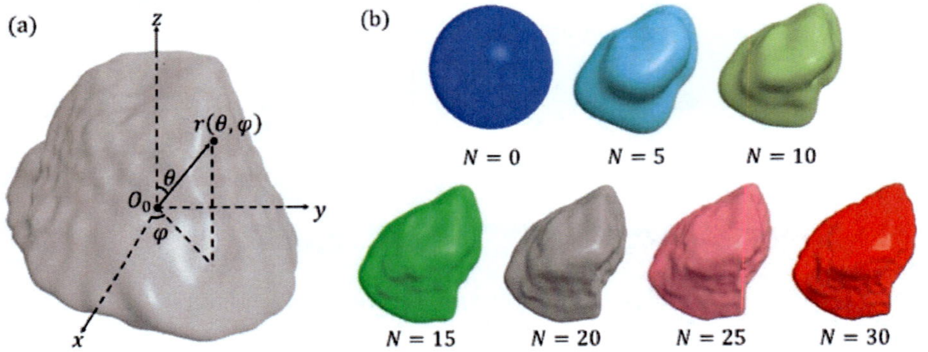

Fig. 5.2 **a** Illustration of the spherical harmonic function in a spherical coordinate system. **b** Particles with different surface roughness determined by the variable parameter N

where $n \in [0, N]$ and $m \in [-n, n]$. As the value of N increases, the particles gradually change from spheres to real particle shapes, and both concave and convex characteristics are observed on the particle surface, as shown in Fig. 5.2b.

5.1.3 Polyhedron Method

The polyhedron method is a mathematical model based on geometric topology (Govender et al. 2018), which can theoretically be used to construct arbitrarily shaped particles. A polyhedron element is composed of several planes, edges, and vertices, as shown in Fig. 5.3.

Fig. 5.3 Arbitrarily shaped particles constructed by polyhedron models

5.2 Contact Algorithm Between Arbitrarily Shaped Particles

In this section, the creation of level set functions and the contact algorithm for arbitrarily shaped particles are introduced in detail, including the creation of zero level set functions, the creation of spatial level set functions, and the calculation of contact points.

5.2.1 Creation of the Zero Level Set Function

Before the DEM simulation, the maximum number of contacts and contact points are predetermined and used to allocate memory for various arrays in the DEM program, expressed as $N_{c,\max} = 20\sigma$ and $N_{p,\max} = \min(0.2N_{zp}, 10^3)$, respectively. $N_{c,\max}$ is the maximum contact number. σ is the aspect ratio of arbitrarily shaped particles, defined as the ratio of the longest axial length to the shortest axial length in the x, y, and z directions. $N_{p,\max}$ is the maximum number of contact points between arbitrarily shaped particles. N_{zp} is the maximum number of zero level set points of differently shaped particles. In future studies, the maximum number of contacts and contact points of granular systems composed of arbitrarily shaped particles need to be further considered to reduce the memory consumption of the DEM program using the level set method.

Owing to the complexity of the contact detection between arbitrarily shaped particles, the bounding sphere and the oriented bounding box can be used to improve the computational efficiency of DEM simulations. First, the bounding sphere is applied as the first rough detection between arbitrarily shaped particles, as shown in Fig. 5.4a. If the sum of the radii of the bounding spheres is greater than the distance between their centroids, the two bounding spheres are in contact. Second, the oriented bounding box is applied as the second rough contact detection between arbitrarily shaped particles, as shown in Fig. 5.4b. This bounding box is based on a separation-axis approach, which can quickly eliminate potential contact pairs and reduce the calculation time. If the projections of the two bounding boxes on the separating axis do not intersect, the two elements are not in contact; otherwise, they are in contact.

Subsequently, each arbitrarily shaped element is discretized into a zero level set function composed of a series of surface points, and the discretized level set function is expressed as $D_p = \psi(\boldsymbol{p})$, where $\boldsymbol{p}$ denotes the surface points of the elements, and D_p denotes the distance between point ($\boldsymbol{p}$) and the surface of the element, which is equal to 0. The superquadric equation and the spherical harmonic function are regarded as the particle shape enveloped by continuous functions, whereas the polyhedron method is regarded as a particle shape enveloped by discretized triangular elements. The methods for creating zero level set functions are introduced below according to the representation characteristics of different DEM models.

(a) Superquadric elements and spherical harmonic functions.

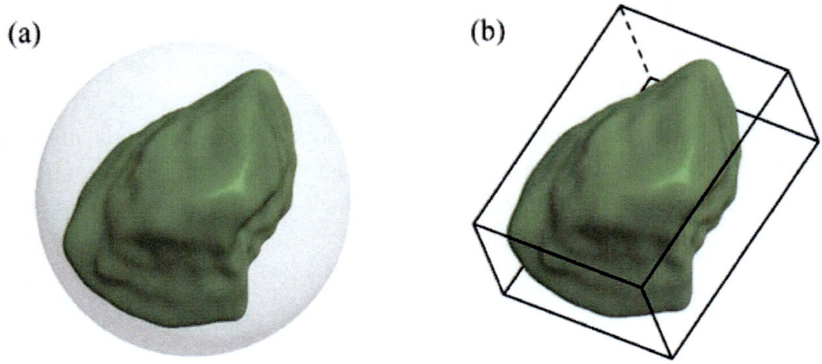

Fig. 5.4 Rough contact detection between arbitrarily shaped particles: **a** bounding sphere and **b** oriented bounding box

To create the zero level set function of the element, the discrete points are evenly distributed on the bounding box of the element, as shown in Fig. 5.5a. $\boldsymbol{O}_{\mathrm{c}}(O_{\mathrm{cx}}, O_{\mathrm{cy}}, O_{\mathrm{cz}})$ are the origin coordinates of the function, $\boldsymbol{q}_i(q_{ix}, q_{iy}, q_{iz})$ are the points on the surface of the bounding box, and $\boldsymbol{p}_i(p_{ix}, p_{iy}, p_{iz})$ are the zero level set points. In the parametric coordinate system, the vector $\boldsymbol{O}_{\mathrm{c}}\boldsymbol{q}_i$ is expressed as:

$$\begin{cases} r_i = \left|\boldsymbol{O}_{\mathrm{c}}\boldsymbol{q}_i\right| \\ \theta_i = \arccos\left(\frac{q_{iz}-O_{\mathrm{cz}}}{r_i}\right) \\ \varphi_i = \arctan\left(\frac{q_{iy}-O_{\mathrm{cy}}}{q_{ix}-O_{\mathrm{cx}}}\right) \end{cases} \tag{5.6}$$

Substituting θ_i and φ_i into the function representing the particle shape, the zero level set points $(\boldsymbol{p}_i)$ in the Cartesian coordinate system are obtained as follows:

$$\begin{cases} p_{ix} = \boldsymbol{f}(\theta_i, \varphi_i)\sin(\theta_i)\cos(\varphi_i) \\ p_{iy} = \boldsymbol{f}(\theta_i, \varphi_i)\sin(\theta_i)\sin(\varphi_i) \\ p_{iz} = \boldsymbol{f}(\theta_i, \varphi_i)\cos(\theta_i) \end{cases} \tag{5.7}$$

where $\boldsymbol{f}(\theta_i, \varphi_i)$ is the parametric equation of the superquadric element or the spherical harmonic function. Considering that there is a one-to-one correspondence between the surface points on the bounding box and the zero level set points, all the zero level set points are obtained by solving the intersection of the surface points of the bounding box and the element surface. Finally, the zero level set points of an arbitrarily shaped element are shown in Fig. 5.5b.

(b) Polyhedral Elements.

To create the zero level set function of the polyhedral element, all the vertices of the polyhedral element are recorded as part of the zero level set points, and the edges and planes are evenly distributed with discrete points. For the points on the edge, a characteristic length is defined in advance, and the zero level set points are evenly

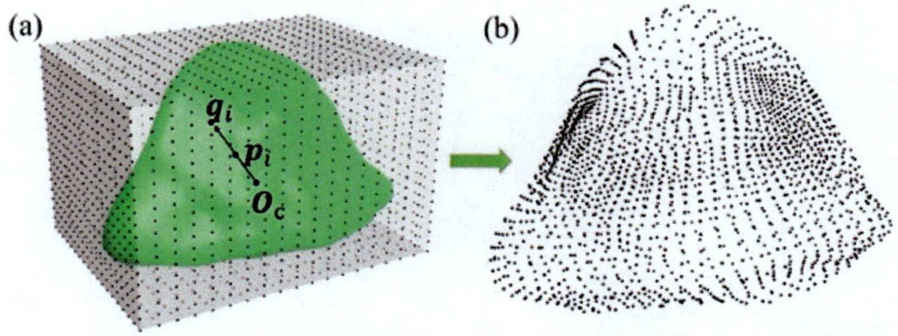

Fig. 5.5 Zero level set function of an arbitrarily shaped element: **a** a series of surface points on the bounding box and **b** a series of zero level set points

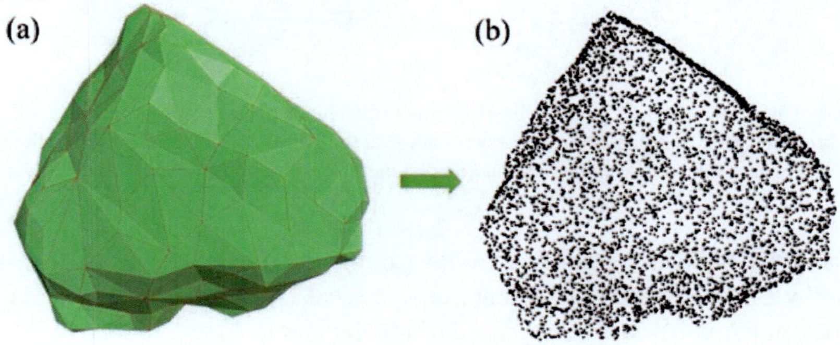

Fig. 5.6 Zero level set function of a polyhedral element: **a** a polyhedral element and **b** a series of zero level set points

distributed according to the ratio of the edge to the characteristic length. The points on the plane are randomly distributed through a random function, and then the effective points are filtered by the characteristic length. As a result, the zero level set points are substantially evenly distributed on the plane. Finally, the zero level set points of the polyhedral elements are shown in Fig. 5.6.

5.2.2 Creation of the Spatial Level Set Function

For different non-spherical DEM models, spatial points are generated and evenly distributed within the bounding box of the element, as shown in Fig. 5.7a. The position of the spatial point i is denoted as $\boldsymbol{e}_i\left(e_{ix}, e_{iy}, e_{iz}\right)$. Meanwhile, the distance (D_i) between this point and the element surface is expressed as $D_i = \psi(\boldsymbol{e}_i)$. If $D_i < 0$, this point is inside the element; if $D_i = 0$, this point is on the surface of the element; if $D_i > 0$, this point is outside the element. For superquadric and spherical harmonic elements, the element surface is easily discretized into a series

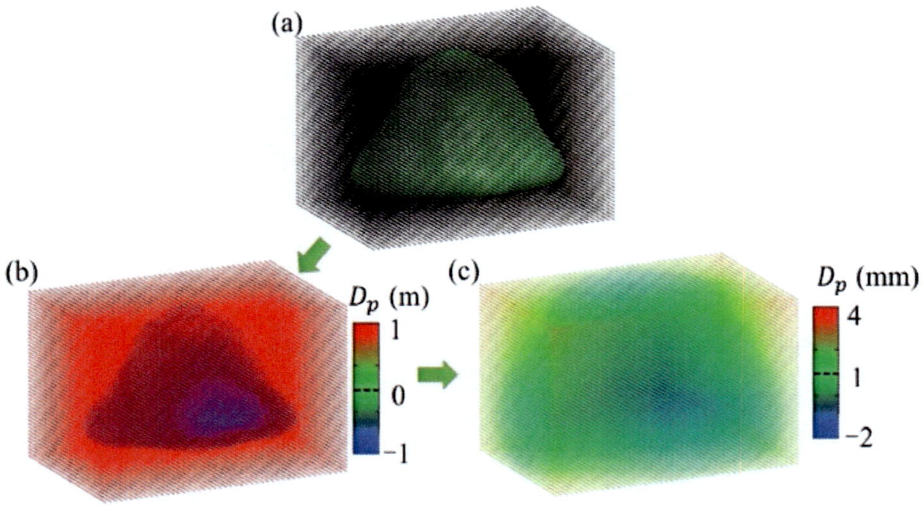

Fig. 5.7 Creation process of a spatial level set function: **a** black discrete points uniformly distributed in the bounding box, **b** red and blue colors indicate that the points are outside and inside the element, respectively, and **c** the color indicates that the distance from the point to the element surface

of triangular elements according to the parametric equation. Therefore, a unified method was developed for different non-spherical DEM models to create a spatial level set function based on discretized triangular elements.

The creation of the spatial level set function consists of two steps: (I) determine whether the spatial point is inside or outside the particle; (II) calculate the shortest distance between the spatial point and the particle surface. First, the positional relationship between the spatial point and particle needs to be determined. An effective method is called the odd–even rule algorithm, which calculates the number of intersections between a ray passing through this point and an arbitrary polyhedral element. If the number of intersection points is odd, this spatial point is inside the element. The distance is expressed as $D_i = -1$, and the color is represented in blue. If the number of intersection points is even, this point is outside the element. The distance is expressed as $D_i = 1$, and the color is represented in red, as exhibited in Fig. 5.7b. It is shown that the element shape composed of the blue points is basically the same as the actual shape.

Second, the shortest distance between the spatial point ($\boldsymbol{e}_i$) and all triangular elements is calculated. A, B, and C represent the three vertices of a triangular element, and the position coordinates of each vertex are denoted as $\boldsymbol{x}_A$, $\boldsymbol{x}_B$, and $\boldsymbol{x}_C$, respectively. $\boldsymbol{a}$, $\boldsymbol{b}$, and $\boldsymbol{c}$ represent the three edge vectors, as shown in Fig. 5.8a. The normal direction of the triangular element is obtained by $\boldsymbol{n}_w = \boldsymbol{a} \times \boldsymbol{b}/|\boldsymbol{a} \times \boldsymbol{b}|$. The projection point of the spatial point onto the plane is denoted as Q, which is obtained by $\boldsymbol{x}_Q(\boldsymbol{x}_{Qx}, \boldsymbol{x}_{Qy}, \boldsymbol{x}_{Qz}) = \boldsymbol{d} + |\boldsymbol{d}\boldsymbol{n}_w| \cdot \boldsymbol{n}_w + \boldsymbol{x}_A$. Subsequently, the positional relationship between point Q and the triangular element must be determined. The position of this point Q is expressed as a function of the unknown parameters α and β with respect to vertex A:

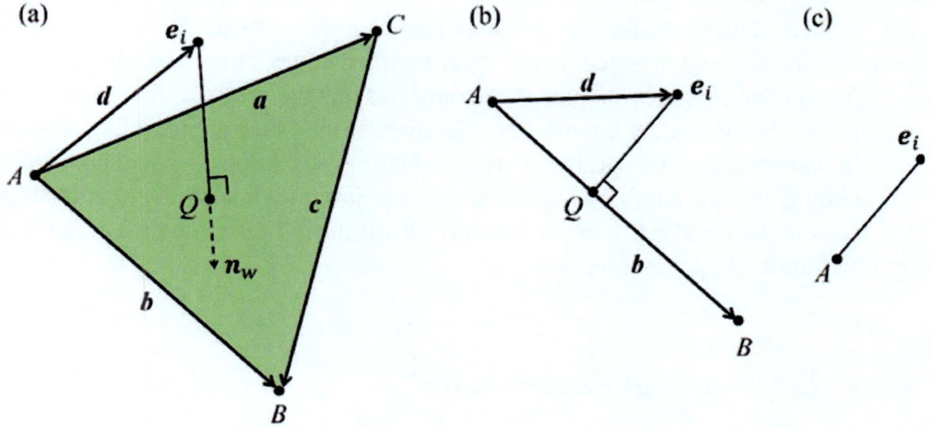

Fig. 5.8 Calculation of the distance between a spatial point and a triangular element: **a** positional relationship between a spatial point and a triangular element, **b** positional relationship between a spatial point and an edge, and **c** positional relationship between a spatial point and a vertex

$$x_Q = \alpha a + \beta b + x_A \tag{5.8}$$

$$\alpha > 0 \cap \beta > 0 \cap \alpha + \beta < 1 \tag{5.9}$$

Here, the values of the parameters α and β are obtained as:

$$\alpha = \frac{(b \cdot b)\big((x_Q - x_A) \cdot a\big) - (b \cdot a)\big((x_Q - x_A) \cdot b\big)}{(a \cdot a)(b \cdot b) - (a \cdot b)(b \cdot a)} \tag{5.10}$$

$$\beta = \frac{(a \cdot a)\big((x_Q - x_A) \cdot b\big) - (a \cdot b)\big((x_Q - x_A) \cdot a\big)}{(a \cdot a)(b \cdot b) - (a \cdot b)(b \cdot a)} \tag{5.11}$$

If the parameters α and β satisfy Eq. (5.9), point Q is located in the triangular element, and the distance $|x_Q e_i|$ is recorded; otherwise, the distance $|x_Q e_i|$ is not recorded.

Third, the shortest distance between the spatial point and all edges is calculated. Taking edge b as an example, the projection of point e_i on edge b is point Q, and the distance AQ is obtained by $AQ = db/|b|$, as shown in Fig. 5.8b. If the distance AQ satisfies $AQ > 0$ and $AQ < |b|$, the distance $|x_Q e_i|$ is obtained by $|x_Q e_i| = |d - AQ \cdot b/|b||$ and recorded; otherwise, the distance $|x_Q e_i|$ is not recorded.

Finally, the shortest distance between the spatial point and all vertices is calculated. Taking vertex A as an example (Fig. 5.8c), the distance $|x_A e_i|$ is recorded. The minimum value of all distances recorded at each spatial point e_i is calculated and denoted as $D_{\min}$. Therefore, the distance (D_i) is obtained by $D_i = D_{\min} \cdot D_i$ and is used for the calculation of the contact points between particles. The spatial level set function of the arbitrarily shaped element is shown in Fig. 5.7c. The color indicates the distance between the level set point and the element surface. The points near the center of the element are blue, while those near the edges of the bounding box are red. The gradual change from blue to red indicates that the spatially discrete level set function is approximately continuous. Meanwhile, the phenomenon of continuous distribution becomes more pronounced for a denser distribution of spatial points. It is worth noting that the number of spatial level set points does not affect the calculation efficiency of the DEM simulations, but only affects the initialization time of the level set functions at the initial moment.

5.2.3 Calculation of Contact Points

Arbitrarily shaped elements with different DEM models are discretized into a zero level set function and a spatial discrete level set function composed of a series of points, and the contact detection between elements is transformed into a solution of two level set functions, as shown in Fig. 5.9. Here, D_p denotes the value of the spatial level set function. In the fixed coordinate system, two adjacent elements are denoted as i and j, respectively. All zero level set points of element j are sequentially brought into the spatial level set function of element i. Considering that the spatial level set function is composed of a series of points uniformly distributed in the bounding box, the distance between these points is the same in the x, y, and z directions, as shown in Fig. 5.10a. A zero level set point of element j is brought into the spatial level set function of element i, and eight level set points of element i around this point are quickly found, as shown in Fig. 5.10b. If the zero level set point is not within the bounding box, this point is outside element i. When all the zero level set points of element j are not within the bounding box of element i, element j is not in contact with element i.

To further determine whether the zero level set point of element j is inside element i, the spacing of the spatial level set points of element i in the three directions are

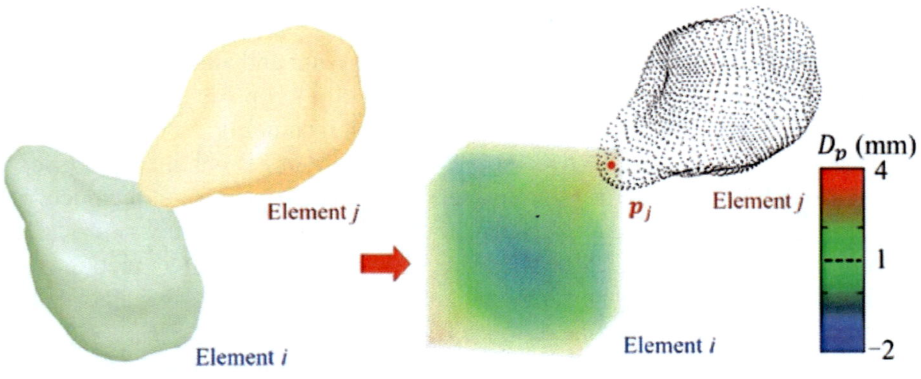

Fig. 5.9 Contact detection between arbitrarily shaped elements based on the level set method

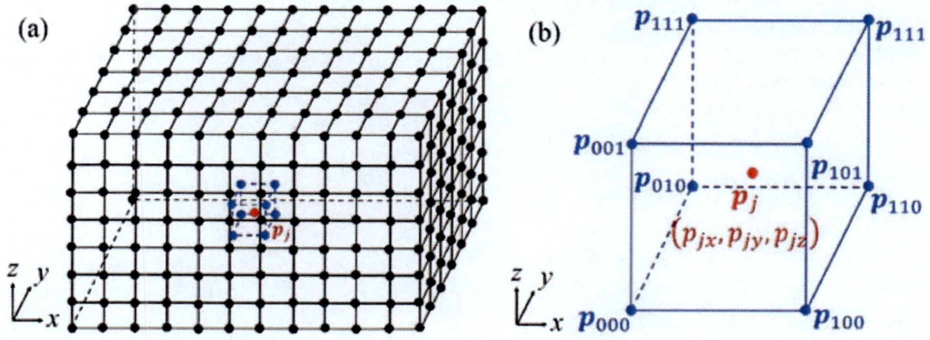

Fig. 5.10 Relationship between a zero level set point ($\boldsymbol{p}_j$) of element j and the spatial level set function of element i: **a** spatial level set function and **b** schematic of point $\boldsymbol{p}_j$ with surrounding level set points

denoted as d_x, d_y, and d_z, respectively. Here, $d_x = d_y = d_z = d_0$. Eight level set points are used to form a hexahedron, and the relative position of point $\boldsymbol{p}_j$ in the hexahedron can be expressed as:

$$\begin{cases} x = \left(p_{jx} - p_{ix,000}\right)/d_0 \\ y = \left(p_{jy} - p_{iy,000}\right)/d_0 \\ z = \left(p_{jz} - p_{iz,000}\right)/d_0 \end{cases} \tag{5.12}$$

The value $\psi\left(\boldsymbol{p}_j\right)$ of the level set function of element i is calculated using the trilinear interpolation method, expressed as Kawamoto et al. (2016):

$$D_{pj} = \psi\left(\boldsymbol{p}_j\right) = \sum_{a=0}^{1}\sum_{b=0}^{1}\sum_{c=0}^{1} \psi_{abc}[(1-a)(1-x)+ax] \\ [(1-b)(1-y)+by][(1-c)(1-z)+cz] \tag{5.13}$$

The gradient value $\nabla\psi\left(\boldsymbol{p}_j\right)$ of the level set function of element i is also calculated using the trilinear interpolation method, expressed as:

$$\frac{\partial\psi\left(\boldsymbol{p}_j\right)}{\partial x} = \sum_{a=0}^{1}\sum_{b=0}^{1}\sum_{c=0}^{1} \psi_{abc}(2a-1)[(1-b)(1-y)+by] \\ [(1-c)(1-z)+cz] \tag{5.14}$$

$$\frac{\partial\psi\left(\boldsymbol{p}_j\right)}{\partial y} = \sum_{a=0}^{1}\sum_{b=0}^{1}\sum_{c=0}^{1} \psi_{abc}[(1-a)(1-x)+ax](2b-1) \\ [(1-c)(1-z)+cz] \tag{5.15}$$

Fig. 5.11 Calculation of multiple contact points between concave elements based on the level set method

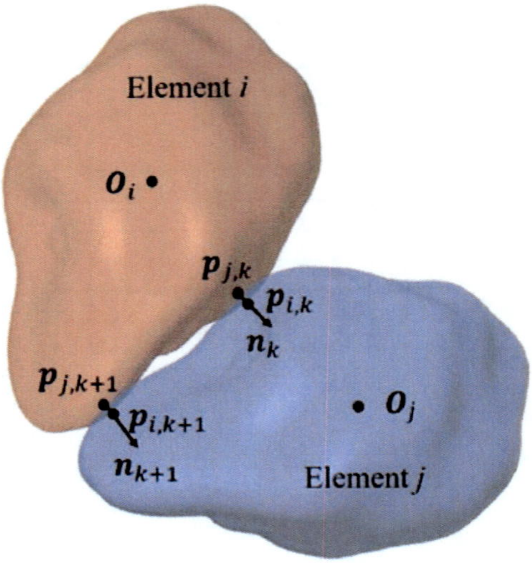

$$\frac{\partial\psi(\boldsymbol{p}_j)}{\partial z}=\sum_{a=0}^{1}\sum_{b=0}^{1}\sum_{c=0}^{1}\psi_{abc}[(1-a)(1-x)+ax]$$
$$[(1-b)(1-y)+by](2c-1) \tag{5.16}$$

If $D_{pj}<0$, point $(\boldsymbol{p}_j)$ is located in element i, and the two elements i and j are in contact. The normal direction $(\boldsymbol{n})$ between elements i and j at point $(\boldsymbol{p}_j)$ is the gradient $(\nabla\psi(\boldsymbol{p}_j))$ of the level set function, and the normal overlap $(\boldsymbol{\delta}_n)$ is obtained by $\boldsymbol{\delta}_n=\psi(\boldsymbol{p}_j)\boldsymbol{n}$, as shown in Fig. 5.11. The multiple contact points between concave particles are calculated for all zero level set points satisfying $D_{pj}<0$, which realizes the calculation of contact points between different non-spherical DEM models.

5.2.4 Calculation of Contact Forces

In the DEM simulations, the total contact force $(\boldsymbol{F})$ of arbitrarily shaped particles includes the normal force $(\boldsymbol{F}_n)$ and the tangential force $(\boldsymbol{F}_t)$, and the total moment $(\boldsymbol{M})$ includes the moments $(\boldsymbol{M}_n$ and $\boldsymbol{M}_t)$ caused by the normal and tangential forces not passing through the centroid of the particle and the rolling friction moment $(\boldsymbol{M}_r)$. The normal force $(\boldsymbol{F}_n)$ includes the normal elastic force $(\boldsymbol{F}_n^e)$ and damping force $(\boldsymbol{F}_n^d)$, expressed as:

$$\boldsymbol{F}_n^e=4/3E^*\sqrt{R^*}\boldsymbol{\delta}_n^{3/2}/N_p \tag{5.17}$$

$$\boldsymbol{F}_n^d=C_n\sqrt{8m^*E^*\sqrt{R^*\boldsymbol{\delta}_n}}\cdot\boldsymbol{v}_{n,ij}/N_p \tag{5.18}$$

where $E^* = \frac{E}{2(1-\nu^2)}$, $R^* = \frac{R_i \cdot R_j}{R_i + R_j}$, and $m^* = \frac{m_i \cdot m_j}{m_i + m_j}$. E, ν, and m are the Young's modulus, Poisson's ratio, and particle mass, respectively. C_n is the normal damping coefficient. $\boldsymbol{v}_{n,ij}$ is the normal relative velocity between particles, and R is the radius of the volume-equivalent sphere of the particle. N_p is the total number of contact points.

The tangential force ($\boldsymbol{F}_t$) includes the tangential elastic force ($\boldsymbol{F}_t^e$) and damping force ($\boldsymbol{F}_t^d$), expressed as:

$$\boldsymbol{F}_t^e = \mu_s |\boldsymbol{F}_n^e| \left(1 - \left(1 - \min(\boldsymbol{\delta}_t, \boldsymbol{\delta}_{t,\max})/\boldsymbol{\delta}_{t,\max}\right)^{3/2}\right) \cdot \boldsymbol{\delta}_t / |\boldsymbol{\delta}_t| \tag{5.19}$$

$$\boldsymbol{F}_t^d = C_t \left(6\mu_s m^* |\boldsymbol{F}_n^e| \sqrt{1 - \min(\boldsymbol{\delta}_t, \boldsymbol{\delta}_{t,\max})/\boldsymbol{\delta}_{t,\max}}\right)^{1/2} \cdot \boldsymbol{v}_{t,ij} \tag{5.20}$$

where μ_s, C_t, and $\boldsymbol{v}_{t,ij}$ are the sliding friction coefficient, tangential damping coefficient, and tangential relative velocity, respectively. $\boldsymbol{\delta}_t$ is the tangential overlap, and $\boldsymbol{\delta}_{t,\max} = \mu_s(2-\nu)/2(1-\nu) \cdot \boldsymbol{\delta}_n$.

When the relative rotation occurs between the elements, the moment caused by the rolling friction is expressed as:

$$\boldsymbol{M}_r = \mu_r R_i |\boldsymbol{F}_n| \widehat{\boldsymbol{\omega}}_{ij} \tag{5.21}$$

where μ_r is the rolling friction coefficient, and $\widehat{\boldsymbol{\omega}}_{ij}$ is the relative rotational velocity, which is obtained by $\widehat{\boldsymbol{\omega}}_{ij} = \boldsymbol{\omega}_{ij} / |\boldsymbol{\omega}_{ij}|$.

Although the DEM time steps of spherical particles have been extensively studied, there are few theoretical formulations for time steps of arbitrarily shaped particles. Therefore, a reasonable range is determined based on the theoretical formula of spherical particles, expressed as Kremmer and Favier (2001):

$$t_{\max} = \frac{\pi r_{\min} \sqrt{\rho / G}}{0.163\nu + 0.8766} \tag{5.22}$$

where $r_{\min}$ is the minimum radius of the volume-equivalent sphere for different particle shapes. ρ and G are the particle density and shear modulus, respectively. In the DEM simulation of a mixed granular flow involving different non-spherical models, $\mathrm{d}t \le 0.02 t_{\max}$ is used as the DEM time step.

5.3 Contact Algorithm Between Particles and Structures

In industrial production, non-spherical granular flows are generally located in complex structures. However, it is still difficult to simulate a mixed granular flow involving multiple DEM models in complex structures. In this section, a contact algorithm between arbitrarily shaped particles and complex structures based on the level set functions is proposed and introduced in detail. Benefiting from the finite element method, complex geometric structures are discretized and meshed through a large number of precisely controllable triangular elements. Hence, the collision force between the particle and the structure is calculated by accumulating multiple contact forces between the particle and several triangular elements. The contact modes between particles and triangular elements include the face, edge, and vertex contacts, which are introduced separately below.

5.3.1 Particle-Plane Contact Detection

$\boldsymbol{x}_A$, $\boldsymbol{x}_B$, and $\boldsymbol{x}_C$ represent the positions of the three vertices of the triangular element, and the vectors of the three edges are denoted as $\boldsymbol{a}$, $\boldsymbol{b}$, and $\boldsymbol{c}$, as shown in Fig. 5.12. The particle centroid is denoted as P, and the projection point of point P on the plane is denoted as Q. The normal direction is obtained by $\boldsymbol{n}_w = \boldsymbol{a} \times \boldsymbol{b}/|\boldsymbol{a} \times \boldsymbol{b}|$, and the position of the projection point is expressed as $\boldsymbol{x}_Q(\boldsymbol{x}_{Qx}, \boldsymbol{x}_{Qy}, \boldsymbol{x}_{Qz}) = \boldsymbol{d} + |\boldsymbol{d}\boldsymbol{n}_w| \cdot \boldsymbol{n}_w + \boldsymbol{x}_A$. The bounding sphere method is also applied for rough detection to improve the calculation efficiency between arbitrarily shaped particles and triangular elements. The distance $|PQ|$ is given by $|PQ| = |\boldsymbol{d}\boldsymbol{n}_w|$. If $|PQ| \leq r_b$, the particle may be in contact with the triangular element. Here, r_b is the radius of the bounding sphere.

Subsequently, all zero level set points satisfying $\boldsymbol{d}_i\boldsymbol{n}_w > 0$ are projected onto the triangular element, as shown in Fig. 5.12b. Taking a level set point $\boldsymbol{p}_i$ as an example,

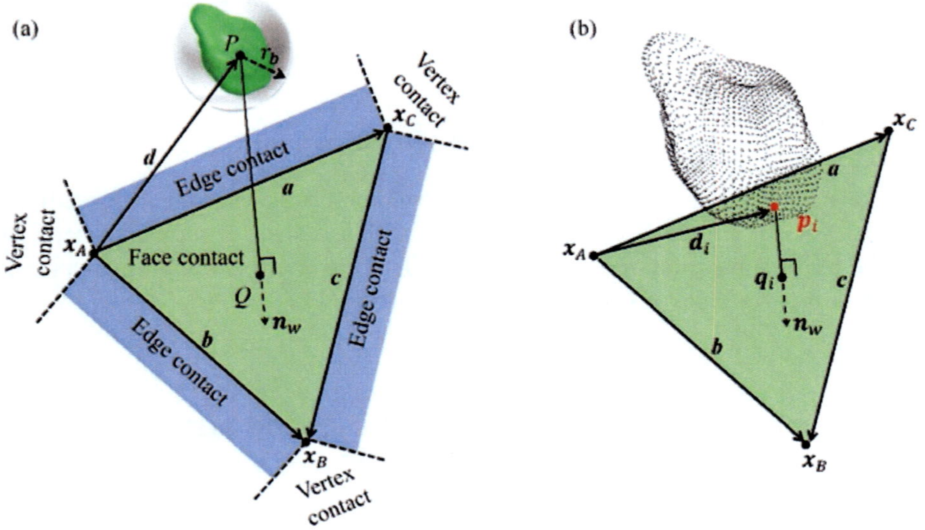

Fig. 5.12 **a** Three contact modes between a particle and a triangular element. **b** Face contact between a particle and a triangular element

vector $\boldsymbol{d}_i$ is directed from vertex $\boldsymbol{x}_A$ to point $\boldsymbol{p}_i$, and point $\boldsymbol{q}_i$ is the projection point of point $\boldsymbol{p}_i$ on the triangular element. The projection point $\boldsymbol{q}_i$ must be further determined according to Eqs. (5.8)–(5.11) whether or not this point is located in the triangular element. If the unknown parameters α and β satisfy Eq. (5.9), point $\boldsymbol{q}_i$ is located in the triangular element, and then it is brought into the spatial level set function of the particles (Fig. 5.10). The value D_{qi} is obtained by solving Eqs. (5.12) and (5.13), respectively. If $D_{qi} < 0$, the particle is in face contact with the triangular element. The overlap is the vector from point $\boldsymbol{p}_i$ to point $\boldsymbol{q}_i$, and the normal direction is expressed as $\boldsymbol{n} = \boldsymbol{n}_w$. The overlap and normal direction of all zero level set points satisfying $D_{qi} < 0$ are recorded to calculate the contact forces between a particle and a triangular element. If all zero level set points satisfy $D_{qi} > 0$, then edge and vertex contacts will be detected.

5.3.2 Particle-Edge Contact Detection

Taking edge $\boldsymbol{b}$ as an example, the projection point of point P on edge $\boldsymbol{b}$ is denoted as Q, as shown in Fig. 5.13. The distance $|PQ|$ is obtained as follows:

$$|PQ| = \left| \boldsymbol{d} - \frac{\boldsymbol{d} \cdot \boldsymbol{b}}{|\boldsymbol{b}|} \cdot \frac{\boldsymbol{b}}{|\boldsymbol{b}|} \right| \tag{5.23}$$

where vector $\boldsymbol{d}$ is directed from vertex A to point P. The prerequisite to proving that the particle is in edge contact with the triangular element can be expressed as

$$|PQ| < r_b \cap 0 < \frac{\boldsymbol{d} \cdot \boldsymbol{b}}{|\boldsymbol{b}|} < |\boldsymbol{b}| \tag{5.24}$$

Next, points $\boldsymbol{x}_{A\min}$ and $\boldsymbol{x}_{B\min}$ can be determined by extending the distances of $\mp r_b$ from point Q along vector $\boldsymbol{b}$, respectively. This indicates that if the particle is in contact with the edge, the contact point is in the range from point $\boldsymbol{x}_{A\min}$ to point $\boldsymbol{x}_{B\min}$. Several points are generated equidistantly between segment $\boldsymbol{x}_{A\min}\boldsymbol{x}_{B\min}$, and all the points ($\boldsymbol{q}_i$) are substituted into the spatial level set function of the particle. Here, the distance (l_0) between points is expressed as $l_0 < r_b/20$, so that a sufficient number of points are in contact with the particle. The value $\left(D_q\right)$ for each point is calculated by solving Eqs. (5.12) and (5.13). If $D_q < 0$, the particle is in edge contact with the triangular element. Subsequently, the normal direction and the overlap are expressed as $\boldsymbol{n} = \nabla\psi\left(\boldsymbol{q}_i\right)$ and $\boldsymbol{\delta}_n = D_q\boldsymbol{n}$, respectively. The overlap and normal direction of all points on the line segment satisfying $D_q < 0$ are recorded to calculate the contact forces between a particle and an edge. If the particle is not in contact with three edges, vertex contacts will be detected.

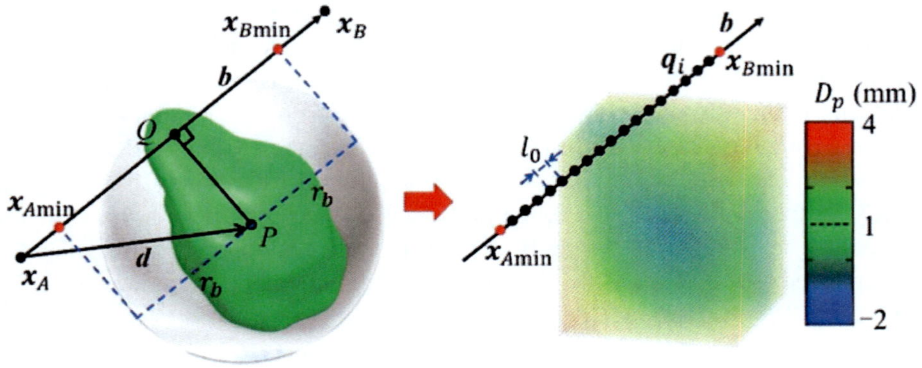

Fig. 5.13 Contact detection between a particle and an edge

5.3.3 Particle-Vertex Contact Detection

Taking vertex $\boldsymbol{x}_A$ as an example, the point is brought into the spatial level set function of the particle, as shown in Fig. 5.14. The value $\left(D_q\right)$ of vertex $\boldsymbol{x}_A$ is calculated by solving Eqs. (5.12) and (5.13). If $D_q < 0$, the particle is in vertex contact with the triangular element. Then, the normal direction and the overlap are expressed as $\boldsymbol{n} = \nabla\psi(\boldsymbol{x}_A)$ and $\boldsymbol{\delta}_n = D_q\boldsymbol{n}$, respectively. The overlap and normal direction of the vertex points satisfying $D_q < 0$ are recorded to calculate the contact forces between a particle and three vertex points. If the particle is not in contact with three vertices, the particle is not in contact with the triangular element.

Moreover, the contact force model between a particle and a triangular element is the same as that between particles. The contact forces are effectively calculated according to the overlap and normal direction between a particle and a triangular element, as described in Sect. 3.4. When a particle is in edge or vertex contact with a triangular element, the calculated contact forces need to be divided by the number of shared edges or vertices. This is because a shared edge or shared vertex is located in several triangular elements, and the contact force between the particle and the shared edge or shared vertex is repeatedly calculated and erroneously accumulated.

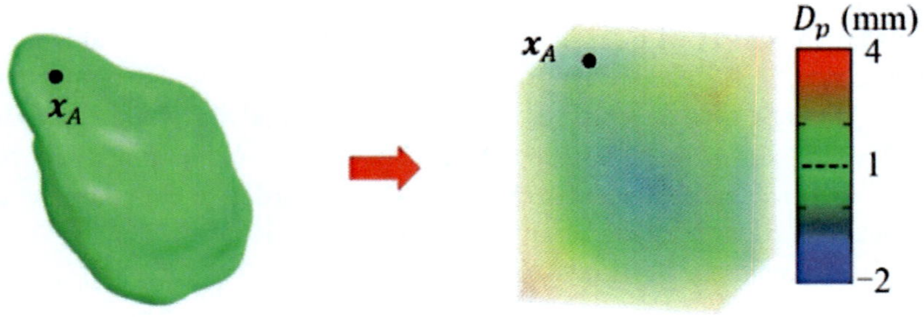

Fig. 5.14 Contact detection between a particle and a vertex

Furthermore, a concave particle may have multiple contact points with a triangular element, and a particle in the complex structure may be in contact with several triangular elements at the same time. Hence, the contact forces of particle i need to be accumulated and expressed as

$$\boldsymbol{F}_w^i = \sum_{k=1}^{N_w^i} \sum_{j=1}^{N_{p,k}^i} \left(\boldsymbol{F}_{n,jk}^i + \boldsymbol{F}_{t,jk}^i \right) \tag{5.25}$$

$$\boldsymbol{M}_w^i = \sum_{k=1}^{N_w^i} \sum_{j=1}^{N_{p,k}^i} \left(\boldsymbol{M}_{n,jk}^i + \boldsymbol{M}_{t,jk}^i + \boldsymbol{M}_{r,jk}^i \right) \tag{5.26}$$

where $\boldsymbol{F}_w^i$ and $\boldsymbol{M}_w^i$ are the total contact force and moment between particle i and the structure, respectively. $\boldsymbol{F}_{n,jk}^i$ and $\boldsymbol{F}_{t,jk}^i$ are the normal and tangential forces contributed by point j of triangular element k. $\boldsymbol{M}_{n,jk}^i$ and $\boldsymbol{M}_{t,jk}^i$ are the torques obtained by $\boldsymbol{F}_{n,jk}^i$ and $\boldsymbol{F}_{t,jk}^i$ when the normal and tangential forces do not pass through the centroid of particle i. $\boldsymbol{M}_{r,jk}^i$ is the rolling friction torque contributed by point j of triangular element k. $N_{p,k}^i$ is the total number of contact points of triangular element k with particle i, and N_w^i is the total number of triangular elements in contact with particle i.

5.4 Numerical Verification and Application

Three sets of numerical tests were compared with the theoretical and experimental results to verify and validate the present DEM model, including its conservation, accuracy, and robustness. These numerical tests were selected in such a manner that they comprise a wide range of applications, including mixed granular flows involving multiple non-spherical DEM models in complex structures.

5.4.1 Collision Between a Single Particle and a Wall

The elastic and inelastic collision of a single particle with a wall was performed as the first benchmark test to verify the ability and conservation of this contact algorithm for different DEM models. Figure 5.15 shows four particle shapes constructed by the sphere, the superquadric equation, the spherical harmonic function, and the polyhedron model. The differently shaped particles have the same volume, and the radius of the volume-equivalent sphere is 0.02 m. The number of zero level set points is ~ 6×10^3, and the number of spatial level set points is ~ 1×10^6. The color of the spatial level set point indicates the distance between the point and the particle

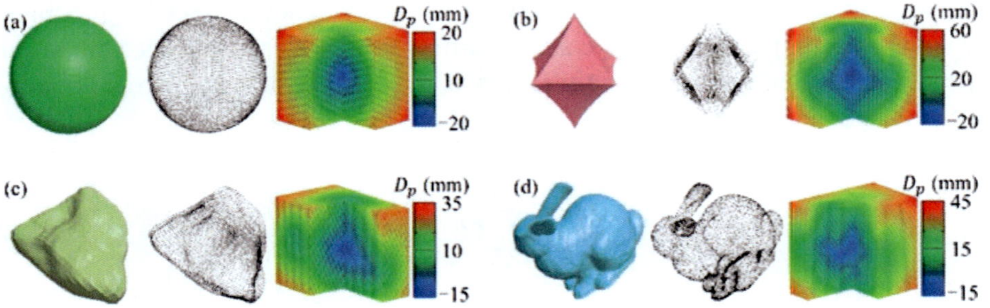

Fig. 5.15 Four types of particle shapes constructed by different DEM models and the corresponding level set functions: **a** sphere, **b** superquadric element, **c** spherical harmonic element, and **d** polyhedral element

surface: blue indicates that the point is close to the center of the particle, and red indicates that the point is far away from the particle surface. The Young's modulus is 5 GPa, Poisson's ratio is 0.3, and the particle density is 2500 kg/m^3. A single particle collides with the wall elastically, and gravity, damping, and friction are not considered during the collision.

The initial translation speed of the particle in the z direction was $-1m/s$, and the particle had a random orientation but no angular velocity. During the collision, the normalized translation, rotation, and total kinetic energy of the particle vary with the normalized time, as shown in Fig. 5.16. The two illustrations show screenshots of a single particle before and after collision with the wall. After the particle collides with the wall, the sphere has no rotational kinetic energy, which is consistent with the theoretical results. Part of the translational energy of a non-spherical particle is converted into rotational kinetic energy. Note that the total kinetic energy of the particles constructed using different DEM models is still conserved.

Moreover, the sensitivity of the energy conservation of the contact algorithm to the number of zero level set points was investigated. Taking the collision of a single spherical harmonic particle with a wall as an example, different numbers of zero level set points are generated, as shown in Fig. 5.17a. The time histories of the normalized total kinetic energy of the particles calculated by the traditional and modified force models are shown in Fig. 5.17b, c. In the traditional force model, the normal elastic and damping forces are not divided by the total number of contact points, whereas in the modified force model, both forces are divided by the total number of contact points. The results show that the kinetic energy calculated by the traditional and modified force models is still conserved after the collision. However, the contact time calculated by the traditional force model decreases as the number of zero level set points increases. This is because the number of contact points between the particle and the wall increases with the number of zero level set points, resulting in higher collision forces and shorter contact times. Applying the modified force model, particles with different numbers of zero level set points have the same contact time. Therefore, the level set algorithm using the modified force model is robust and insensitive to the number of zero level set points representing general shapes.

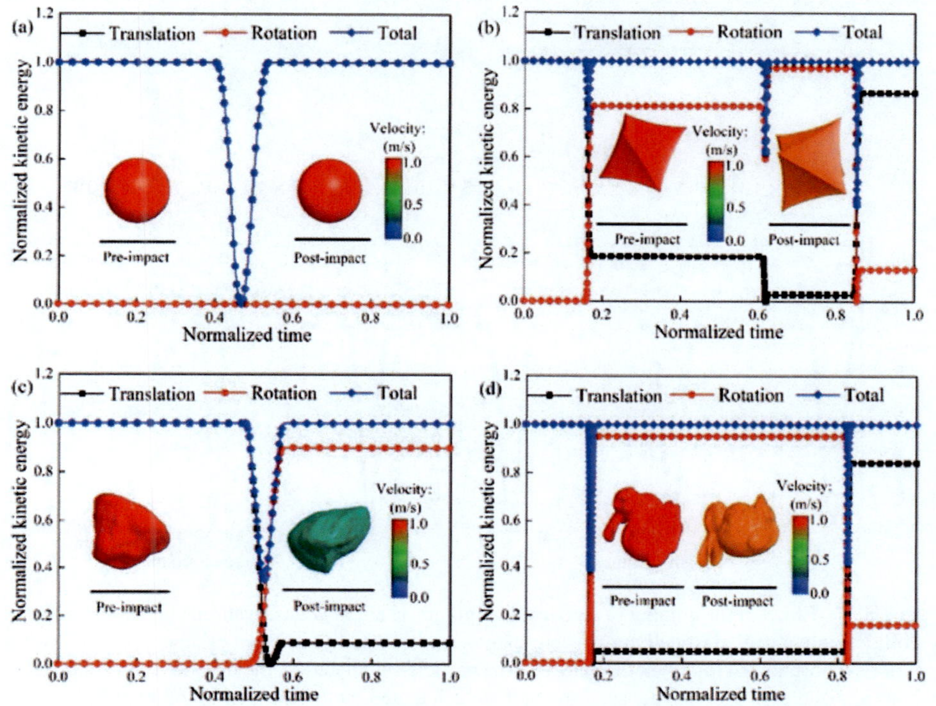

Fig. 5.16 Time history of the normalized kinetic energy when a single particle collides with a wall elastically: **a** spheres, **b** superquadric elements, **c** spherical harmonic elements, and **d** polyhedral elements

To further verify the level set method, the numerical results of a single particle impacting a wall were compared with the theoretical results. First, the superquadric equation is used to represent a cylindrical particle, and the rebound velocity and rebound angular velocity of a single cylinder impacting a wall are compared with the theoretical results (Kodam et al. 2010). A cylinder has the following superquadric parameters: $2a = 2b = 2c = 2R_p = L_p = 0.04\text{m}, n_1 = 80$, and $n_2 = 2$. The corresponding level set functions are shown in Fig. 5.18a. The number of zero level set points is ~ 6×10^3, and the number of spatial level set points is ~ 1×10^6. Young's modulus, Poisson's ratio, and particle density are 1.0 GPa, 0.3, and 2500 kg/m^3, respectively. The normal damping coefficient is 0.05, which approximately corresponds to the restitution coefficient $\varepsilon = 0.85$. Gravity, tangential damping, sliding and rolling friction were not considered during collisions. The rebound angular velocity (w_y^+) and translational velocity (v_z^+) were theoretically obtained by Kodam et al. (2010)

$$w_y^+ = mv_z^-(1+\varepsilon)r_0\cos(\varphi+\theta_0)/\left(I_{yy} + mr_0{}^2\cos^2(\varphi+\theta_0)\right) \tag{5.27}$$

$$v_z^+ = w_y^+ r_0\cos(\varphi+\theta_0) - \varepsilon v_z^- \tag{5.28}$$

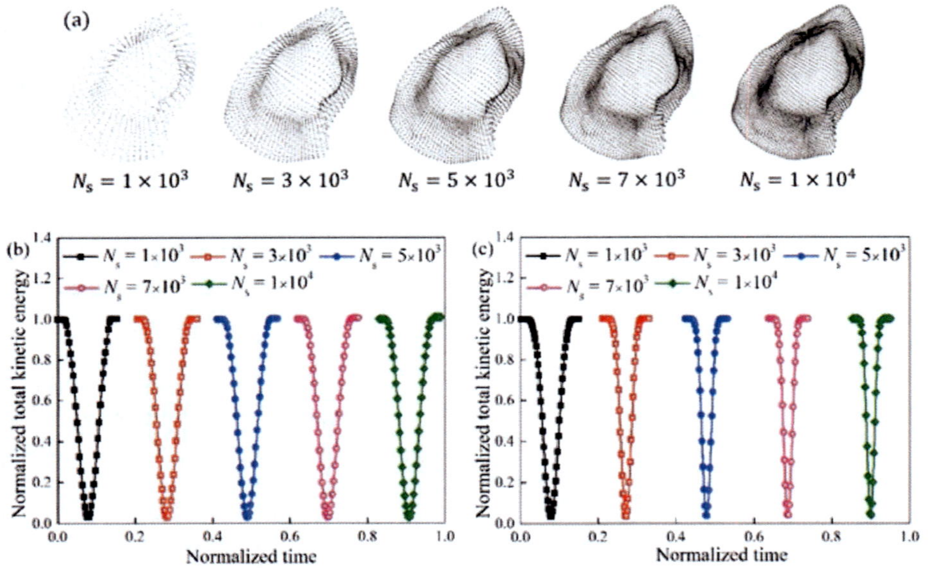

Fig. 5.17 Effect of the number of zero level set points on energy conservation when a single particle collides with a wall elastically: **a** particle shapes represented by different numbers of zero level set points, **b** normalized total kinetic energy of the particle calculated by the modified force model, and **c** normalized total kinetic energy of the particle calculated by the traditional force model

where m and I_{yy} are the particle mass and moment of inertia in the y direction, respectively. φ is the angle between the particle surface and the line connecting the contact point and the particle centroid. For different impact angles (θ_0), the pre-impact translational speed $\left(v_z^-\right)$ is always 1m/s. The rebound coefficient (ε) is always 0.85, which approximately corresponds to the normal damping coefficient ($C_n = 0.05$) between a particle and the wall in DEM simulations. When the impact angle varies from 0° to 90°, the dimensionless rebound velocity and rebound angular velocity are calculated and compared with the theoretical results (Kodam et al. 2010), as shown in Fig. 5.18b, c. The numerical results are slightly different from the theoretical results. This difference is mainly because the zero level set points are randomly distributed on the particle surface. The cylinder and the plane are in edge contact, and the zero level set point close to the edge may be the contact point, resulting in a slight difference in rebound velocity.

5.4.2 Packing Process of Multiple Particles

A mixed packing simulation involving multiple DEM models was performed as the third test to verify the ability and conservation of the level set method. Further, the numerical results of the final accumulation of spherical harmonic particles were compared with the experimental results to verify that the level set method is applicable

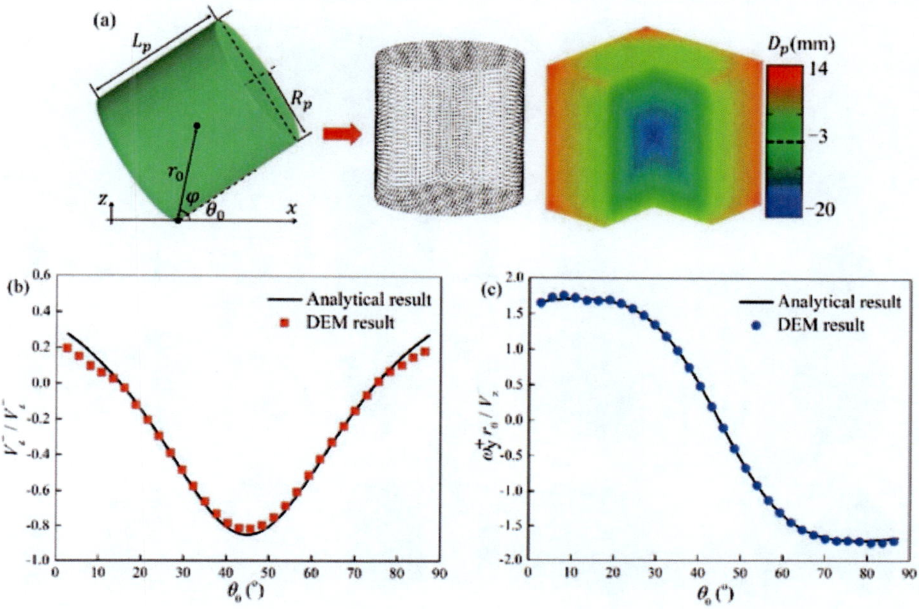

Fig. 5.18 Comparison of analytical results (Kodam et al. 2010) and numerical results for cylinder-wall impact: **a** schematics of the cylinder-wall contact and the corresponding level set functions, **b** dimensionless rebound translational velocity, v_z^+/v_z^-, and **c** dimensionless rebound angular velocity, $w_y^+ r_0/v_z^-$

to spherical harmonic functions. In the mixed packing simulations, the particle shapes include a sphere, four convex and concave particles constructed by the superquadric equations, four convex and concave particles constructed by the polyhedron models, and four concave particles constructed by the spherical harmonic functions. A total of 13 particle shapes and corresponding level set functions are shown in Fig. 5.19. The differently shaped particles have the same volume, and the radius of the volume-equivalent sphere is 0.02 m. The number of zero level set points is ~ 1×10^3, and the number of spatial level set points is ~ 1×10^6. Young's modulus, Poisson's ratio, and particle density are 0.1 GPa, 0.3, and 2500 kg/m^3, respectively. Both normal and tangential damping coefficients are 0.3, and both normal and tangential friction coefficients are 0.3. The rolling friction coefficient is 0.001.

A rabbit-shaped particle was constructed using a polyhedron model, and the corresponding level set functions are shown in Fig. 5.15d. Initially, two thousand particles had random positions and orientations. All the rabbit-shaped particles were dropped into a cubic container to form a stable granular bed under gravity, as shown in Fig. 5.20a. Two particles are in contact with each other if there is at least one contact point between them. Rabbit-shaped particles located in the central region of the stable granular bed are used to count the contact number, and the maximum and average contact numbers are 8 and 6.7, respectively. The maximum and average

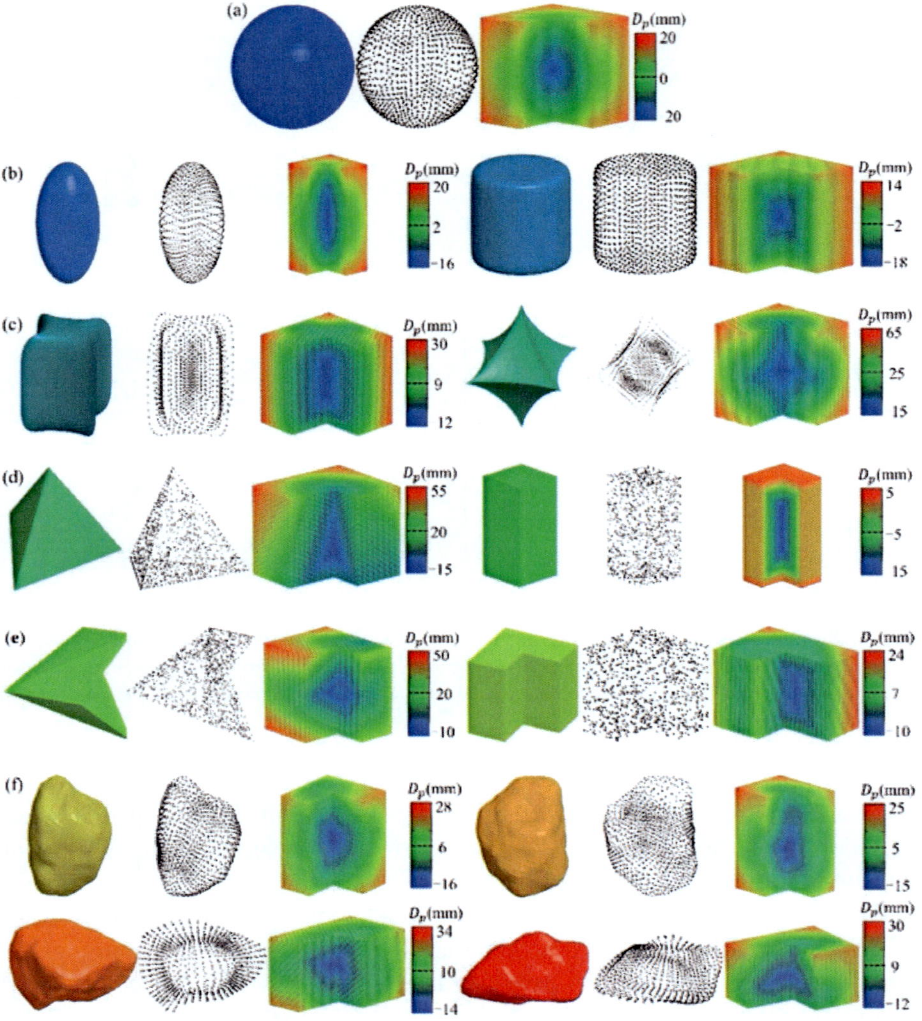

Fig. 5.19 Various particle shapes constructed by different DEM models and the corresponding level set functions: **a** spheres, **b** convex particles constructed by superquadric equations, **c** concave particles constructed by superquadric equations, **d** convex particles constructed by polyhedron models, **e** concave particles constructed by polyhedron models, and **f** concave particles constructed by spherical harmonic functions

numbers of contact points between particles are 577 and 282, respectively. Meanwhile, the normalized energy of the granular system during the falling and stabilization processes was calculated, as shown in Fig. 5.20b. As time increases, the kinetic energy of the granular system first increases and then decreases, while the potential energy and total energy both decrease. Finally, the total kinetic energy of the granular material tends to zero, and the total energy is equal to the total potential energy.

Moreover, 13 particle shapes were randomly generated in the cubic container, and these particles had random positions and orientations. Two thousand particles fell

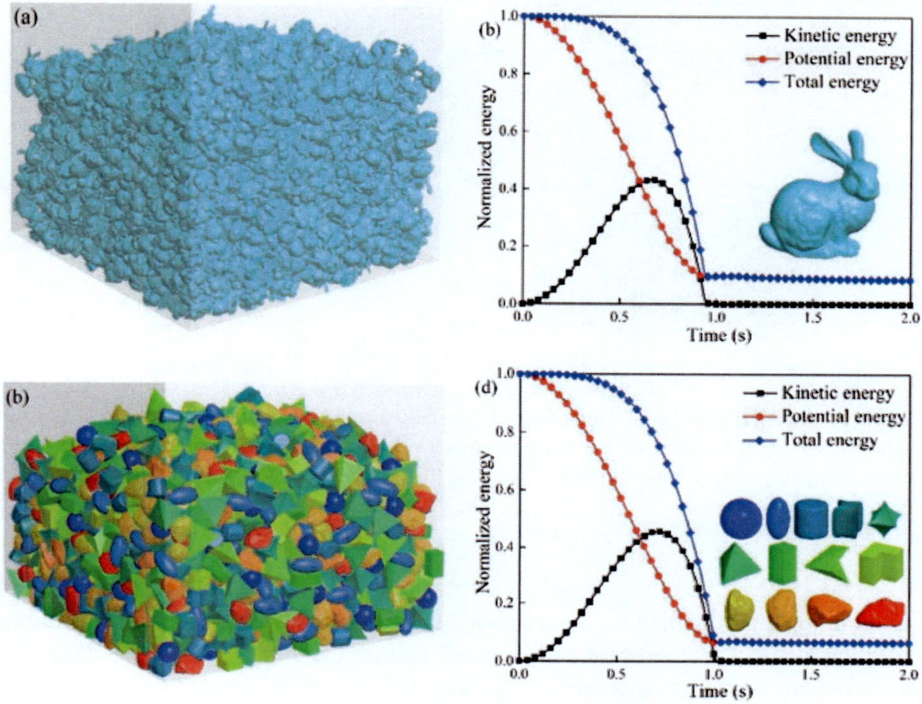

Fig. 5.20 Packing process of differently shaped particles and time history of the normalized energy of the granular system: **a** stable granular system composed of rabbit-shaped particles, **b** time history of the normalized energy of rabbit-shaped granular systems, **c** stable mixed granular system involving different DEM models, and **d** time history of the normalized energy of mixed granular systems

and formed a stable mixed granular system under gravity, as shown in Fig. 5.20c. In this mixed granular system, the maximum and average contact numbers are 8 and 6.1, respectively, and the maximum and average numbers of contact points between particles are 109 and 46, respectively. Meanwhile, the time history of the normalized energy of the mixed granular system was calculated, as shown in Fig. 5.20d. Similar to the granular system composed of rabbit-shaped particles, the total kinetic energy of the mixed granular system tends to zero, and the total energy is the same as the total potential energy. Therefore, the level set method also has good stability for simulating the dynamic process of a mixed granular system involving multiple DEM models.

Furthermore, a tablet-shaped particle was constructed using the spherical harmonic function, and the corresponding level set functions are shown in Fig. 5.21a. The size of the tablet-shaped particle in the simulation was the same as that used in the experiment (Kodam et al. 2012). The tablet diameter was 11.35 mm, and the band thickness and total thickness were 4 and 6.46 mm, respectively. The number of zero level set points is ~ 6×10^3, and the number of spatial level set points is ~ 1×10^6. The Young's modulus, Poisson's ratio, and particle density are 2.99 GPa,

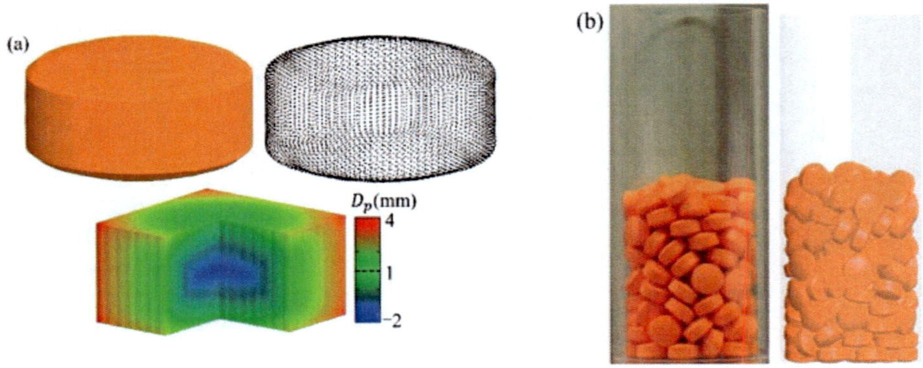

Fig. 5.21 Packing process of tablet-shaped particles: **a** tablet-shaped particle constructed by a spherical harmonic function and the corresponding level set functions, and **b** comparison between the experimental result (Kodam et al. 2012) (left) and the simulation result (right) for the accumulation of tablet particles

0.3, and 1192 kg/m^3, respectively, corresponding to the material properties of the acetaminophen tablets in the experiments (Kodam et al. 2012). The sliding friction coefficients between particles and between particles and walls were measured using a pin-on-disk test, and these parameters were 0.22 and 0.38, respectively. The restitution coefficient was 0.6, and it has been reported that the restitution coefficient has no significant effect on the motion behaviors of bi-convex tablets (Kalbag et al. 2008). In DEM simulations, the sliding friction coefficients between particles and between particles and walls are 0.22 and 0.38, respectively, which are the same as those measured experimentally. Both the normal and tangential damping coefficients are set to 0.3, which approximately corresponds to the restitution coefficient. The rolling friction coefficient is 0.001. A total of 150 tablet-shaped particles were generated in a cylindrical container, and these particles had random positions and orientations. The tablet particles were dropped into the container under gravity and formed a stable granular bed. The screenshot of the final granular bed was compared with the experimental results, as shown in Fig. 5.21b. In the DEM simulation, the filling height and volume fraction of the granular bed are 68.2 mm and 0.58, respectively. In the experiment (Kodam et al. 2012), the filling height and volume fraction are 66.8 ± 2.0 mm and 0.59 ± 0.02, respectively. Although there are small differences between the simulation and experimental results, the numerical results are within a reasonable range of the experimental results. Therefore, the level set method can also reflect the macroscopic behavior of spherical harmonic particles.

5.4.3 Engineering Applications of Mixed Granular Materials

To examine the robustness of the present contact algorithm in solving the contact problem between a mixed granular system and a complex structure, the stabilization process on a terrain surface, the mixing behavior in a rotating drum, and the conveying motion in a screw conveyor were simulated. Young's modulus, Poisson's ratio, and particle density are 0.1 GPa, 0.3, and 2500 kg/m^3, respectively. The damping and friction coefficients in the normal and tangential directions are 0.3, and the rolling friction coefficient is 0.001.

First, a terrain surface with complex mountains and valleys, which is widely used in the prediction of geological disasters such as landslides and mudslides, was simulated. The terrain surface was composed of 6438 nodes and 12,872 triangular elements. A total of 2000 mixed particles involving different DEM models were generated above the terrain surface, as shown in Fig. 5.22a. The mixed particles had initial random orientations, and these particles fell to the terrain surface under gravity, as shown in Fig. 5.22b. The color indicates the value of the particle velocity, and blue and red denote the lowest and highest velocities, respectively. The friction coefficient between the particles and the terrain surface is 0.1, which is conducive to the sliding and rolling of particles on the terrain surface. Under the action of gravity and the collision force between particles, the mixed particles on the terrain surface move to valleys with lower heights, as shown in Fig. 5.22c, d. Finally, all particles have no velocity, and a stable granular bed is formed on the terrain surface.

Second, rotating drums have a wide range of applications in industrial production and processing, such as separation, mixing, drying, and coating. This rotating drum was composed of 72 nodes and 84 triangular elements, and it had a rotating speed of 30 rpm. Initially, two thousand particles containing 13 particle shapes were generated with random orientations, and these particles formed a stable mixed granular system in the drum under gravity, as shown in Fig. 5.23a. As the mixed particles were repeatedly rotated, the particles close to the drum were continuously lifted and then fell to the free surface, as shown in Fig. 5.23b–d. It can be found that the entire granular system is divided into three parts: the flowing layer near the drum, the static layer at the center of the granular bed, and the collapsed layer on the free surface. This is similar to the velocity distribution of the granular system observed in a previous study (Ji et al. 2020).

Third, screw conveyors are commonly used in agriculture, food, mining, pharmaceuticals, and plastic processing for short to medium distance lifting and/or transportation of granular materials. The screw conveyor was composed of 2334 nodes and 4602 triangular elements, and the rotation speed of the screw conveyor was 30 rpm. Mixed particles composed of 13 particle shapes were generated with random orientations and formed a stable mixed granular system in the hopper under gravity, as shown in Fig. 5.24a. When the hopper orifice was opened and the screw conveyor started to rotate, the mixed particles fell into the screw conveyor under gravity and moved forward along the axis of the trough by the thrust of the screw thread or flight, as shown in Fig. 5.24b–c. The color of the particles located in the hopper and screw conveyor is blue, indicating that the particles have the lowest velocity. The color of the particles is red only at the opening of the hopper, indicating that the particles move at the highest velocity. This velocity distribution phenomenon is similar to that observed in previous studies (Fernandez et al. 2011).

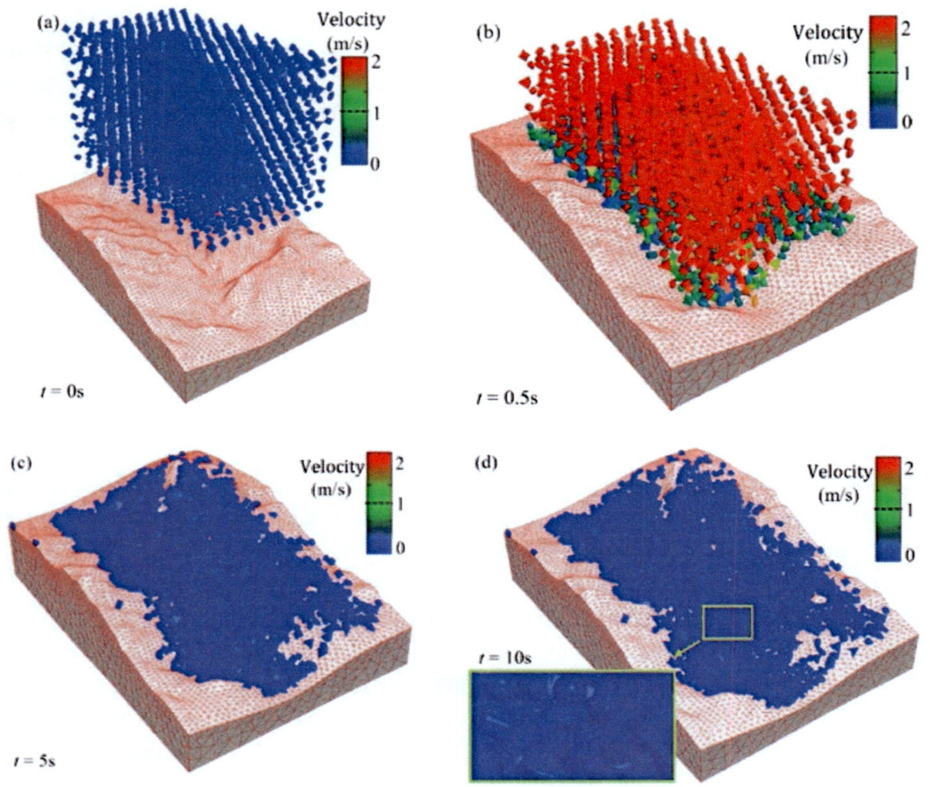

Fig. 5.22 Stabilization process of mixed granular systems on a terrain surface: **a** $t = 0$s, **b** $t = 0.5$s, **c** $t = 5$s, and **d** $t = 10$s

For the above three examples, no numerical instability is observed in the simulations. The proposed numerical procedures related to the level set contact algorithm are robust and can effectively simulate mixed granular flows involving multiple DEM models in moving or static complex structures.

5.5 Summary

In this chapter, a unified level set method is presented to simulate the flow behaviors of a mixed granular system involving multiple DEM models in complex structures. In this model, a unified method for creating level set functions of arbitrarily shaped particles constructed by different non-spherical DEM models was developed, and the contact points between arbitrarily shaped particles were determined by the level set method. Because the collision between particles and complex structures is still challenging for engineering applications, a contact algorithm between arbitrarily shaped particles and complex structures was proposed based on the level

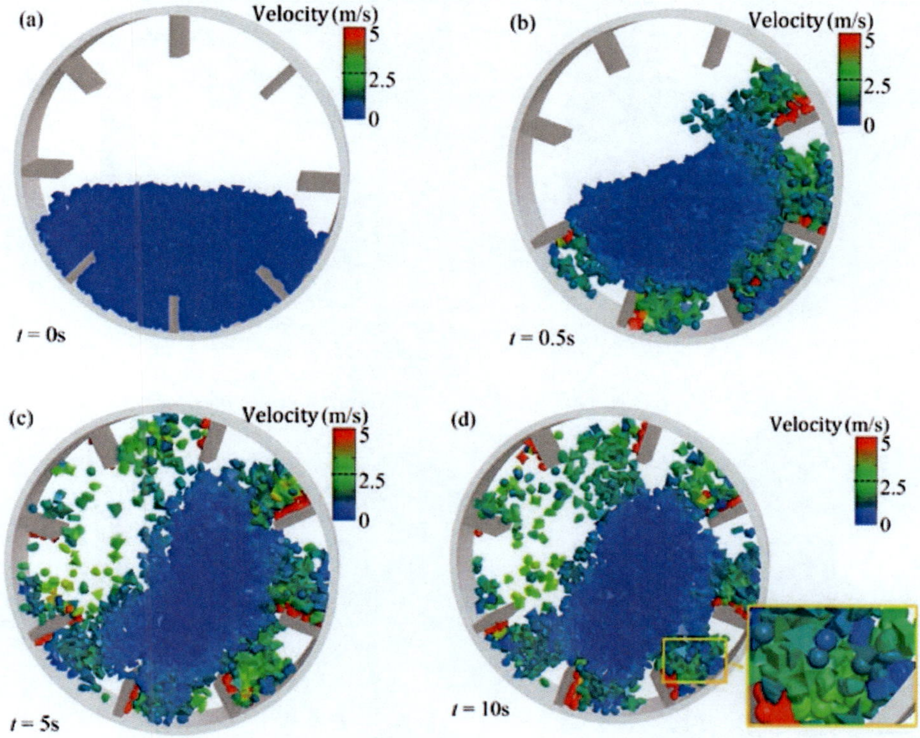

Fig. 5.23 Mixing behavior of the mixed granular system in a rotating drum: **a** $t = 0$s, **b** $t = 0.5$s, **c** $t = 5$s, and **d** $t = 10$s

set functions. Applying this contact algorithm, a mixed granular system containing multiple DEM models can realize granular flow in moving or static complex structures. The numerical results are in good agreement with the theoretical and experimental results, indicating that the level set method is applicable to arbitrarily shaped particles constructed by spheres, superquadric equations, spherical harmonic functions, and polyhedral models. In particular, this level set method is suitable and robust for simulating the flow characteristics of a mixed granular system involving multiple non-spherical DEM models in moving or static complex structures. This model has a wide range of applications for mixed granular flows.

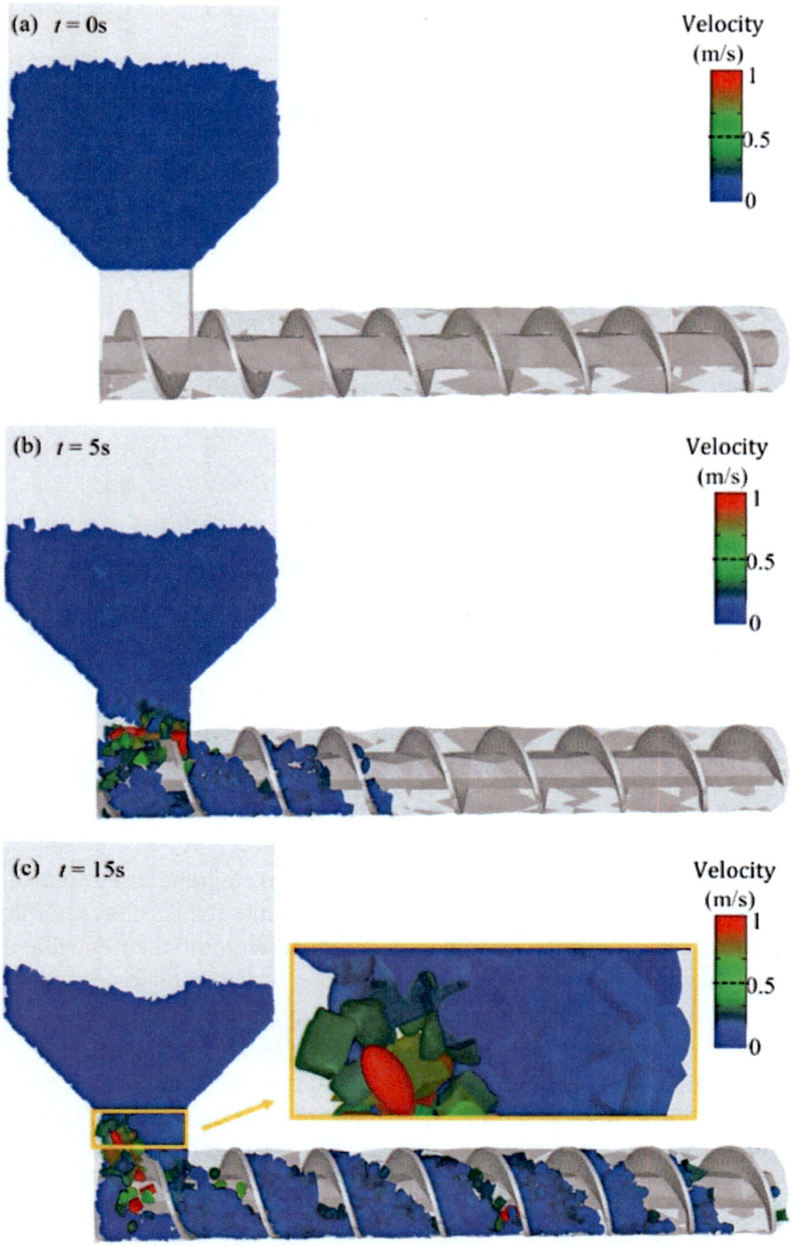

Fig. 5.24 Conveying motions of mixed granular systems in a screw conveyor: **a** t = 0s, **b** t = 5s, and **c** t = 10s

References

Bar AH (1981) Superquadrics and angle-preserving transformations. IEEE Comput Graph Appl 1:11–23

Bhattacharya D, Kawamoto R, Karapiperis K, Andrade JE, Prashant A (2021) Mechanical behaviour of granular media in flexible boundary plane strain conditions: experiment and level-set discrete element modelling. Acta Geotech 16:113–132

Caselles V, Catté F, Coll T, Dibos F (1993) A geometric model for active contours in image processing. Numer Math 66:1–31

Duriez J, Bonelli S (2021) Precision and computational costs of level set-discrete element method (LS-DEM) with respect to DEM. Comput Geotech 134:104033

Duriez J, Galusinski C (2021) A level set-discrete element method in YADE for numerical, micro-scale, geomechanics with refined grain shapes. Comput Geosci 157:104936

Fernandez JW, Cleary PW, McBride W (2011) Effect of screw design on hopper drawdown of spherical particles in a horizontal screw feeder. Chem Eng Sci 66:5585–5601

Govender N, Wilke DN, Pizette P, Abriak NE (2018) A study of shape non-uniformity and poly-dispersity in hopper discharge of spherical and polyhedral particle systems using the Blaze-DEM GPU code. Appl Math Comput 319:318–336

Harmon JM, Arthur D, Andrade JE (2020) Level set splitting in DEM for modeling breakage mechanics. Comput Methods Appl Mech Eng 365:112961

Harmon JM, Karapiperis K, Li L, Moreland S, Andrade JE (2021) Modeling connected granular media: particle bonding within the level set discrete element method. Comput Methods Appl Mech Eng 373:113486

Hettich T, Ramm E (2006) Interface material failure modeled by the extended finite-element method and level sets. Comp Methods Appl Mech Eng 195:4753–4767

Hettich T, Hund A, Ramm E (2008) Modeling of failure in composites by X-FEM and level sets within a multiscale framework. Comput Methods Appl Mech Eng 197:414–424

Jerves AX, Kawamoto RY, Andrade JE (2016) Effects of grain morphology on critical state: a computational analysis. Acta Geotech 11:493–503

Jerves AX, Kawamoto RY, Andrade JE (2017) A geometry-based algorithm for cloning real grains. Granular Matter 19:30

Ji S, Wang S, Zhou Z (2020) Influence of particle shape on mixing rate in rotating drums based on super-quadric DEM simulations. Adv Powder Technol 31:3540–3550

Kalbag A, Wassgren C, Penumetcha SS, Pérez-Ramos JD (2008) Inter-tablet coating variability: residence times in a horizontal pan coater. Chem Eng Sci 63:2881–2894

Kawamoto R, Andò E, Viggiani G, Andrade JE (2016) Level set discrete element method for three-dimensional computations with triaxial case study. J Mech Phys Solids 91:1–13

Kawamoto R, Andò E, Viggiani G, Andrade JE (2018) All you need is shape: predicting shear banding in sand with LS-DEM. J Mech Phys Solids 111:375–392

Kodam M, Bharadwaj R, Curtis J, Hancock B, Wassgren C (2010) Cylindrical object contact detection for use in discrete element method simulations, Part II—experimental validation. Chem Eng Sci 65:5863–5871

Kodam M, Curtis J, Hancock B, Wassgren C (2012) Discrete element method modeling of bi-convex pharmaceutical tablets: contact detection algorithms and validation. Chem Eng Sci 69:587–601

Kremmer M, Favier JF (2001) A method for representing boundaries in discrete element modelling—part II: kinematics. Int J Numer Methods Eng 51:1423–1436

Legrain G, Cartraud P, Perreard I, Moës N (2011) An X-FEM and level set computational approach for image-based modelling: application to homogenization. Int J Numer Methods Eng 86:915–934

Li L, Marteau E, Andrade JE (2019) Capturing the inter-particle force distribution in granular material using LS-DEM. Granular Matter 21:43

Lim KW, Kawamoto R, Andò E, Viggiani G, Andrade JE (2016) Multiscale characterization and modeling of granular materials through a computational mechanics avatar: a case study with experiment. Acta Geotech 11:243–253

Macedo RBD, Marshall JP, Andrade JE (2018) Granular object morphological generation with genetic algorithms for discrete element simulations. Granular Matter 20:73

Osher S, Fedkiw RP (2001) Level set methods: an overview and some recent results. J Comput Phys 169:463–502

Osher S, Sethian JA (1988) Front propagating with curvature dependent speed: algorithms based on Hamilton-Jacobi formulations. J Comput Phys 78:12–49

Sethian JA (2001) Evolution, implementation, and application of level set and fast marching methods for advancing fronts. J Comput Phys 169:503–555

Stafford DS, Jackson TL (2010) Using level sets for creating virtual random packs of non-spherical convex shapes. J Comput Phys 229:3295–3315

Sukumar N, Chopp DL, Moës N, Belytschko T (2001) Modeling holes and inclusions by level sets in the extended finite-element method. Comput Methods Appl Mech Eng 190:6183–6200

Tahmasebi P (2018) Packing of discrete and irregular particles. Comput Geotech 100:52–61

Tran AB, Yvonnet J, He QC, Toulemonde C, Sanahuja J (2011) A multiple level set approach to prevent numerical artefacts in complex microstructures with nearby inclusions within XFEM. Int J Numer Methods Eng 85:1436–1459

Vlahini I, Andò E, Viggiani G, Andrade JE (2014) Towards a more accurate characterization of granular media: extracting quantitative descriptors from tomographic images. Granular Matter 16:9–21

Vlahinic I, Kawamoto R, Ando E, Viggiani G, Andrade JE (2017) From computed tomography to mechanics of granular materials via level set bridge. Acta Geotech 12:85–95

Wang SY, Lim KM, Khoo BC, Wang MY (2007) An extended level set method for shape and topology optimization. J Comput Phys 221:395–421

Wang X, Yin ZY, Xiong H, Su D, Feng YT (2021) A spherical-harmonic-based approach to discrete element modeling of 3D irregular particles. Int J Numer Methods Eng 122:5626–5655

Zhong W, Yu A, Liu X, Tong Z, Zhang H (2016) DEM/CFD-DEM modelling of non-spherical particulate systems: theoretical developments and applications. Powder Technol 302:108–152

Chapter 6
High Performance Computation and DEM Software Development

The high-performance parallel computing approaches mainly include the open multi-processing method (OpenMP) (Amritkar et al. 2014), the message passing interface (MPI) (Kačianauskas et al. 2010), and the MPI-OpenMP algorithm (Berger et al. 2015). The above three methods use the CPU as the computing core, and the cores cooperate with each other to improve the calculation efficiency (Kačeniauskas et al. 2011). In addition, the parallel algorithms based on a graphics processing unit (GPU) have been successfully used for large-scale computations in the discrete element method (Zhang et al. 2013). Compared to a CPU, the number of computing cores in a GPU is significantly increased, and the hardware costs of the computer are reduced, which makes the GPU have more powerful parallel computing capabilities (Xu et al. 2011). The GPU parallel computing method for spheres was applied to simulate large-scale granular systems with a wide range of sizes (He et al. 2018). Another GPU parallel algorithm for the multi-sphere model was used to analyze the effects of the process parameters on large-scale tablet coating processes (Boehling et al. 2016). Moreover, considering the computational requirements of non-spherical DEM simulations, the ellipsoidal GPU parallel approach was developed and used to compare the speedup ratio of a GPU to a CPU (Gan et al. 2016). The polyhedral GPU parallel method was validated and successfully applied to the hopper discharge (Govender et al. 2014, 2018). The coating process of 20 million biconvex tablets in a rotating drum was simulated by a GPU parallel algorithm (Kureck et al. 2019). However, thus far, few efforts have been made to contribute to the detailed description of GPU parallel algorithms for superquadric elements.

S. Wang and S. Ji, *Computational Mechanics of Arbitrarily Shaped Granular Materials*,
Springer Tracts in Mechanical Engineering,
https://doi.org/10.1007/978-981-99-9927-9_6

This chapter introduces the parallel algorithm for superquadric elements using the CUDA-GPU architecture. This method is based on the parallel algorithm of spherical elements in shared-memory systems (Nishiura and Sakaguchi 2011) and adds the quaternion method, the bounding sphere approach, the oriented bounding box, and the Newton iterative algorithm for superquadric elements. Then, this parallel algorithm is applied to the generation of a large-scale non-spherical granular bed, the granular flow process, the dynamic hopper discharge, and the mixing behavior within the rotating drum. The corresponding numerical results are verified by the experimental results, and the speedup ratio of the GPU to the CPU is investigated. Finally, the influences of the particle shape on the computational efficiency and the flow process are analyzed.

6.1 Modeling Methodology of Arbitrarily Shaped Elements

In this section, superquadric equations are used to construct spherical and non-spherical particles, and the particle motion is governed by Newton's second law of motion. Subsequently, the quaternion method is used to establish the transformation matrices of the superquadric particles in the local and global coordinate systems, which realizes the rotational motion of the non-spherical particles in three-dimensional space.

6.1.1 Non-spherical Particles Represented by Superquadric Equations

The superquadric equation is one of the most effective methods for representing the shape of non-spherical elements. It is reported that the superquadric equation can be used to describe 80% of the solid elements in nature (Williams and Pentland 1992). The superquadric equation is expressed as follows (Bar 1981):

$$\left(\left|\frac{x}{a}\right|^{n_2}+\left|\frac{y}{b}\right|^{n_2}\right)^{n_1/n_2}+\left|\frac{z}{c}\right|^{n_1}-1=0, \tag{6.1}$$

where a, b, and c are the half-axis lengths of the superquadric element in the directions of the x, y, and z axes, respectively, and n_1 and n_2 are used to determine the basic shape and surface roughness of the elements. Figure 6.1 shows the differently shaped elements that are constructed by changing the parameters in Eq. (6.1). A sphere or an ellipsoid is obtained if $n_1 = n_2 = 2$, a cylinder-like element is obtained if $n_1 > 2$ and $n_2 = 2$, and a cube-like element is obtained if $n_1 = n_2 > 2$.

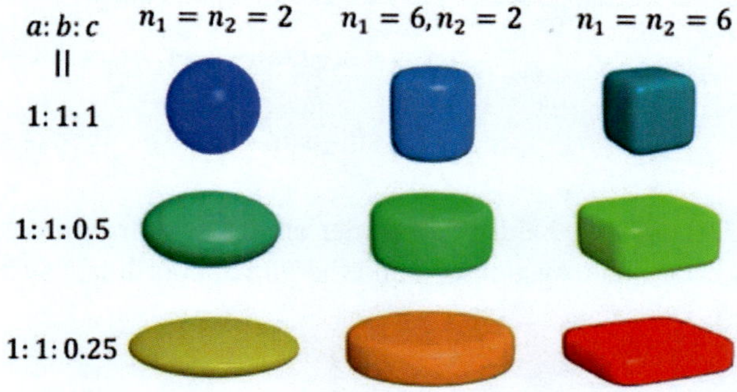

Fig. 6.1 Three-dimensional samples of differently shaped elements

6.1.2 *Motion Equation of Superquadric Particles*

In granular systems, a given element has translational and rotational motion, which is determined by Newton's second law of motion:

$$m_i \frac{\mathrm{d}\boldsymbol{v}_i}{\mathrm{d}t} = \sum_{j=1}^{N_p} \left(\boldsymbol{F}_{n,ij} + \boldsymbol{F}_{t,ij}\right) + m_i \boldsymbol{g}, \tag{6.2}$$

$$\boldsymbol{I}_i \frac{\mathrm{d}\boldsymbol{\omega}_i}{\mathrm{d}t} = \sum_{j=1}^{N_p} \left(\boldsymbol{M}_{n,ij} + \boldsymbol{M}_{t,ij} + \boldsymbol{M}_{r,ij}\right), \tag{6.3}$$

where m_i and $\boldsymbol{I}_i$ are the mass and the moment of inertia, respectively. $\boldsymbol{v}_i$ and $\boldsymbol{\omega}_i$ are the translational and rotational velocities of elements, respectively. $\boldsymbol{F}_{n,ij}$ and $\boldsymbol{M}_{n,ij}$ are the normal contact force and the torque caused by the normal force not passing through the centroid of the element, respectively. $\boldsymbol{F}_{t,ij}$ and $\boldsymbol{M}_{t,ij}$ are the tangential contact force and the torque caused by the tangential force, respectively. $\boldsymbol{M}_{r,ij}$ is the rolling friction torque. N_p is the total number of elements in contact with element i.

In the non-spherical granular system, the contact detection and calculation of contact forces between particles depend on the orientation of the particles, and the orientation of a non-spherical particle can usually be expressed as a rotation of the coordinate system. The quaternion is one of the most effective tools for determining the transition relationship between the local coordinate system and the global coordinate system. Therefore, the transformation matrix $\boldsymbol{T}$ can be obtained by quaternions $\boldsymbol{Q}(q_0, q_1, q_2, q_3)$ and expressed as (Fritzer 2001):

$$\boldsymbol{T} = \begin{bmatrix} q_0^2 + q_1^2 - q_2^2 - q_3^2 & 2(q_1 q_2 + q_0 q_3) & 2(q_1 q_3 - q_0 q_2) \\ 2(q_1 q_2 - q_0 q_3) & q_0^2 - q_1^2 + q_2^2 - q_3^2 & 2(q_2 q_3 + q_0 q_1) \\ 2(q_1 q_3 + q_0 q_2) & 2(q_2 q_3 - q_0 q_1) & q_0^2 - q_1^2 - q_2^2 + q_3^2 \end{bmatrix}, \tag{6.4}$$

where ${q_0}^2 + {q_1}^2 + {q_2}^2 + {q_3}^2 = 1$. Moreover, the relationship between quaternions and Euler angles can be expressed as:

$$q_0 = \cos\frac{\emptyset}{2}\cos\frac{\varphi+\psi}{2}, q_1 = \sin\frac{\emptyset}{2}\cos\frac{\varphi-\psi}{2},$$
$$q_2 = \sin\frac{\emptyset}{2}\sin\frac{\varphi-\psi}{2}, q_3 = \cos\frac{\emptyset}{2}\sin\frac{\varphi+\psi}{2}, \quad (6.5)$$

where $\varnothing$, φ, and ψ are Euler angles. The transition relationship between the local coordinate system ($\boldsymbol{v}^l$) and the global coordinate system ($\boldsymbol{v}^g$) can be expressed as:

$$\boldsymbol{v}^l = \boldsymbol{T} \cdot \boldsymbol{v}^g. \quad (6.6)$$

Moreover, the transformation matrix $\boldsymbol{T}$ has the property of $\boldsymbol{T}^T = \boldsymbol{T}^{-1}$. More descriptions of quaternions can be found in Ref. (Fritzer 2001).

6.2 GPU Parallel Computing Based on CUDA Architecture

In this section, the DEM simulation is mainly divided into three parts: grid partitioning and creating the neighbor list, creating the rough contact list and the Newton iteration list, and calculating the contact force and updating the element information. The flowchart of the superquadric DEM simulation under the GPU parallel architecture is shown in Fig. 6.2.

It is worth noting that the data transfer between CPU memory and GPU memory significantly affects the computational efficiency of DEM simulations. Therefore, parallelized DEM simulation code is developed specifically for GPU architectures, and all the data used for DEM is stored on GPU memory. The current parallelized DEM code is based on the parallel-vector algorithm of spherical elements, and the shared-memory systems achieve better load balancing for large-scale granular materials (Nishiura and Sakaguchi 2011). When the DEM results need to be output and further analyzed, the data on GPU memory will be copied to the CPU memory. However, this operation is not required at each DEM time step.

6.2.1 Spatial Grid Partitioning and Neighbor List Generation

First, the calculation region is divided into a large number of cubic grids whose length is 1.5 times the maximum diameter of the bounding sphere of all superquadric elements. Meanwhile, the maximum number of neighbors of the particle needs to be given in advance, which depends on the ratio of the maximum diameter to the minimum diameter of the bounding sphere of the particle in the multi-sized granular system. As the ratio of the maximum size to the minimum size of the particle increases, the grid size and the maximum number of the particle neighbors increase. As a result, the calculation efficiency of the DEM simulation decreases and the consumption of memory increases.

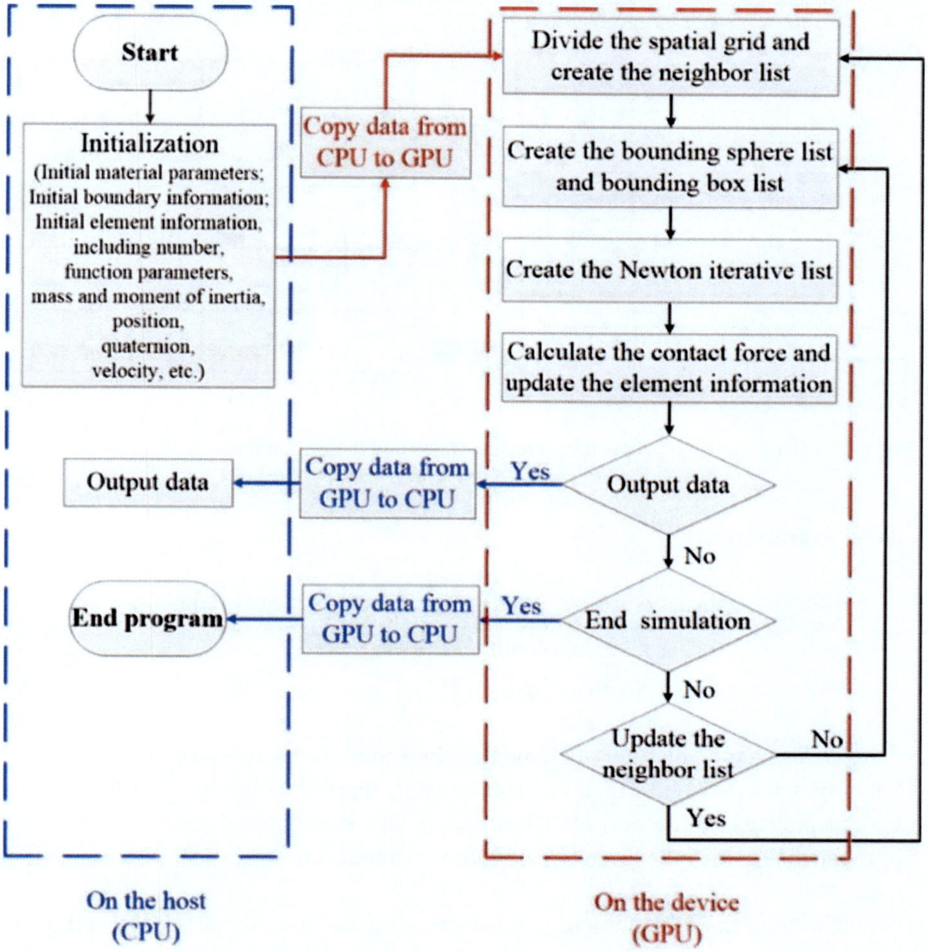

Fig. 6.2 CUDA-GPU architecture for the superquadric DEM simulation

The element index (i) is considered as the parallel basis, which means that one element index corresponds to one thread. The relationship between the element index and the thread index is expressed as:

$$i = \text{blockIdx.x} \times \text{blockDim.x} + \text{threadIdx.x} \tag{6.7}$$

The relationship between the element index and the grid index is determined by the cell list method, as shown in Fig. 6.3. The grid index is shown at the bottom-left corner of each grid, and the element index is shown in each particle. The grid index corresponding to the element index is stored in the array $P_g[i]$, and the element index (i) before sorting is stored in the array $P_e[i]$. Then, the element indexes are reordered according to the grid index. Related operations are performed directly on the GPU by using a kernel function named "sort_by_key" in the CUDA Thrust library and

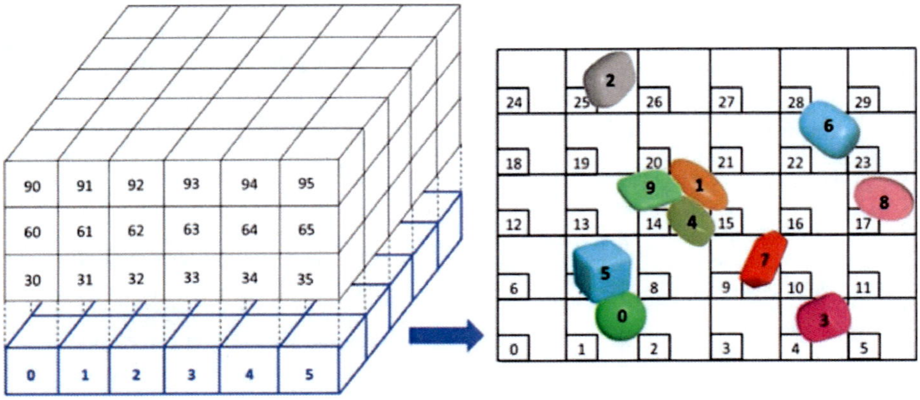

Fig. 6.3 The relationship between the element index and the grid index

can be expressed as:

$$
\begin{aligned}
&\text{thrust::sort\_by\_key}\left(\text{thrust::device\_ptr}\langle\text{int}\rangle\left(P_g\right),\right.\\
&\text{thrust::device\_ptr}\langle\text{int}\rangle\left(P_g + N_e\right),\\
&\left.\text{thrust::device\_ptr}\langle\text{int}\rangle\left(P_e\right)\right)
\end{aligned}
\tag{6.8}
$$

where N_e is the total number of elements. Therefore, the reordered element index (i^*) is obtained, as shown in Fig. 6.4a. The array P_e represents the relationship between element indexes before and after reordering and is expressed as $P_e[i^*] = i$. This approach makes the indexes of two neighbor elements closer in the three-dimensional space.

Considering the new element index as the parallel basis, the array S_g is created in shared memory and used to store the grid index, i.e., $S_g[\text{threadIdx.x} + 1] = P_g[i^*]$. For the first thread in each block, the array S_g satisfies $S_g[0] = P_g[i^* - 1], i^* > 0$. The data between different threads in the block is synchronized by using a kernel

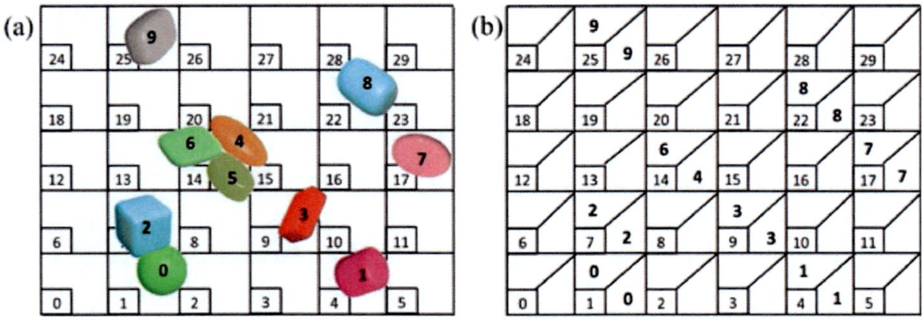

Fig. 6.4 The relationship between the reordered element index and the grid index: **a** The sorted element index and **b** the maximum and minimum element indexes in each grid

function named "__syncthreads" in the CUDA library, which ensures that all threads in the block are executed to the same location. Although the kernel function "__syncthreads" is used to coordinate communication between different threads in the same block, it forces some threads to be idle and reduces the efficiency of parallel programs. Subsequently, the arrays I_{gmax} and I_{gmin} are used to store the maximum and minimum values of the new element index in each grid, respectively. If the grid index corresponding to the element index is not equal to the grid index stored in the array S_g, i.e., $P_g[i^*] \neq S_g[\text{threadIdx.x}]$, the current element index is the minimum index in this grid, while it is the maximum index in the previous grid. Therefore, the element index (i^*) is stored in arrays I_{gmin} and I_{gmax}, respectively, which is expressed as $I_{gmin}\left[P_g[i^*]\right] = i^*$ and $I_{gmax}\left[S_g[\text{threadIdx.x}]\right] = i^* - 1$. The whole procedure is summarized in Algorithm 6.1, and the minimum and maximum values of the element index within each grid are shown in Fig. 6.4b. The minimum element index is shown at the bottom-right corner, and the maximum index is shown at the top-left corner of each grid. This operation is to quickly create the neighbor list on the GPU device (Nishiura and Sakaguchi 2011).

Algorithm 6.1 C program for determining the maximum and minimum element indexes in the grid.

extern __shared__ int S_g[]; // The array S_g is created in shared memory

$i^* = \text{blockIdx.x} * \text{blockDim.x} + \text{threadIdx.x}; (i^* < N_e)$ // Each thread index corresponds to an element index

$S_g[\text{threadIdx.x} + 1] = P_g\left[i^*\right]$; // The grid index is stored in shared memory

If ($i^* > 0$ && threadIdx.x == 0) $S_g[0] = P_g\left[i^* - 1\right]$; // The grid index of the neighbor element is stored

__syncthreads(); // Different threads in the same block are synchronized

If (i^* == 0 || $P_g\left[i^*\right]$! = S_g[threadIdx.x])

{

  $I_{gmin}\left[P_g\left[i^*\right]\right] = i^*$; // The minimum index of the elements in each grid is stored

  If ($i^* > 0$) $I_{gmax}\left[S_g[\text{threadIdx.x}]\right] = i^* - 1$; // The maximum index of the elements in each grid is stored

}

If ($i^* == N_e - 1$) $I_{gmax}\left[P_g\left[i^*\right]\right] = i^*$; // The maximum index of all elements is the maximum index in this grid

Considering the element index as the parallel basis, the first rough contact detection is performed on the elements located in the same or adjacent grids to create the neighbor list. The bounding sphere method is used as the first rough detection. If the distance between the centroids of the two elements is less than the sum of their radii, then the elements are in contact, and a neighbor list $A_{nei}\left[i^*, I_{grid}\right]$ corresponding to each element index (i^*) is created. I_{grid} records the index of adjacent elements, which ranges from 0 to $C_{max} - 1$. Here, C_{max} is the maximum number of contacts between superquadric elements. In this paper, the differently shaped elements are constructed, and the aspect ratio of elements is expressed as σ. It is assumed that $C_{\max} = 24\sigma$. Moreover, if the index (j^*) of the neighboring element is greater than i^*, j^* will be stored from the left, starting from the smallest I_{grid}, and the total number of neighboring elements with $j^* > i^*$ is stored separately in the array $N_{\text{g}}[i^*]$; otherwise, j^* will be stored from the right, starting from the largest I_{grid}, and the total number of neighboring elements with $j^* < i^*$ is stored separately in the array $N_{\text{l}}[i^*]$. The results of the particle contact shown in Fig. 6.4 are listed in Table 6.1 by the above statistical method.

The fast prefix summation of the array $N_{\text{g}}[i^*]$ is done directly on the GPU by using a kernel function named "inclusive_scan" in the CUDA Thrust library, i.e., $N_{\text{sum}}[i^*] = \sum_{k=0}^{i^*} N_{\text{g}}[i^*]$. This kernel function can be expressed as

$$\begin{aligned}
&\text{thrust::inclusive_scan}\left(\text{thrust::device_ptr}\langle\text{int}\rangle\left(N_{\text{g}}\right),\right.\\
&\text{thrust::device_ptr}\langle\text{int}\rangle\left(N_{\text{g}} + N_e\right),\\
&\left.\text{thrust::device_ptr}\langle\text{int}\rangle(N_{\text{sum}})\right),
\end{aligned} \tag{6.9}$$

where N_e is the total number of elements. The results of the prefix summation of the array $N_{\text{g}}[i^*]$ are stored in the array $N_{\text{sum}}[i^*]$, which is listed in Table 6.1.

Considering the element index as the parallel basis, the contact pair list and reference list are created simultaneously. The index of each contact pair can be obtained by $N_{\text{list}} = N_{\text{sum}}[i^* - 1] + I_{\text{grid}}$, $I_{\text{grid}} \in \left[0, N_{\text{g}}[i^*] - 1\right]$. The neighbor element j^* of the element index i^* is computed by cycling through I_{grid} from 0 to $N_{\text{g}}[i^*] - 1$. Thus, the arrays $L_{\text{pa}}[N_{\text{list}}]$ and $L_{\text{pb}}[N_{\text{list}}]$ of the contact pairs created by the neighbor list $A_{\text{nei}}\left[i^*, I_{\text{grid}}\right]$ in Table 6.1 are listed in Table 6.2. Moreover, the index (N_{list})

Table 6.1 Neighbor element and the number of contacts for element index i^*

i^*	$A_{\text{nei}}\left[i^*, I_{\text{grid}}\right]$							$N_{\text{g}}\left[i^*\right]$	$N_{\text{l}}\left[i^*\right]$	$N_{\text{sum}}\left[i^*\right]$
	$I_{\text{grid}} =$ 0	1	2	…	$C_{\max} - 3$	$C_{\max} - 2$	$C_{\max} - 1$			
0	$j^* = 2$			…				1	0	1
1				…				0	0	1
2				…			$j^* = 0$	0	1	1
3				…				0	0	1
4	$j^* = 5$	$j^* =$ 6		…				2	0	3
5	$j^* = 6$			…			$j^* = 4$	1	1	4
6				…		$j^* = 5$	$j^* = 4$	0	2	4
7				…				0	0	4
8				…				0	0	4
9				…				0	0	4

of the contact pair is stored in the reference list $A_{\mathrm{ref}}\left[i^{*}, I_{\mathrm{grid}}\right]$. I_{grid} records the index of the contact pair, and the storage location is the same as in the neighbor list $A_{\mathrm{nei}}\left[i^{*}, I_{\mathrm{grid}}\right]$. Figure 6.5 shows the calculation process of creating the reference lists for two neighbor particles $i^{*} = 4$ and $j^{*} = 5$. The creation of the reference list $A_{\mathrm{ref}}\left[i^{*}, I_{\mathrm{grid}}\right]$ consists of four steps (Nishiura and Sakaguchi 2011): (I) according to the correspondence between the neighbor list $A_{\mathrm{nei}}\left[i^{*}, I_{\mathrm{grid}}\right]$ and the reference list $A_{\mathrm{ref}}\left[i^{*}, I_{\mathrm{grid}}\right]$, the contact pair index N_{list} is stored in the reference list of the element i^{*}; (II) the storage location I_{grid} of the element index i^{*} in the neighbor list of the element j^{*} is searched from right to left; (III) according to the correspondence between the neighbor list and the reference list, the storage location of the contact pair index in the reference list of the element j^{*} is found; (IV) the contact pair index obtained in the first step is stored in the corresponding location obtained in the third step. The above procedure is summarized in Algorithm 6.2. The reference list of each element facilitates the rapid superposition of the contact force and moments of each element on the GPU device.

Table 6.2 List of contact pairs obtained from the neighbor list

i^{*}	0	1	2	3	4	5	6	7	8	9	
$j^{*}(> i^{*})$	2				5	6	6				
$N_{\mathrm{g}}\left[i^{*}\right]$	1	0	0	0	2	1	0	0	0	0	
$N_{\mathrm{sum}}\left[i^{*}\right]$	1	1	1	1	3	4	4	4	4	4	
N_{list}	0				1	2	3				
$L_{\mathrm{pa}}[N_{\mathrm{list}}] = i^{*}$	0				4	4	5				
$L_{\mathrm{pb}}[N_{\mathrm{list}}] = j^{*}$	2				5	6	6				

Fig. 6.5 Process of creating the reference list ($A_{\mathrm{ref}}\left[i^{*}, I_{\mathrm{grid}}\right]$) from the neighbor list ($A_{\mathrm{nei}}\left[i^{*}, I_{\mathrm{grid}}\right]$)

Algorithm 6.2 C program for creating the contact pair list and reference list on the GPU.

$i^* = \text{blockIdx.x} * \text{blockDim.x} + \text{threadIdx.x}; (i^* < N_e)$ // Each thread index corresponds to an element index
for $(I_{\text{grid}} = 0;\ I_{\text{grid}} < N_g[i^*];\ I_{\text{grid}} + +)$ // Loop over neighbor list
{
$N_{\text{list}} = N_{\text{sum}}[i^* - 1] + I_{\text{grid}};$ // The contact pair index is obtained
  $A_{\text{ref}}[i^*, I_{\text{grid}}] = N_{\text{list}};$ // Step (I) of Fig. 6.5
  $j^* = A_{\text{nei}}[i^*, I_{\text{grid}}];$ // The neighbor element index j^* of element i^* is obtained
  $L_{\text{pa}}[N_{\text{list}}] = i^*;\ L_{\text{pb}}[N_{\text{list}}] = j^*;$ // The contact pair list is created
  for $(\text{k} = 0;\ \text{k} < N_1[j^*];\ \text{k} + +)$ // Step (II) of Fig. 6.5
  {
if $(i^* == A_{\text{nei}}[j^*, C_{\max} - 1 - \text{k}])\ A_{\text{ref}}[j^*, C_{\max} - 1 - \text{k}] = N_{\text{list}};$ // Steps (III) and (IV) of Fig. 6.5
  }
}

6.2.2 *Bounding Box and Newton Iterative List Generation*

Considering the index of the contact pair as the parallel basis, the bounding box list $L_{\text{box}}[N_{\text{list}}]$ corresponding to the contact pair list is created. This facilitates the rapid elimination of potential contact pairs that are unlikely to collide with each other, thereby improving the computational efficiency of parallel programs. More descriptions of the oriented bounding box algorithm can be found in Ref. (Portal et al. 2010). If the bounding boxes of the two non-spherical elements are in contact, "1" is stored in the array $L_{\text{box}}[N_{\text{list}}]$; otherwise, "0" is stored in the array $L_{\text{box}}[N_{\text{list}}]$.

If $L_{\text{box}}[N_{\text{list}}] = 1$ is satisfied, the Newton iterative method is used to accurately calculate the overlap between the superquadric elements. Here, $L_{\text{pa}}[N_{\text{list}}] = i^*$ and $L_{\text{pb}}[N_{\text{list}}] = j^*$. According to the relationship between the old and new element indexes, the element indexes before reordering can be expressed as $i = P_e[i^*]$ and $j = P_e[j^*]$. The nonlinear equations for the midway point $\boldsymbol{X}_0$ can be expressed as (Cleary et al. 1997):

$$\begin{cases} \nabla F_i(\boldsymbol{X}) + \alpha^2 \nabla F_j(\boldsymbol{X}) = 0 \\ F_i(\boldsymbol{X}) - F_j(\boldsymbol{X}) = 0 \end{cases}, \tag{6.10}$$

where $\boldsymbol{X} = (x, y, z)^{\text{T}}$ and α is a Lagrange multiplier. F_i and F_j are superquadric equations for elements i and j, respectively. The Newton iterative equation for

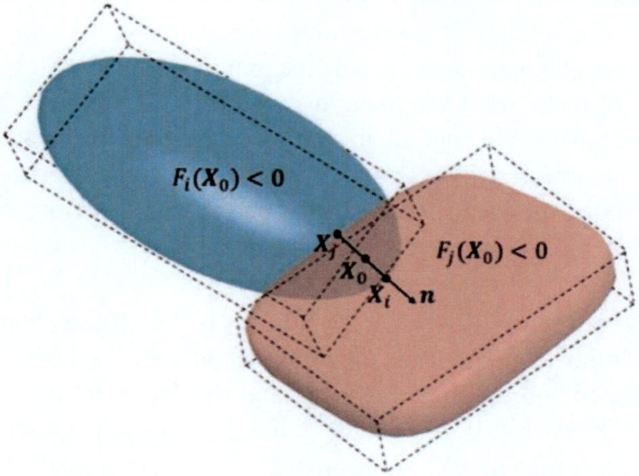

Fig. 6.6 Contact detection between two superquadric elements

Eq. (6.10) can be expressed as (Podlozhnyuk et al. 2016):

$$\begin{pmatrix} \nabla^2 F_i(\boldsymbol{X}) + \alpha^2 \nabla^2 F_j(\boldsymbol{X}) & 2\alpha \nabla F_j(\boldsymbol{X}) \\ \nabla F_i(\boldsymbol{X}) - \nabla F_j(\boldsymbol{X}) & 0 \end{pmatrix} \begin{pmatrix} \mathrm{d}X \\ \mathrm{d}\alpha \end{pmatrix} = - \begin{pmatrix} \nabla F_i(\boldsymbol{X}) + \alpha^2 \nabla F_j(\boldsymbol{X}) \\ F_i(\boldsymbol{X}) - F_j(\boldsymbol{X}) \end{pmatrix}, \tag{6.11}$$

where $\boldsymbol{X}^{(k+1)} = \boldsymbol{X}^{(k)} + \mathrm{d}\boldsymbol{X}^{(k)}$ and $\alpha^{(k+1)} = \alpha^{(k)} + \mathrm{d}\alpha^{(k)}$. If $F_i(\boldsymbol{X}_0) < 0$ and $F_j(\boldsymbol{X}_0) < 0$ are satisfied, element i is in contact with element j, and the normal direction is expressed as $\boldsymbol{n} = \nabla F_i(\boldsymbol{X})/\|\nabla F_i(\boldsymbol{X})\|$, as shown in Fig. 6.6. Then, the surface points $\boldsymbol{X}_i$ and $\boldsymbol{X}_j$ can be expressed as $\boldsymbol{X}_i = \boldsymbol{X}_0 + \beta\boldsymbol{n}$ and $\boldsymbol{X}_j = \boldsymbol{X}_0 + \gamma\boldsymbol{n}$, respectively. The unknown parameters β and γ can be obtained by the Newton iterative method (Podlozhnyuk et al. 2016): $\beta^{(k+1)} = \beta^{(k)} - F_i(\boldsymbol{X}_i^{(k)})/(\nabla F_i(\boldsymbol{X}_i^{(k)}) \cdot \boldsymbol{n})$ and $\gamma^{(k+1)} = \gamma^{(k)} - F_j(\boldsymbol{X}_j^{(k)})/(\nabla F_j(\boldsymbol{X}_j^{(k)}) \cdot \boldsymbol{n})$. The normal overlap is $\boldsymbol{\delta}_n = \boldsymbol{X}_i - \boldsymbol{X}_j$.

To accelerate the computational efficiency of numerical iterations, the midway point $\boldsymbol{X}_0$ is stored in the Newton iterative list $L_\mathrm{n}[N_\mathrm{list}]$ and used for the initial predicted value of the Newton iteration at the next DEM time step. Moreover, the tangential force of the particles is calculated based on the tangential displacement, and the tangential displacement of each element is superimposed in an incremental form. Here, the calculation of the tangential force in the DEM model will be given in Sect. 6.3. It means that if two neighbor elements are in contact at the previous time step and remain in contact at the current step, then the tangential displacement of the previous time step must be inherited and superimposed at the current step. Therefore, it is crucial for the superquadric DEM model to determine whether the current contact pair is the contact pair of the previous time step.

At the current time step, the index of the contact pair is considered as the parallel basis. i and i^* represent the element index before and after reordering, respectively. At the previous time step, $i^{\mathrm{pre}*}$ represent the element index after reordering, and the array $P_e^{\mathrm{pre}}[i]$ is used to store the correspondence between the new and old element indexes. Moreover, the array $N_{\mathrm{sum}}^{\mathrm{pre}}[i^{\mathrm{pre}*}]$ is used to store the prefix summation of the number of neighbors at the previous time step, and the array $L_{\text{pb}}^{\text{pre}}\left[N_{\text{list}}^{\text{pre}}\right]$ is used to store the neighbor element index in the previous contact pair list. The tangential displacements of the current and previous time steps are stored in arrays $D_t[N_{\text{list}}]$ and $D_t^{\text{pre}}\left[N_{\text{list}}^{\text{pre}}\right]$, respectively, and the midway point of the current and previous time steps are stored in arrays $L_{\text{n}}[N_{\text{list}}]$ and $L_n^{\text{pre}}\left[N_{\text{list}}^{\text{pre}}\right]$, respectively. The C program for determining the correspondence between the current and previous contact pairs and inheriting the iteration point and previous tangential displacements is shown in Algorithm 6.3. If $j^{\text{pre}*} > i^{\text{pre}*}$, the previous contact pair list $L_{\text{pb}}^{\text{pre}}\left[N_{\text{list}}^{\text{pre}}\right]$ is searched in the range of $N_{\text{list}}^{\text{pre}}$ between $N_{\text{sum}}^{\text{pre}}[i^{\text{pre}*}-1]$ and $N_{\text{sum}}^{\text{pre}}[i^{\text{pre}*}]$. If the element index $j^{\text{pre}*}$ is matched, the midway point of the previous step is used for the initial iteration point of the current step and the previous tangential displacement is superimposed at the current step. Otherwise, if $j^{\text{pre}*} < i^{\text{pre}*}$, the previous contact pair list $L_{\text{pb}}^{\text{pre}}\left[N_{\text{list}}^{\text{pre}}\right]$ is searched in the range of $N_{\text{list}}^{\text{pre}}$ between $N_{\text{sum}}^{\text{pre}}[j^{\text{pre}*}-1]$ and $N_{\text{sum}}^{\text{pre}}[j^{\text{pre}*}]$. If the element index $i^{\text{pre}*}$ is matched, the tangential displacement and midway point of the previous step are inherited at the current step.

It is worth noting that the element index is reordered at each DEM time step, but the element index before reordering at the current time step is the same as the element index before reordering at the previous time step. In other words, the information update at each time step in the DEM simulation is based on the original element index. In addition, the tangential displacements are vectors extending from the element with a smaller index to that with a larger index for $j^{\text{pre}*} > i^{\text{pre}*}$. Therefore, when $j^{\text{pre}*} < i^{\text{pre}*}$, the tangential displacement vector with the opposite sign at the previous step is superimposed at the current step. Finally, the contact forces acting on elements i^* and j^* are stored in arrays $\boldsymbol{F}_i^{ij}[N_{\text{list}}]$ and $\boldsymbol{F}_j^{ij}[N_{\text{list}}]$, respectively, and the moments acting on elements i^* and j^* are stored in arrays $\boldsymbol{M}_i^{ij}[N_{\text{list}}]$ and $\boldsymbol{M}_j^{ij}[N_{\text{list}}]$, respectively.

Algorithm 6.3 C program for determining the correspondence between current and previous contact pairs.

N_{list} = blockIdx.x * blockDim.x + threadIdx.x;// Each thread index corresponds to a contact pair index

$i^* = L_{\text{pa}}[N_{\text{list}}]$; $j^* = L_{\text{pb}}[N_{\text{list}}]$; // Two neighbor element indexes are obtained at the current time step

$i^{\text{pre}*} = P_e^{\text{pre}}\left[P_e\left[i^*\right]\right]$; $j^{\text{pre}*} = P_e^{\text{pre}}\left[P_e\left[j^*\right]\right]$; // Two element indexes are obtained at the previous time step

if ($j^{\text{pre}*} > i^{\text{pre}*}$)

{

  for ($N_{\text{list}}^{\text{pre}} = N_{\text{sum}}^{\text{pre}}\left[i^{\text{pre}*}-1\right]$; $N_{\text{list}}^{\text{pre}} < N_{\text{sum}}^{\text{pre}}\left[i^{\text{pre}*}\right]$; $N_{\text{list}}^{\text{pre}}++$) // Loop over contact pair list

  {

    if ($j^{\text{pre}*} == L_{\text{pb}}^{\text{pre}}\left[N_{\text{list}}^{\text{pre}}\right]$) // The current contact pair is the contact pair of the previous time step

{ // The tangential displacement and midway point at the previous time step are inherited

$D_t[N_{\text{list}}] = D_t^{\text{pre}}\left[N_{\text{list}}^{\text{pre}}\right]$; $L_{\text{n}}[N_{\text{list}}] = L_n^{\text{pre}}\left[N_{\text{list}}^{\text{pre}}\right]$;

}

}

}

else if $(j^{\text{pre}*} < i^{\text{pre}*})$

{

for $(N_{\text{list}}^{\text{pre}} = N_{\text{sum}}^{\text{pre}}\left[j^{\text{pre}*} - 1\right];\ N_{\text{list}}^{\text{pre}} < N_{\text{sum}}^{\text{pre}}\left[j^{\text{pre}*}\right];\ N_{\text{list}}^{\text{pre}} + +)$ // Loop over contact pair list

{

if $(i^{\text{pre}*} == L_{\text{pb}}^{\text{pre}}\left[N_{\text{list}}^{\text{pre}}\right])$ // The current contact pair is the contact pair of the previous time step

{ // The tangential displacement and midway point at the previous time step are inherited

$D_t[N_{\text{list}}] = -1.0 \times D_t^{\text{pre}}\left[N_{\text{list}}^{\text{pre}}\right]$; $L_{\text{n}}[N_{\text{list}}] = L_n^{\text{pre}}\left[N_{\text{list}}^{\text{pre}}\right]$;

}

}

}

6.2.3 Contact Force Calculation and Element Information Updating

Considering the element index as the parallel basis, the contact forces and moments of several neighbor elements j^* acting on the element index i^* are cumulatively summed. The reference list $A_{\text{ref}}\left[i^*, I_{\text{grid}}\right]$ obtained in Algorithm 6.2 is used to sum the contact force and moment of each element on the GPU, which is expressed as (Nishiura and Sakaguchi 2011):

$$\boldsymbol{F}\left[i^*\right] = \sum_{I_{\text{grid}}=0}^{N_{\text{g}}[i^*]-1} \boldsymbol{F}_i^{ij}\left[A_{\text{ref}}\left[i^*, I_{\text{grid}}\right]\right] + \sum_{I_{\text{grid}}=C_{\max}-N_{\text{l}}[i^*]}^{C_{\max}-1} \boldsymbol{F}_j^{ij}\left[A_{\text{ref}}\left[i^*, I_{\text{grid}}\right]\right], \quad (6.12)$$

$$\boldsymbol{M}\left[i^*\right] = \sum_{I_{\text{grid}}=0}^{N_{\text{g}}[i^*]-1} \boldsymbol{M}_i^{ij}\left[A_{\text{ref}}\left[i^*, I_{\text{grid}}\right]\right] + \sum_{I_{\text{grid}}=C_{\max}-N_{\text{l}}[i^*]}^{C_{\max}-1} \boldsymbol{M}_j^{ij}\left[A_{\text{ref}}\left[i^*, I_{\text{grid}}\right]\right], \quad (6.13)$$

where $\boldsymbol{F}[i^*]$ and $\boldsymbol{M}[i^*]$ represent the resultant force and moment acting on the element index i^*, respectively. In the reference list $A_{\text{ref}}\left[i^*, I_{\text{grid}}\right]$, I_{grid} from 0 to

$N_g[i^*] - 1$ represents the contact force or moment of element j^* acting on element i^*, and I_{grid} from $C_{max} - N_1[i^*]$ to $C_{max} - 1$ represents the contact force or moment of element i^* acting on element j^*. Therefore, the cumulative sum of the contact force and moment of each element is independent for each thread by using the reference list.

Considering the element index as the parallel basis, the information update of each element is performed independently in each thread. The original element index i is obtained by $i = P_e[i^*]$, and all information updates of the element are based on the original element index. The translational velocity and position of the element are obtained by solving the Eq. (6.2) with the leap-frog algorithm. However, the solution of the rotation velocity of a single particle is relatively difficult. It is mainly because the rotation of a single particle in the global coordinate system causes a change in the moment of inertia, which means that the moment of inertia varies with time. Therefore, a local coordinate system is established for each element, and a transformation matrix is used to determine the relationship between the local and global coordinate systems. Here, the transformation matrix $\boldsymbol{T}$ obtained by the quaternion is expressed as Eq. (6.4), and the total moment acting on the particle i in the global coordinate system is denoted as $\boldsymbol{M}^g$. Moreover, the total moment $\boldsymbol{M}^l$ acting on the particle i in the local coordinate system can be obtained by $\boldsymbol{M}^l = \boldsymbol{T} \cdot \boldsymbol{M}^g$. The rotational motion of a single particle in the local coordinate system can be expressed as

$$I_{xx}^l \frac{d\omega_x^l}{dt} + \left(I_{zz}^l - I_{yy}^l\right)\omega_y^l \omega_z^l = M_x^l, \tag{6.14}$$

$$I_{yy}^l \frac{d\omega_y^l}{dt} + \left(I_{xx}^l - I_{zz}^l\right)\omega_z^l \omega_x^l = M_y^l, \tag{6.15}$$

$$I_{zz}^l \frac{d\omega_z^l}{dt} + \left(I_{yy}^l - I_{xx}^l\right)\omega_x^l \omega_y^l = M_z^l, \tag{6.16}$$

where I_{xx}^l, I_{yy}^l, and I_{zz}^l are the diagonal elements of the inertial tensor of particle i. The subscripts x, y, and z denote the components of the vector in the x, y, and z directions, respectively. Subsequently, the local angular velocity $\left(\boldsymbol{\omega}^l\right)$ of particle i is obtained by solving Eqs. (6.14–6.16), and the updated quaternion is expressed as

$$\begin{bmatrix} q_0 \\ q_1 \\ q_2 \\ q_3 \end{bmatrix} = \begin{bmatrix} q_0 \\ q_1 \\ q_2 \\ q_3 \end{bmatrix} + 0.5dt \cdot \begin{bmatrix} q_0 & -q_1 & -q_2 & -q_3 \\ q_1 & q_0 & -q_3 & q_2 \\ q_2 & q_3 & q_0 & -q_1 \\ q_3 & -q_2 & q_1 & q_0 \end{bmatrix} \begin{bmatrix} 0 \\ \omega_x^l \\ \omega_y^l \\ \omega_z^l \end{bmatrix}. \tag{6.17}$$

Here, a more detailed description for quaternion updates can be found in Refs. (Kosenko 1998; Miller et al. 2002). The angular velocity $(\boldsymbol{\omega}^g)$ of particle i can be obtained using the transformation matrix $\boldsymbol{T}$ and expressed as $\boldsymbol{\omega}^g = \boldsymbol{T}^{-1} \cdot \boldsymbol{\omega}^l$. It is worth noting that these operations for solving equations of motion and updating information are parallelized for each element index. Finally, the basic information of each element (such as position, translational velocity, global angular velocity, local angular velocity, and quaternion) is used for the DEM simulation of the next time step.

6.3 Domestic and Foreign DEM Software Development

In this section, foreign and domestic discrete element computational analysis software are introduced in detail. Domestic software include GDEM, DEMms, MatDEM, StreamDEM, DEMSLab, CoSim, SudoSim, and SDEM. Foreign software include EDEM, PFC3D, 3DEC, Rocky, Yade, ESyS-Particle, LAMMPS, and LIGGGHTS.

6.3.1 Development of Foreign DEM Software

High-performance discrete element algorithms have been widely used in different engineering fields, such as geotechnical, mining, environmental, chemical, and cold regions, and computational analysis software based on high-performance discrete element methods has been rapidly developed and successfully applied in industrial fields.

Software EDEM is a commercially global CAE simulation and analysis software based on the discrete element method, which can perform industrial-scale discrete element simulations of granular materials through CPU parallelism and GPU parallelism. The computational scale can be up to tens of millions of complex granular systems, as shown in Fig. 6.7. The software is used in several industries, such as off-road, mining, steelmaking, manufacturing to predict the specific behavior of granular materials and to optimize and evaluate equipment performance. In the software EDEM, Particle Factory can automatically generate multiple types of particles with different sizes, material parameters, and initial states of motion, and the particle shapes are spheres and polyhedrons. Moreover, EDEM can be coupled with other software such as FEA (Finite Element Analysis), CFD (Computational Fluid Dynamics), and MBD (Multi-Body Dynamics) to realize the simulation of the granular materials coupled with the fluid, mechanical structure, and electromagnetic field.

Software PFC (Particle Flow Code) is a computational software based on the discrete element method developed by Itasca, and this software is divided into two-dimensional (PFC2D) and three-dimensional software (PFC3D) (Fig. 6.8), which are discrete element programs based on two-dimensional disk and three-dimensional spheres, respectively. The software was initially developed for geotechnical engineering and has been widely used in geotechnical, mining, petroleum, mechanical,

Fig. 6.7 Flow processes of granular materials simulated by the software EDEM

and other industrial production fields. The advantage of PFC lies in the nodal characterization of discontinuous features, such as structural surfaces and internal defects of geotechnical rocks. It has a more mature method for simulating the mechanical behavior of geotechnical rocks and soils in the process of damage. With the upgrading of the software version, the application of PFC has been enriched and expanded gradually, such as the three-dimensional fissure network simulation for the structural surface of the rock.

UDEC and 3DEC are other discrete element calculation software developed by Itasca. UDEC is a 2D version of 3DEC. The software is mainly used for geotechnical analysis of soil, rock, groundwater, structural support, and masonry. The software provides users with several analysis modules, including dynamic analysis, and temperature analysis. It also can realize the simulation of various coupling scenarios represented by heat-force coupling (Fig. 6.9).

Rocky DEM is a powerful discrete element simulation software developed by Granular Dynamics International, LLC, and Engineering Simulation and Scientific Software Company (ESSS). Rocky DEM has been widely used in the fields of mining equipment, engineering and agricultural machinery, chemical industry, steel, food

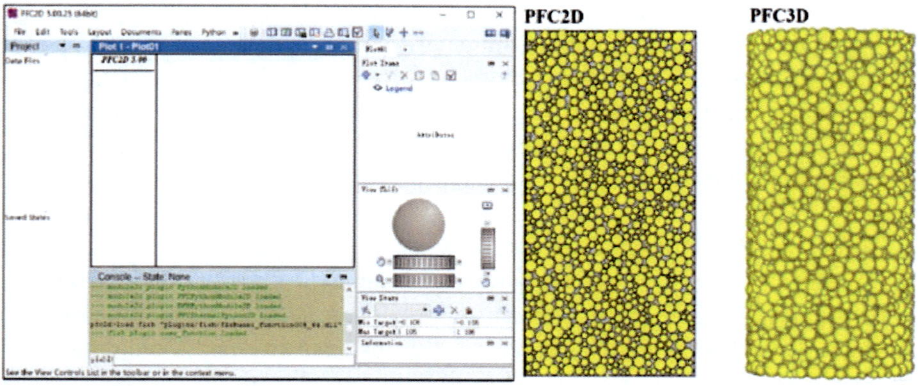

Fig. 6.8 PFC 5.0 graphical user interface and 2D/3D particle modeling

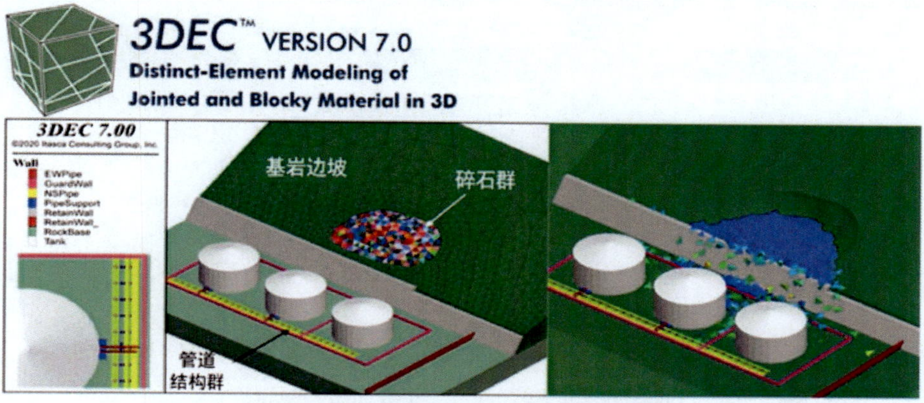

(a) DEM simulation of debris roll-off.

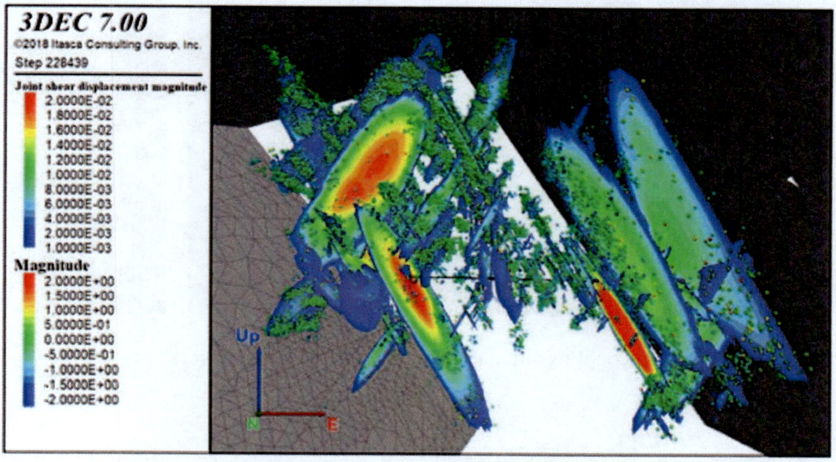

(b) Shear displacement of geotechnical surfaces under thermal-force coupling.

Fig. 6.9 Different application scenarios for the software 3DEC

and medicine, etc. Rocky DEM can realize large-scale GPU parallel computation, and it can also realize the modeling and simulation of complex morphology particles. Rocky DEM has been fully integrated into the Ansys Workbench environment (Fig. 6.10), and can be coupled with Fluent software (Fig. 6.11). Furthermore, the software has a powerful secondary development function.

In addition to commercially available DEM software, some open-source DEM frameworks such as Yade, ESyS-Particle, LAMMPS, and LIGGGHTS have also developed rapidly, as shown in Fig. 6.12. These software also realize large-scale DEM simulations with multiple GPUs or GPU clusters.

ESyS-Particle is primarily used in geotechnical applications such as rock fragmentation, landslides, and earthquakes. ESyS-Particle includes a Python scripting interface that provides flexibility in the simulation setup and real-time data analysis.

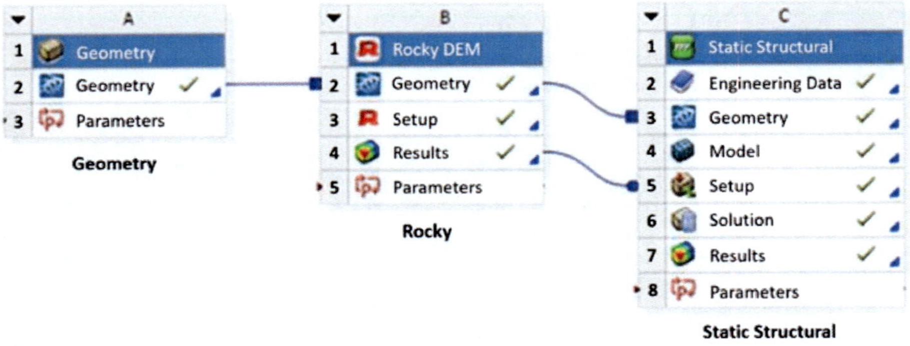

Fig. 6.10 Rocky DEM integration in Ansys workbench environment

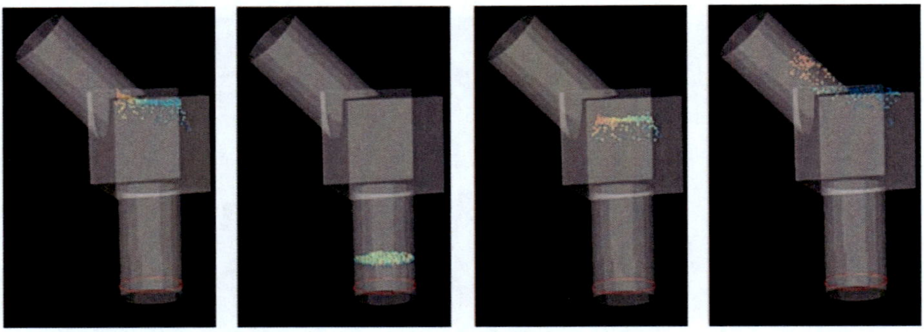

Fig. 6.11 Unidirectional coupling simulation with Rocky-Fluent software

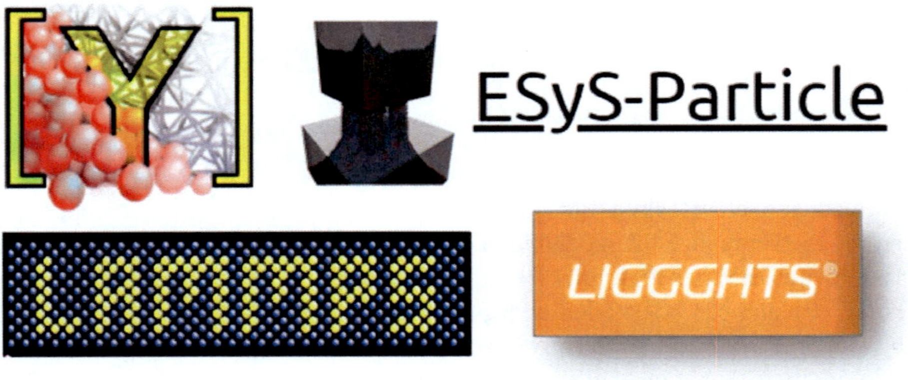

Fig. 6.12 Signs of some open-source DEM platforms

ESyS-Particle's DEM computation engine is written in C++ and parallelized using MPI to support the simulation of millions of particles on clusters or workstations.

Yade, a community-driven open-source computational platform for granular materials, was first developed by Prof. Frederic in France and named SDEC (Spherical Discrete Element Code). In 2004, PhD student Olivier started the development of the Yade project and rewrote SDEC using C++ and Python. Then, PhD student Janek led the development of Yade. Yade has developed into a large open-source software platform, which is still being updated and is used in a variety of applications. It is worth noting that the Yade project aims to provide users with a more comfortable interactive environment, such as running on Windows systems, while providing commercial customization services.

LAMMPS is a molecular dynamics numerical simulation software developed by Sandia National Laboratories in the USA. It is released under GPL license, and the open-source code can be obtained free of charge. It means that the users can modify the source code according to their own needs. LAMMPS can support the numerical simulation of atomic and molecular systems of millions of particles, and can provide a variety of potential functions. Moreover, LAMMPS has good parallel scalability, and CFDEM coupling can be used to realize CFD-DEM coupling calculation.

6.3.2 Development of Domestic DEM Software

The development of the discrete element method and related program software in China originated from the application in geotechnical mechanics in the 1980s. The Joint Laboratory of Discontinuous Medium Mechanics and Engineering Disasters of the Chinese Academy of Sciences (CAS) and Jidao Chengran Technology Co., Ltd. Have jointly developed the large-scale commercial software of discrete element methods (GDEM). GDEM was developed by Prof. Shihai Li and Prof. Chun Feng's group. The core algorithm of this software is CDEM (Continuum-based Discrete Element Method), which is suitable for simulating the discontinuous deformation and progressive damage of materials under static and dynamic loads, and has been widely used in many fields, such as geotechnical engineering, mining engineering, structural engineering, and water conservancy and hydroelectricity engineering, as shown in Fig. 6.13. GDEM couples the finite element method (FEM) and discrete element methods (DEM), and carries out FEM calculations inside the block and DEM calculations at the boundary of the block. GDEM can not only simulate the deformation and motion characteristics of the material in the continuous and discontinuous state, but also realize the gradual destruction process of the material from the continuum to the discontinuous state. GDEM also adopts the high-performance computing technology of the GPU, which can realize large-scale and high-precision calculations.

Software DEMms is a research and engineering software developed by Prof. Wei Ge and Prof. Limin Wang's group at the Institute of Process Engineering, Chinese Academy of Sciences, for large-scale high-performance simulations of granular materials and multiphase systems. DEMms can efficiently utilize various computational resources such as CPUs and GPUs. This software couples a unique

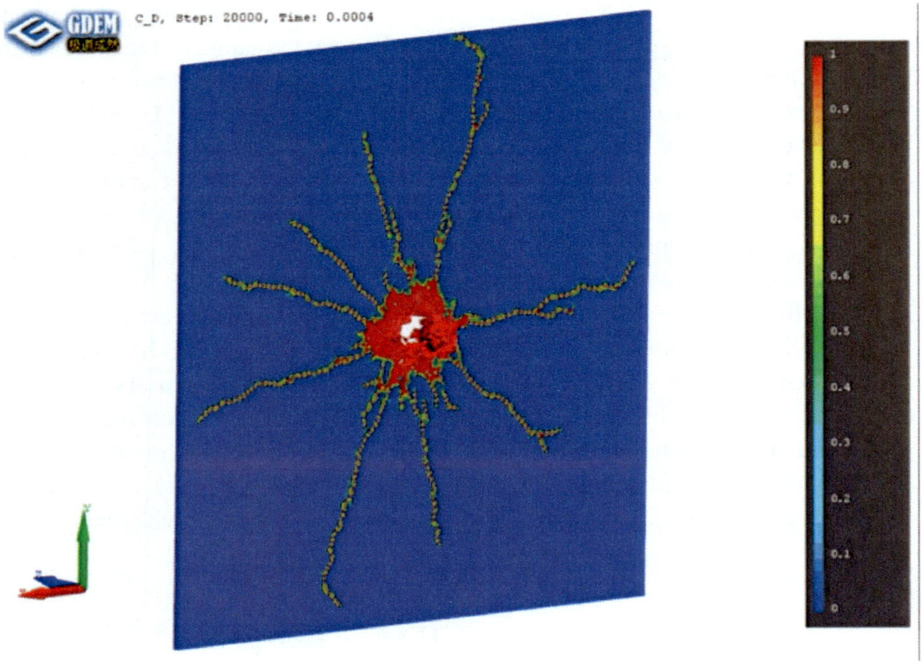

Fig. 6.13 Typical examples of software GDEM

particle coarse-graining model with a fluid–solid coupling method, which can efficiently interface with a variety of open-source flow solvers, and has the ability to simulate industrial processes with wide particle size distributions and reaction-transfer coupling in a long time or in quasi-real time. It can also construct complex particle shapes, such as non-spherical and deformed particles, as well as simulate the multiphase transfer reaction and other complex processes, as shown in Fig. 6.14.

Software MatDEM is a high-performance DEM software developed by Prof. Chun Liu's group at Nanjing University, as shown in Fig. 6.15. This software adopts the matrix DEM calculation method, which can effectively simulate the large deformation and damage processes in the field of geological and geotechnical engineering. It can also realize GPU parallel computation with millions of particles. In addition, this software supports automatic packing modeling, layered material assignment, and load setting, and also realizes multi-field coupling analysis based on the MATLAB language. MatDEM is widely used in many engineering fields, such as landslides, rockburst, impact damages, pile-soil action, hobbing rock breakage, hydraulic fracturing, and so on.

Software DEMSLab is a large-scale commercial software developed by Prof. Yongzhi Zhao's group at Zhejiang University to simulate granular systems. DEMSLab is based on the non-spherical particles to realize the simulation of large-scale granular systems on an industrial scale. DEMSLab contains three parts: the pre-processor, solver, and post-processor (Fig. 6.16). The pre-processor of DEMSLab

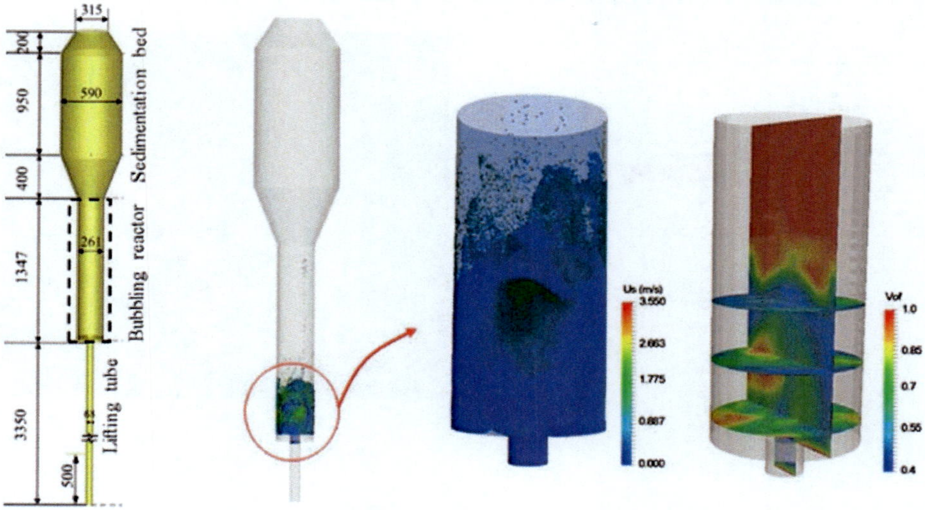

Fig. 6.14 Example of DEMms software

Fig. 6.15 Interface of software MatDEM

can perform complex geometric modeling of the device (through standard 3D software such as UG NX, Pro/E, SolidWorks, and CATIA). The particle generator is based on the adaptive generation of particles (such as spheres, combined spheres, bonded spheres, micro-droplets, super-ellipsoids, combined super-ellipsoids, convex

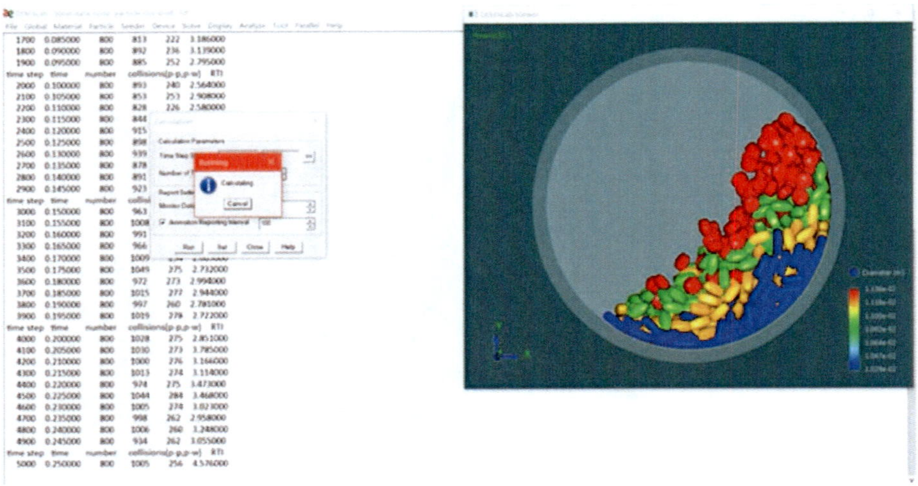

Fig. 6.16 Post-processing module of the software DEMSLab

polyhedrons, and concave polyhedrons). The DEMSLab solver contains the contact and non-contact force models (such as Van der Waals force, capillary force), and has been designed in parallel with OpenMP technology. It can simulate spherical and non-spherical granular systems of ten million scales, and support complex structural and kinematic boundary conditions, periodic boundary conditions, etc., as well as a powerful API for secondary development.

Prof. Wenjie Xu's group at Tsinghua University has developed the DEM software CoSim, which is centered on the unique multiphase, multi-process, and multi-scale characteristics of geotechnical/geological engineering, and based on the physical and mechanical mechanisms of deformation and damage process of geotechnical/geological engineering. It uses different numerical methods such as Finite Element Model (FEM), Discrete Element Model (DEM) for spheres and polyhedrons, Matter Point Method (MPM), Smooth Particle Hollow (SPH), Lattice Boolean Method (LBM), and Finite Volume Method (FVM). It is supported by the parallel acceleration of GPUs, and the different numerical computation and analysis methods are developed by utilizing the advantages of different numerical methods in the aspects of solids, fluids, continuum and discontinuity, fineness, and macroscopicity. Figure 6.17 shows the main functions of CoSim, which has been widely used in the multi-scale mechanics of geotechnical engineering (e.g., rocks, sands, soil-stone mixtures, etc.), stability analysis of slopes, dynamic assessment of landslide hazards, analysis of landslide surges, and wave chain hazards in reservoirs.

Prof. Jidong Zhao's group at the Hong Kong University of Science and Technology (HKUST) developed the DEM software SudoSim. This software is a computational platform that integrates all kinds of meshless numerical methods (including DEM, MPM, LBM, PD, SPH, etc.) and inherits the basic framework of the open-source DEM code, YADE. The core program of SudoSim is written in C++, CUDA, and Python. SudoSim GUI is the visualization tool of SudoSim (Fig. 6.18), mainly

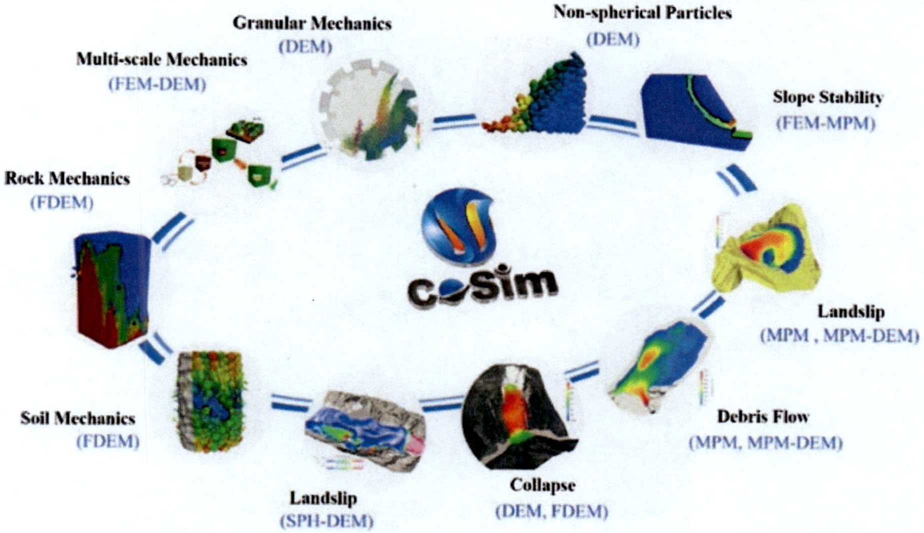

Fig. 6.17 Main functions of software CoSim

used for modeling and debugging. SudoSim supports the three operating systems of Linux, Windows, and MacOS, but the MacOS version does not provide the local GPU computation function. SudoSim provides Client/Server mode to realize cross-platform remote computation and real-time operation.

Prof. Jianqun Yu's group at Jilin University developed the software AgriDEM (Agricultural DEM), and this software is a 3D CAE software focusing on the field of agricultural engineering. The software adopts the multi-sphere method, lysimeter aggregator method, and mass-spring model to establish crop seed, cob, and plant models, and realizes numerical simulations of the working process of agricultural machinery components based on the Discrete Element Method (DEM), Computational Fluid Dynamics (CFD), and Planar Multi-Rigid Body Dynamics/Kinematics (MBD/MBK).

6.4 Computational Analysis Software SDEM

In this section, the software SDEM developed by Prof. Shunying Ji is introduced in detail. Then, four tests are performed to examine the applicability of the GPU parallel algorithm for superquadric elements, including the generation of a large-scale non-spherical granular bed, the flow process of the granular column, the dynamic hopper discharge, and the mixing behavior within a horizontal drum. Then, the numerical results are compared with the experimental results, and the speedup ratios of the GPU to the CPU for differently shaped elements are investigated.

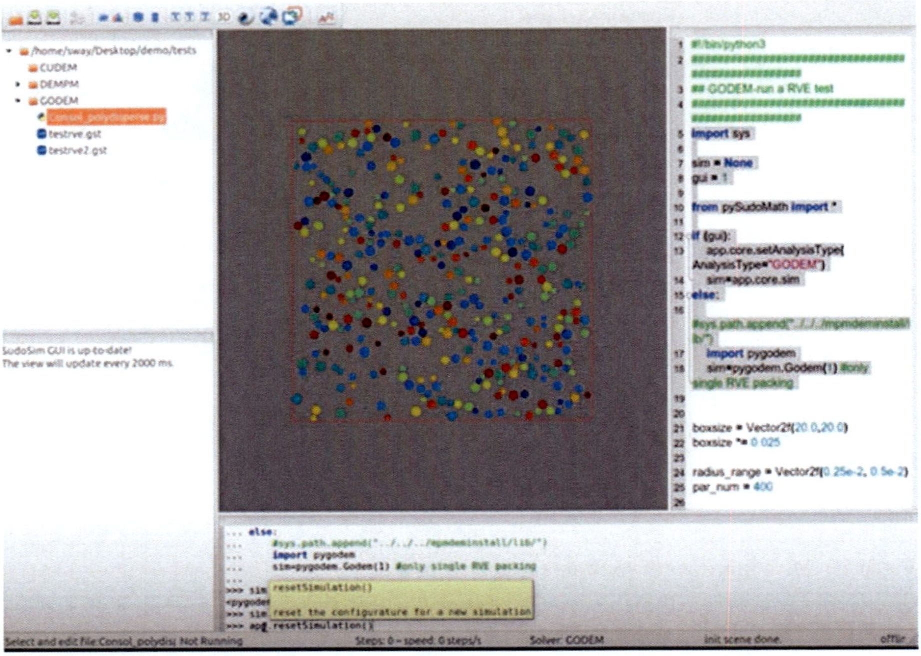

Fig. 6.18 SudoSim GUI and the visualization tool of software SudoSim

6.4.1 *Introduction to the Software SDEM*

Prof. Shunying Ji's group at Dalian University of Technology has developed the software SDEM, which is a high-performance DEM computational analysis software (Fig. 6.19). The software has numerical simulation functions such as multi-particle morphology, multi-media, multi-scale, and can realize GPU parallel computing.

Software SDEM provides a powerful library, covering the various discrete element methods of arbitrarily shaped particles. The types of particle shapes include the sphere, superquadric equation, polyhedron, dilated polyhedron, spherical harmonic function, bonding element, dilated disk, level set function, and large deformation element (Fig. 6.20). The spherical bonding model can be used for effective simulations of the damage process of brittle materials, such as sea ice and rock. The dilated polyhedral model based on the Minkowski sum algorithm can realize the simulation of non-regular particles. In addition, the level set algorithm and large deformation model greatly enrich the DEM simulations.

Software SDEM provides various special coupling methods, including DEM-FEM (Fig. 6.21), DEM-SPH (Fig. 6.22), DEM-FEM-SPH, and other algorithms. The numerical algorithms have a wide range of applications in the fields of dynamic analysis of regular/non-regular granular materials, ocean engineering in cold regions, evolution of geologic hazards, mechanical response of ballasted railroad beds, and the landing process of landers. To improve the calculation speed of DEM simulations, this software also includes the local grid algorithm, local cell refinement technology,

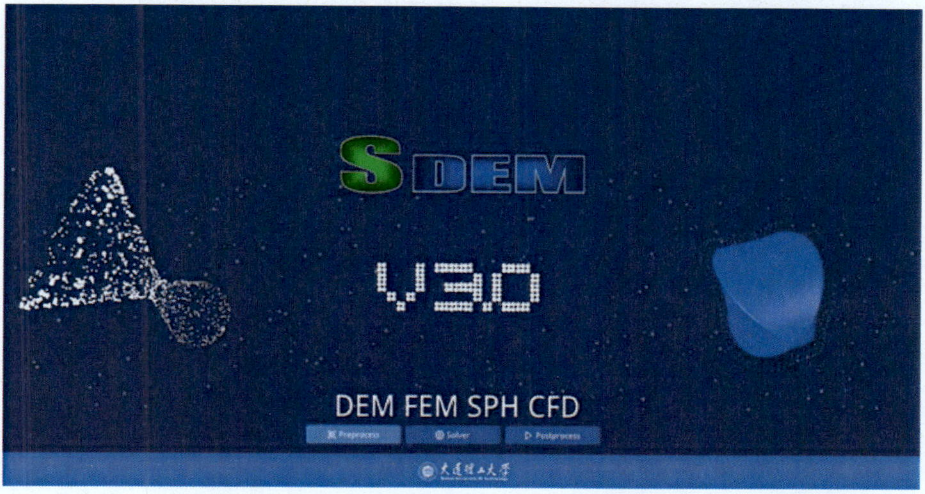

Fig. 6.19 Pre-processing, solver, and post-processing displays of software SDEM

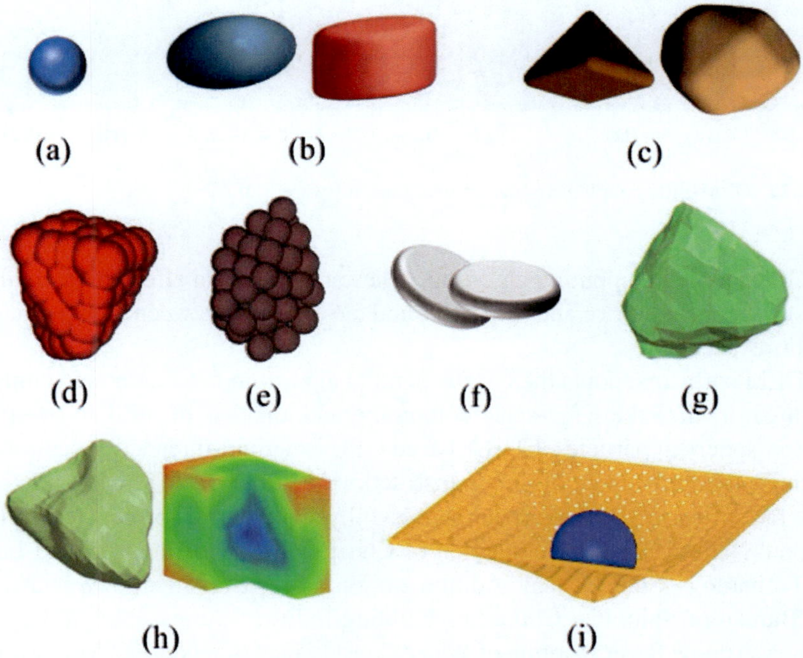

Fig. 6.20 Particle shapes in software SDEM: **a** spherical model; **b** superquadric model; **c** dilated polyhedron; **d** multi-spheres; **e** bonded spheres; **f** dilated disk; **g** polyhedral model; **h** level set function; **i** large deformable model

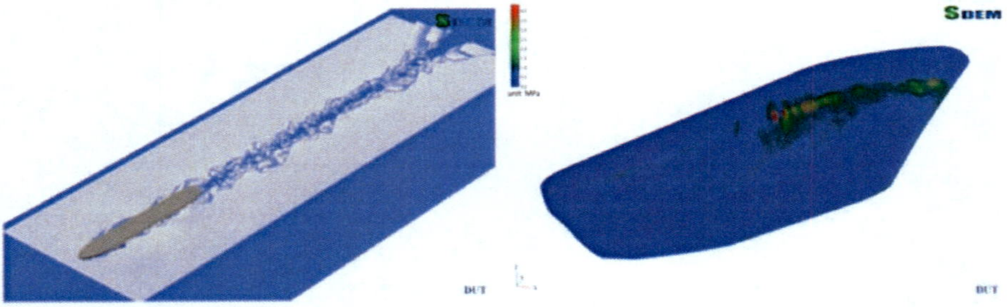

(a) Ice navigation simulation of an icebreaker (b) Stress cloud of the hull structure

Fig. 6.21 DEM-FEM coupling simulations using the software SDEM

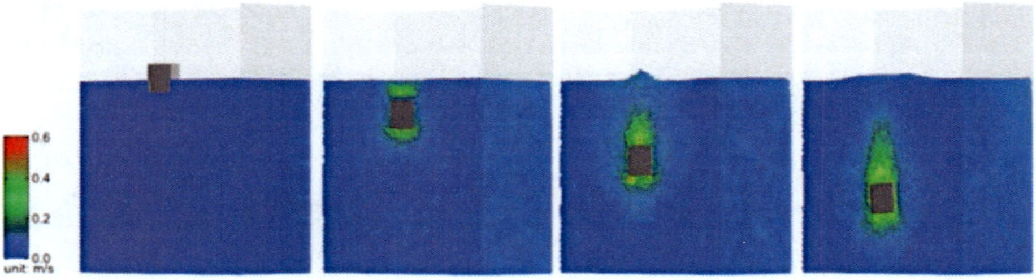

(a) $t = 0\text{s}$ (b) $t = 0.1\text{s}$ (c) $t = 0.2\text{s}$ (d) $t = 0.28\text{s}$

Fig. 6.22 DEM-SPH coupled simulation using the software SDEM

and GPU parallel computing algorithm. These methods can effectively simulate the collision between sea ice and ship hulls, and the collision between sea ice and marine platforms, etc.

SDEM software adopts the CUDA parallel algorithm for accelerated computing, which can realize the large-scale computing and analysis of millions of spherical and non-spherical particles. CUDA-based GPU is a multi-threaded execution mode, which is an improvement of Single Instruction Multiple Data (SIMD).

So far, software SDEM has been successfully applied in the project research of the National Natural Science Foundation of China, Ministry of Industry and Information Technology, State Oceanic Administration, China Classification Society, American Bureau of Shipping, China Shipbuilding Industry, China Shipbuilding Heavy Industry, Yellow River Institute of Water Conservancy Science, etc. This software is one of the most utilized DEM software in China. The latest version of the SDEM software is V3.0, and the team of computational particle mechanics at the Dalian University of Technology is still making unremitting efforts to enrich the software with particle morphology, large deformation particles/structures, fracture crushing, and DEM-FEM-CFD coupling algorithms.

6.4.2 Million-Scale Granular Materials

The generation of a large-scale non-spherical granular bed is used to verify the applicability of GPU parallel algorithms for superquadric elements. First, a sequential packing approach based on the advancing front concept is used to generate a large-scale spherical granular bed (Li and Ji 2018), as shown in Fig. 6.23a. Here, the sphere has a diameter of 1.25 mm, and the number of elements is 2 million. The main simulation parameters are listed in Table 6.3. The cuboid container has a length (L_0) of 270 mm, a width (W_0) of 90 mm, and a height (H_0) of 160 mm. The initial packing density of the spherical granular bed is 0.53, and there is no overlap between particles in the granular system. Then, a randomly oriented sphere, cube, or cylinder is generated inside the basic sphere, and the position of the sphere is replaced by a superquadric element, as shown in Fig. 6.23b. Therefore, the initial non-spherical granular system is generated and the packing density is 0.28, as shown in Fig. 6.24a. A total of 2 million non-spherical elements with arbitrary orientations are dropped into this container to form a granular bed under gravity, as shown in Fig. 6.24b. Finally, the packing density is 0.64. Particles located in the central region of the stable granular bed are used to count the contact number, and the average contact number is 10.4. A stable granular bed is observed, which also confirms that the GPU parallel algorithm is capable of generating a large-scale non-spherical granular system.

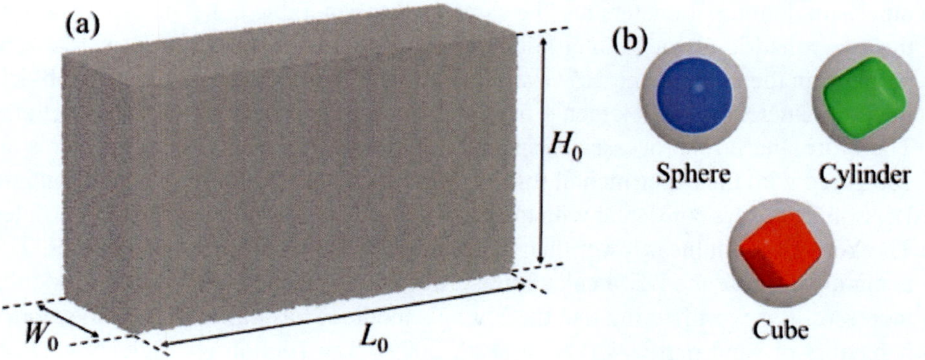

Fig. 6.23 Granular bed composed of 2 million spheres: **a** initial state and **b** a basic sphere replaced by a superquadric element

Table 6.3 Major computational parameters of granular systems

Definition	Value	Definition	Value
Density (kg/m^3)	2500	Normal damping coefficient	0.4
Young's modulus (Pa)	1×10^8	Tangential damping coefficient	0.4
Poisson's ratio	0.2	Sliding friction coefficient	0.4

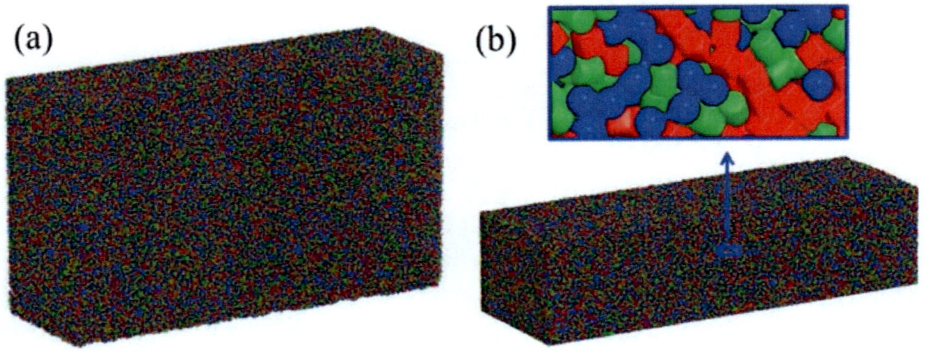

Fig. 6.24 Granular bed composed of 2 million non-spherical elements: **a** initial state and **b** stable state

6.4.3 Flow Process of Large-Scale Granular Materials

The flow process of the granular column is used to verify the validity of GPU parallel algorithms and the numerical results are compared with the experimental results (Owen et al. 2009). The superquadric elements have parameters of $2a = 0.6 \sim 0.71$mm, $b/a = 0.6 \sim 0.85$, $c/a = 0.75 \sim 1.0$, and $n_1 = n_2 = 2 \sim 3.5$. Young's modulus, the element density, and the coefficient of friction are 1×10^7Pa, 1520 kg/m^3, and 0.52, respectively. The rolling friction coefficient is 0.005. The other simulation parameters are the same as those in Table 6.3. The aspect ratio of the superquadric elements is a fair representation of the aspect ratio of real sand particles in the experiment. However, it is difficult to effectively estimate the blockiness parameters of the elements from the surface sharpness of the sand particles. Therefore, the flow processes of spherical and non-spherical granular systems are compared with the experimental results. Moreover, both the particle density and the Poisson's ratio are consistent with the basic material properties of real sand particles. The Young's modulus is lower than the Young's modulus of real sand particles. This is mainly because the DEM calculation efficiency decrease as the Young's modulus increases. It is worth noting that the Young's modulus has little effect on the motion behaviors of sand particles (Owen et al. 2009). The friction coefficients between particles and between particles and walls are calibrated based on the angle of repose of the experiments (Owen et al. 2009). The damping coefficients and the rolling friction coefficient are within reasonable ranges (Zhou et al. 2011). However, it is not sufficient to calibrate the simulation parameters based on the experimental data, and further analysis is needed in future research.

A total of 155,000 elements with arbitrary orientations and positions are generated in a rectangular container, and there is no overlap between all elements at the initial moment. Meanwhile, the initial granular system is layered by color, which helps to show the flow patterns of the granular materials at different moments. The elements fall into the container under gravity and form a stable granular bed, as shown in Fig. 6.25. The final granular bed has a height of 60 mm, a length of 40 mm, and

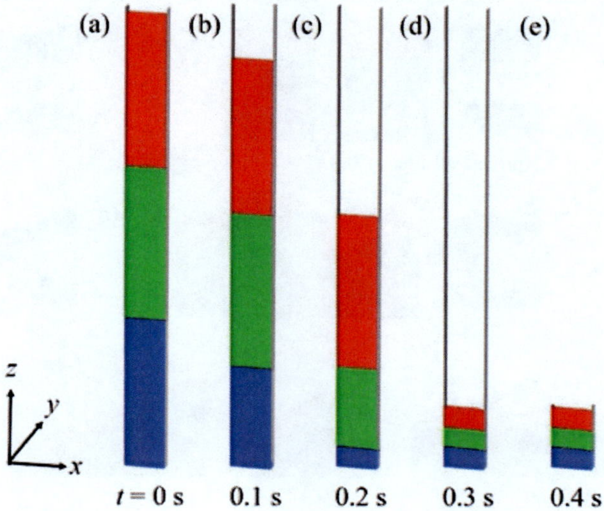

Fig. 6.25 Snapshots of the generation process of the granular bed at different moments: **a** $t = 0\text{s}$, **b** $t = 0.1\text{s}$, **c** $t = 0.2\text{s}$, **d** $t = 0.3\text{s}$, and **e** $t = 0.4\text{s}$

a width of 12 mm. The left wall is fixed, and the right wall moves at 6.11 mm/s. Figure 6.26 shows the comparison of the flow process of the superquadric elements with the experimental results (Owen et al. 2009) at different normalized moments. The granular bed slowly forms an angle of repose under gravity. Moreover, the flow process of the granular bed is simulated by spherical elements, and the heights of the granular bed at the fixed wall and the moving wall are compared with the experimental results, as shown in Fig. 6.27. Although the heights of the non-spherical granular bed at the fixed wall and the moving wall are in good agreement with the experimental results, the flow patterns of the non-spherical granular bed are slightly different from the flow processes of real sand particles. This is mainly because the flow process of sand particles involves deformation of the microstructure and failure of the granular material, resulting in complex flow patterns. In addition, the simulation results of the non-spherical elements are closer to the experimental results than the spherical elements. Spherical granular beds are too weak, which leads to continuous collapse and deformation of the entire system. Therefore, the particle shape plays an essential role in determining the flow characteristics of the granular materials, especially the failure and continuous deformation of the granular bed, and the shear strength of the material. Generally, a good agreement is observed in terms of the macroscopic flow characteristics of granular materials, which verifies that the current GPU parallel algorithm is able to capture the main flow features of large-scale non-spherical systems.

6.4.4 Comparison of GPU and CPU Computational Speed

The discharging process of granular systems is used to calculate the speedup ratio of the GPU to the CPU. The length (L_h) and width (W_h) of the hopper container

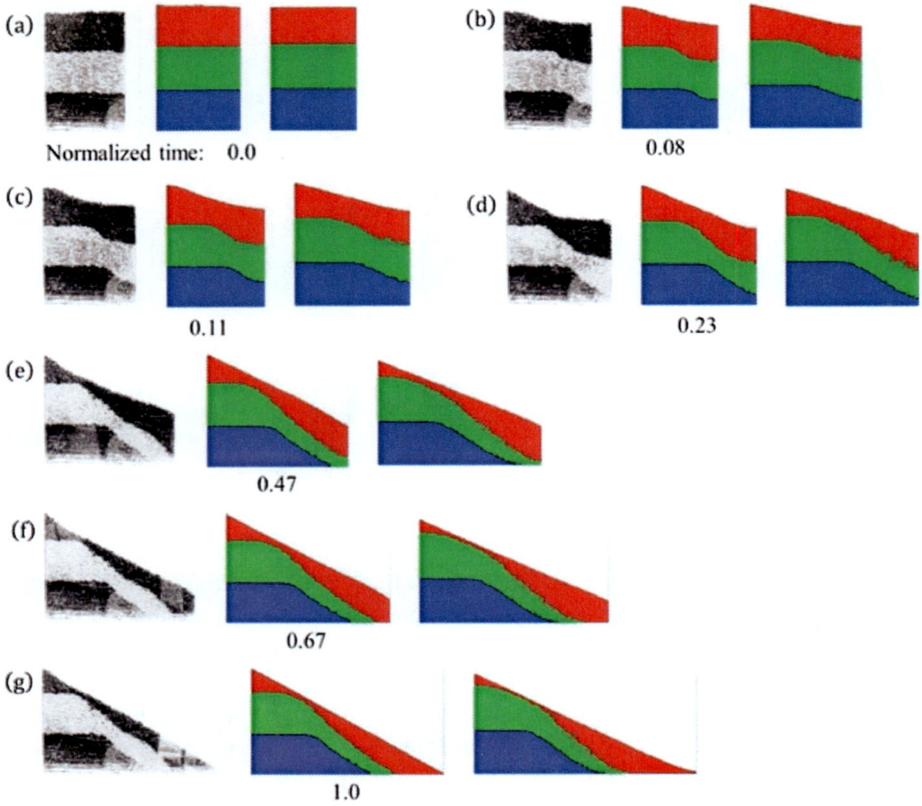

Fig. 6.26 Comparison of snapshots of the fall patterns between experiments (Owen et al. 2009) (left) and DEM simulations of non-spherical (middle) and spherical (right) granular systems at different normalized moments

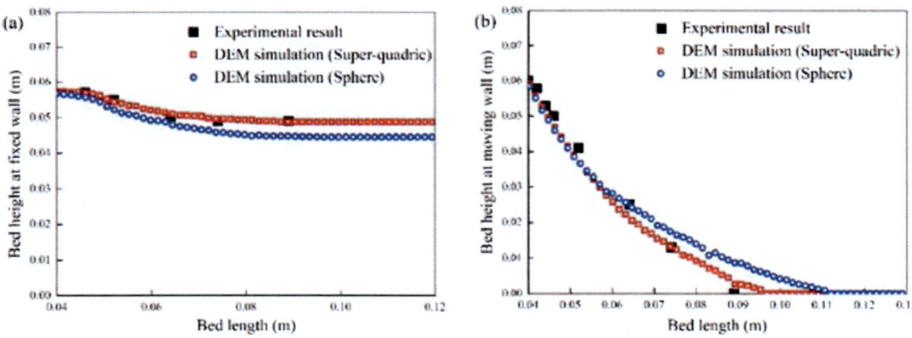

Fig. 6.27 The height of the granular bed varying with its length: **a** The height of the granular bed at the fixed wall and **b** the height of the granular bed at the moving wall

are both 40 mm, and the length (L_r) and width (W_r) of the orifice are both 16 mm. The hopper angle (θ_0) is 60°, as shown in Fig. 6.28a. The superquadric elements have parameters of $a = b$ and the aspect ratio is given as $\sigma = c/a$. A sphere is obtained if $n_1 = n_2 = 2$, a cylinder is obtained if $n_1 = 8$ and $n_2 = 2$, and a cube is obtained if $n_1 = n_2 = 8$. Different shapes of elements have the same mass and the radius of the volume-equivalent sphere is 1 mm. Both the normal and tangential damping coefficients are 0.2 and the friction coefficient is 0.1. The other parameters are the same as those in Table 6.3. The CPU is an Intel(R) Xeon(R) Silver 4114, and only one CPU core is used for DEM numerical simulations. In addition, the GPU is an NVIDIA Quadro GV100. Figure 6.28b shows the relationship between the actual consumption times of the GPU and CPU and the simulation time during the discharging process. Here, the actual consumption time is the wall-clock time. The number of elements is approximately 31,620. It is shown that the computational efficiency of the GPU is clearly higher than that of the CPU and the computational efficiency of differently shaped elements is significantly different.

When all the elements flow out of the hopper, the total consumption times of the CPU code and GPU code are denoted as t_c and t_g, respectively. The speedup ratio is defined as $\tau = t_c/t_g$. Thus, t_c, t_g, and τ obtained under different particle shapes and numbers are listed in Table 6.4. Meanwhile, the relationship between the speedup ratio and the number of elements is shown in Fig. 6.29a. The speedup ratio increases as the number of elements increases. When the number of elements is > 30,000, the speedup ratio of the cylinders is > 300, and the cylinders have the highest speedup ratio; further, the speedup ratio of the spheres is > 150, and the spheres have the lowest speedup ratio. In addition, the influence of particle shapes on the computational efficiency is shown in Fig. 6.29b. Here, the number of elements is 10,000, and the total consumption time of the spheres is denoted as t_0. It is shown that the calculation

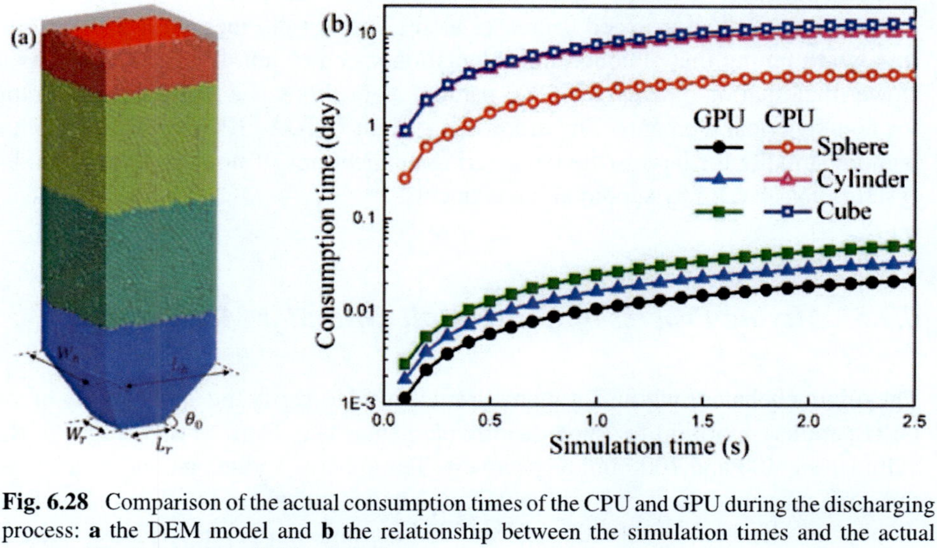

Fig. 6.28 Comparison of the actual consumption times of the CPU and GPU during the discharging process: **a** the DEM model and **b** the relationship between the simulation times and the actual consumption times of the GPU and CPU

Table 6.4 Total consumption times of the CPU code and GPU code

Number of particles ($\times 10^3$)	Sphere			Cylinder			Cube		
	t_c(s)	t_g(s)	t_c/t_g	t_c(s)	t_g(s)	t_c/t_g	t_c(s)	t_g(s)	t_c/t_g
1.0	331	39	8.5	1078	53	20.3	1288	71	18.1
3.16	3458	93	37.2	10,855	143	75.9	13,820	249	55.5
10	27,963	363	77.0	87,431	576	151.8	115,949	939	123.5
31.62	315,104	1938	162.6	939,495	3022	310.9	1,197,062	5212	229.7

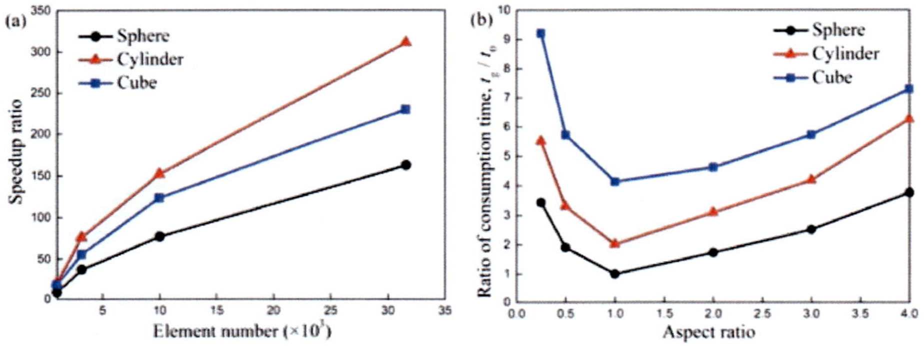

Fig. 6.29 Effects of the particle shapes and the number of elements on the computational efficiency: **a** The speedup ratio as a function of the number of elements and **b** the relationship between the calculation speed and the particle shapes

speed of the cubes is the slowest, and the spheres have the fastest calculation speed. Further, the calculation speed decreases as the aspect ratio increases or decreases. It is worth noting that although the calculation speed of non-spherical elements is slower than that of spheres, the GPU parallel algorithm has a higher speedup ratio for non-spherical elements. Therefore, the current CUDA-GPU parallel algorithm is more suitable for large-scale engineering applications of non-spherical granular systems constructed by superquadric elements.

6.4.5 *Mixing Process of Large-Scale Granular Materials*

The mixing behavior within a rotating drum is used to verify the applicability of the GPU parallel algorithm for superquadric elements. The diameter and length of the cylinder are 400 and 100 mm, respectively. The spheres, cylinders, and cubes have the same mass and the radius of the volume-equivalent sphere is 1.5 mm. The total number of elements is 200,000, and the main simulation parameters are listed in Table 6.3. The formation of a stable granular bed in the rotating drum includes three steps, as shown in Fig. 6.30. (I) A total of 300,000 elements with random positions and orientations are generated in a rectangular container at the initial moment. Here, the container has a height of 1300 mm, a length of 400 mm, and a width of 100 mm. All elements fall into the container under gravity and form a stable granular bed. When $t = 0.6$s, all information of the elements located in the horizontal drum is extracted and used in the subsequent DEM simulation. (II) The elements in the horizontal drum form a new granular bed under gravity. When $t = 0.5$s, all information of 200,000 elements is extracted from low to high according to the height of each element. Meanwhile, these elements are colored red and blue according to the coordinates in the x direction. (III) The initial granular bed is generated in the horizontally rotating drum, and the drum does not begin to rotate until all the elements remain motionless.

Figure 6.31 shows the different mixing patterns for the spheres, cylinders, and cubes at different moments. Here, the rotation speed is 30 rpm. The S-shaped surface of the granular systems is observed. Additionally, the red and blue elements exhibit a spiral and distinct layered pattern. This pattern gradually disappears as the number of revolutions increases. Moreover, the non-spherical elements mix faster than the spherical elements. Finally, the influence of the rotating speeds on the mixing processes of the cylinders is shown in Fig. 6.32. The granular system is divided into three parts: the collapse layer, the static layer, and the ascending layer. As the rotation speed increases, both the areas of the collapse layer and the ascending layer increase, while the area of the static layer that is located in the middle of the granular system decreases.

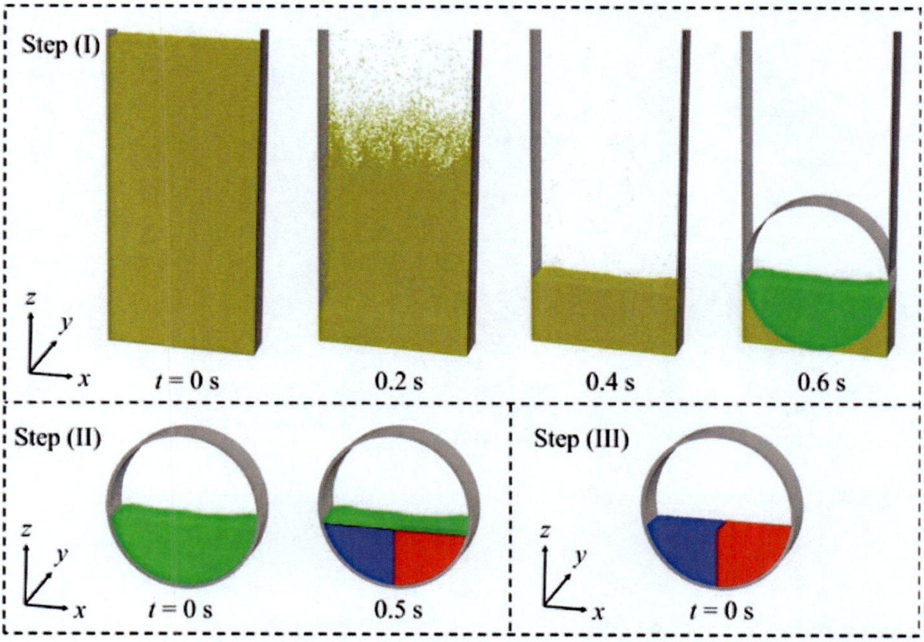

Fig. 6.30 Snapshots of the generation process of the initial granular bed in the horizontal drum

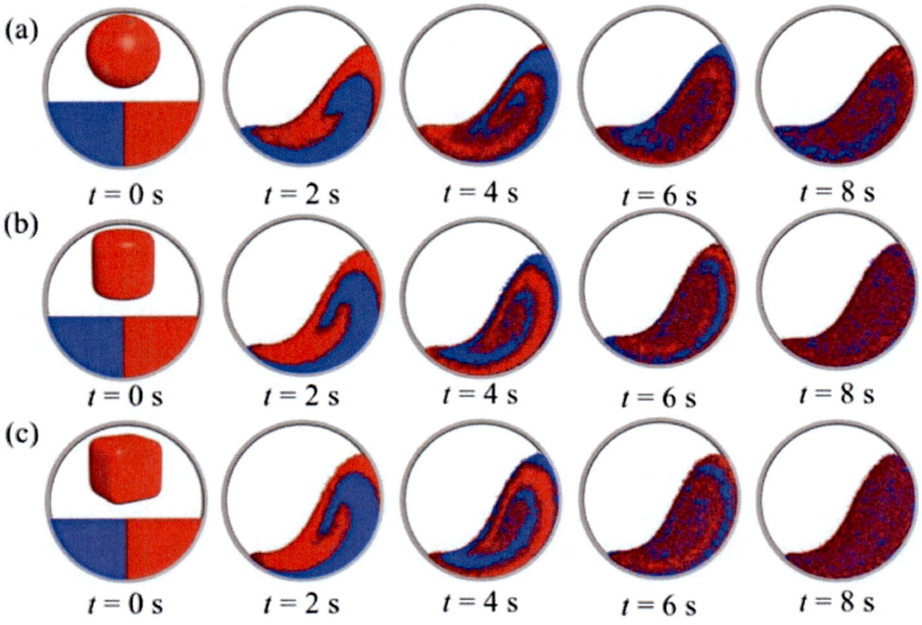

Fig. 6.31 Mixing patterns of differently shaped elements at different moments: **a** spheres, **b** cylinders, and **c** cubes

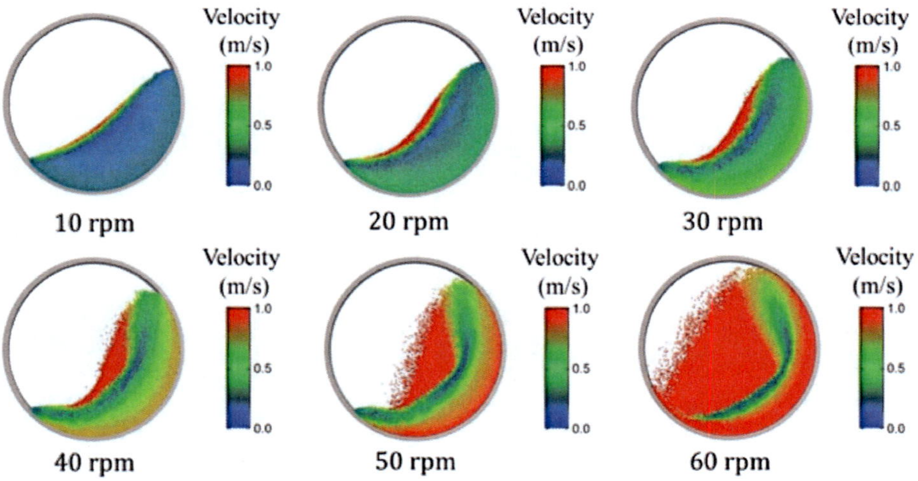

Fig. 6.32 Velocity distribution of cylindrical granular systems at different rotating speeds

6.5 Summary

In this chapter, the CUDA-GPU parallel algorithm for superquadric elements has been developed. This method is based on the spherical parallel-vector concept, and the bounding box list and the Newton iteration list are added to improve the computational efficiency of non-spherical granular systems. The validation tests consist of the generation of a large-scale non-spherical granular bed, the flow process of the granular column, the dynamic hopper discharge, and the mixing behavior within the horizontal drum. The simulated results are in good agreement with the previous experimental results, which verifies the applicability and reliability of current GPU parallel algorithms. Then, the effects of the particle shape on the computational efficiency are studied. The results show that the calculation speed of the cubes is the slowest, and the spheres have the fastest calculation speed. Additionally, the calculation speed becomes slower as the aspect ratio increases or decreases. Moreover, the speedup ratio of the GPU to the CPU increases as the number of elements increases. Non-spherical elements have a higher speedup ratio than spherical elements. Therefore, the current CUDA-GPU parallel algorithm is suitable for the large-scale engineering applications of non-spherical granular systems constructed by superquadric elements. In future work, multiple GPUs based on message passing interface (MPI) will be further applied to accelerate the computational efficiency of the non-spherical systems. Meanwhile, the computing efficiency of CPU parallel code and GPU parallel code needs to be further compared.

References

Amritkar A, Deb S, Tafti D (2014) Efficient parallel CFD-DEM simulations using OpenMP. J Comput Phys 256:501–519

Bar AH (1981) Superquadrics and angle-preserving transformations. IEEE Comput Graph Appl 1:11–23

Berger R, Kloss C, Kohlmeyer A, Pirker S (2015) Hybrid parallelization of the LIGGGHTS open-source DEM code. Powder Technol 278:234–247

Boehling P, Toschkoff G, Knop K, Kleinebudde P, Just S, Funke A, Rehbaum H, Khinast JG (2016) Analysis of large-scale tablet coating: modeling, simulation and experiments. Eur J Pharm Sci off J Eur Feder Pharm Sci 90:14–24

Cleary PW, Stokes N, Hurley J (1997) Efficient collision detection for three dimensional super-ellipsoidal particles. Comput Techn Appl CTAC

Fritzer HP (2001) Molecular symmetry with quaternions. Spectrochim Acta A 57:1919–1930

Gan JQ, Zhou ZY, Yu AB (2016) A GPU-based DEM approach for modelling of particulate systems. Powder Technol 301:1172–1182

Govender N, Wilke DN, Kok S, Els R (2014) Development of a convex polyhedral discrete element simulation framework for NVIDIA Kepler based GPUs. J Comput Appl Math 270:386–400

Govender N, Wilke DN, Pizette P, Abriak NE (2018) A study of shape non-uniformity and poly-dispersity in hopper discharge of spherical and polyhedral particle systems using the Blaze-DEM GPU code. Appl Math Comput 319:318–336

He Y, Evans TJ, Yu AB, Yang RY (2018) A GPU-based DEM for modelling large scale powder compaction with wide size distributions. Powder Technol 333:219–228

Kačeniauskas A, Kačianauskas R, Maknickas A, Markauskas D (2011) Computation and visualization of discrete particle systems on gLite-based grid. Adv Eng Softw 42:237–246
Kačianauskas R, Maknickas A, Kačeniauskas A, Markauskas D, Balevičius R (2010) Parallel discrete element simulation of poly-dispersed granular material. Adv Eng Softw 41:52–63
Kosenko II (1998) Integration of the equations of a rotational motion of a rigid body in quaternion algebra: the Euler case J. Appl Math Mech 62:193–200
Kureck H, Govender N, Siegmann E, Boehling P, Radeke C, Khinast JG (2019) Industrial scale simulations of tablet coating using GPU based DEM: a validation study. Chem Eng Sci 202:462–480
Li Y, Ji S (2018) A geometric algorithm based on the advancing front approach for sequential sphere packing. Granular Matter 20:59
Miller TF, Eleftheriou M, Pattnaik P, Ndirango A, Newns D, Martynaa GJ (2002) Symplectic quaternion scheme for biophysical molecular dynamics. J Chem Phys 116:8649–8659
Nishiura D, Sakaguchi H (2011) Parallel-vector algorithms for particle simulations on shared-memory multiprocessors. J Comput Phys 230:1923–1938
Owen PJ, Cleary PW, Mériaux C (2009) Quasi-static fall of planar granular columns: comparison of 2D and 3D discrete element modelling with laboratory experiments. Geomech Geoeng 4:55–77
Podlozhnyuk A, Pirker S, Kloss C (2016) Efficient implementation of superquadric particles in discrete element method within an open-source framework. Comput Part Mech 4:101–118
Portal R, Dias J, de Sousa L (2010) Contact detection between convex superquadric surfaces. Arch Mech Eng 57:165–186
Williams JR, Pentland AP (1992) Superquadrics and modal dynamics for discrete elements in interactive design. Eng Comput 9:115–127
Xu J, Qi H, Fang X, Lu L, Ge W, Wang X, Xu M, Chen F, He X, Li J (2011) Quasi-real-time simulation of rotating drum using discrete element method with parallel GPU computing. Particuology 9:446–450
Zhang L, Quigley SF, Chan AHC (2013) A fast scalable implementation of the two-dimensional triangular discrete element method on a GPU platform. Adv Eng Softw 60–61:70–80
Zhou ZY, Yu AB, Choi SK (2011) Numerical simulation of the liquid-induced erosion in a weakly bonded sand assembly. Powder Technol 211:237–249

Chapter 7
DEM Analysis of Flow Characteristics of Non-spherical Particles

The flow characteristics of granular materials have attracted extensive attention, and the research focus has gradually expanded from spherical granular materials to irregular granular materials. Zeng et al. (2017) used the combined sphere method to construct rice-shaped particles, and also analyzed the pulsation behavior of granular materials during the flow process through the total contact force and velocity fluctuation. Govender et al. (2018) compared the flow process of polyhedral and spherical particles in hoppers, and found that the polyhedral shapes significantly affected the flow rate of granular materials. Besides, granular materials exhibit more unique mechanical behaviors under external driving force. Liu and Nagel (1998) firstly proposed the phase transition mode of granular materials, which is achieved by applying external stresses and then transforming the granular materials from blockage to flow. Meanwhile, the external driving force changes the structural form of the granular material, leading to more strain intensification and expansion effects. In localized regions, the granular materials exhibit fluid-like properties and incomplete Taylor vortices (Huang et al. 2014). In addition, numerical simulation of the mixing process of granular materials in a rotating drum is another important research to analyze the granular flow, and the mechanistic study of its mixing and segregation characteristics is beneficial for the application of granular materials in industrial production. Particle friction and rotational speed significantly affect the flow characteristics of particles inside the rotating drum, and the dynamic angle of repose of the granular material increased with the increase of friction coefficient and rotational speed (Chou et al. 2016). The mixing behavior of polyhedral particles inside a two-dimensional rotating drum was simulated by the SIPHPM method, and the results showed that triangular, quadrilateral, and hexagonal particles had higher mixing degrees compared to spherical particles (Gui et al. 2017). Superquadric equations were used to construct ellipsoidal particles for investigating the effects of different aspect ratios on the mixing process of non-regular granular materials (You and Zhao 2018). The results showed that the main axis direction of particles with high aspect ratios was basically parallel to the flow direction of the particles.

S. Wang and S. Ji, *Computational Mechanics of Arbitrarily Shaped Granular Materials*, Springer Tracts in Mechanical Engineering,
https://doi.org/10.1007/978-981-99-9927-9_7

This chapter introduces spherical and irregular particle morphologies based on superquadric equations, where irregular particles are ellipsoidal, cylindrical, and cubic particles with different aspect ratios and surface sharpness. These particles have the basic characteristics of irregular particles, and these particles constructed by superquadric equations are characterized by high operational efficiency and large computational scale compared to concave particles. CUDA-GPU parallel algorithm is used to simulate the buffering performance of granular materials under impact load, the flow characteristics of granular materials under gravity field and external pressure field, and the mixing and segregation process of particles in a rotating drum, respectively. The numerical results are verified by comparing them with the experimental results, and the microscopic mechanisms of the flow pattern transformation, mixing, and segregation characteristics of granular materials are further analyzed.

7.1 Analysis of Buffering Performance of Irregular Granular Materials

In this section, the non-spherical particles are constructed based on the superquadric equation. The numerical model of granular bed filled with differently shaped particles under the action of spherical impactor is established by the discrete element method, and the computational model is verified by the experimental results of a large sphere impacting a spherical granular bed. The effects of layer thickness, aspect ratio, and surface sharpness of particles on the buffering properties of non-spherical granular beds are investigated. Subsequently, the intrinsic mechanism of the buffering performance of non-spherical particles is revealed by the initial packing fraction of the granular material.

7.1.1 Experimental Verification of the Impact Process of Granular Materials

In order to verify the reliability of the DEM model of superquadric particles, the numerical results of the impact force at the bottom of the container are compared with the experimental results (Ji et al. 2012). The layer thicknesses of the spherical granular bed are 0.5 and 2.0 cm. The particle sizes of the spherical particle in the DEM simulation are randomly distributed within the range of [4.0, 5.0 mm] according to the uniform probability density function, and the mean value is 4.5 mm. The main DEM parameters are listed in Table 7.1.

When the layer thickness of the granular bed is 0.5 and 2.0 cm, respectively, the time histories of the impact force on the bottom of the container are shown in Fig. 7.1. When the layer thickness is 0.5 cm, the peak value of the bottom force and the duration of the impact are basically consistent with the experimental results.

Table 7.1 DEM simulation parameters of sphere impact

Definition	Value	Definition	Value
Cylinder diameter D_1 (m)	0.19	Elastic modulus E (GPa)	1.0
Cylinder height H_1 (m)	0.3	Poisson's ratio ν, (–)	0.2
Diameter of impactor D_P (m)	0.05	Sliding friction coefficient μ_s (–)	0.5
Impact velocity V_0 (m/s)	0.0	Normal damping coefficient C_n (–)	0.06
Particle size d (mm)	4.0–5.0	Tangential damping coefficient C_t (–)	0.1
Particle density ρ (kg/m^3)	2551.6	DEM Time step dt (–)	2×10^{-6}

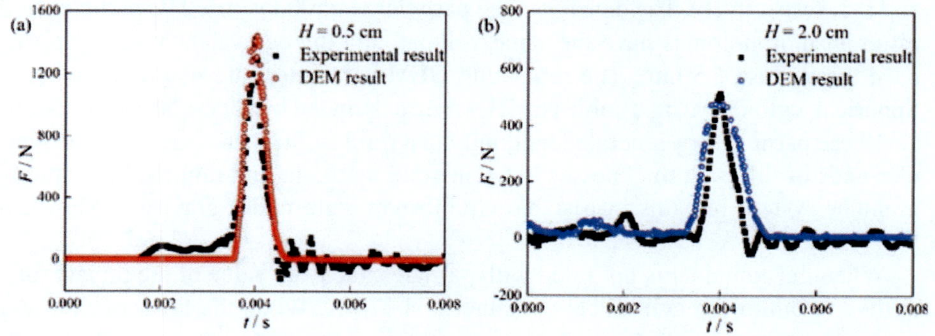

Fig. 7.1 Comparison of impact loads between experiment results and DEM results under different granular thicknesses: **a** 0.5 cm; **b** 2.0 cm

When the layer thickness is 2.0 cm, the peak value of the bottom force is lower than the experimental result, and the duration of the impact is larger than the experimental result. Besides, the impact force shows several fluctuations during the experiment, while the numerical results of the impact force obtained from the DEM are more stable. This is because the spherical impactor causes vibration of the container bottom during the experimental measurements, which results in a noticeable fluctuation phenomenon. However, the container bottom is set as a rigid plate in the DEM simulation, which results in little oscillations of the impact. Although there is some deviation between the DEM results and the experimental results, the variation of the impact force is consistent. This indicates that the discrete element method based on the superquadric equation can reasonably analyze the buffering characteristics of the granular system during the impact process.

7.1.2 Effect of Granular Layer Thickness on the Buffering Performance

The layer thickness of the granular bed and the particle shape are important factors affecting the buffering characteristics. In previous studies of experimental and DEM simulations of the buffering properties of particles, the peak value of the impact force at the bottom of the container decreased with increasing layer thickness of the spherical granular bed. When the layer thickness of the granular bed is greater than the critical thickness, the peak value of the impact force at the bottom of the container remained essentially constant as the layer thickness increased. Moreover, the non-spherical particles showed special properties different from spherical particles such as low mobility and high interlocking (Zhong et al. 2016).

In order to obtain a better impact process between particles of different shapes, the initial velocity of the impactor is 5 m/s, and the initial height of the impactor on the bed surface is 30 cm. The friction and damping coefficients between particles are 0.4 and 0.1, respectively. The density of the particles is 2500 kg/m^3. The particles with different morphologies have the same volume, and the equivalent volume spheres have the radius of 5 mm. The remaining DEM parameters are listed in Table 7.1. Spherical, cylindrical, and cubic particles are constructed by superquadric equations, and these particles are generated randomly in a rigid cylindrical container. The granular beds of different thicknesses are generated by the falling rain method, and the granular system remains in a stable equilibrium state under gravity as shown in Fig. 7.2.

When the container is not filled with particles, the peak value of the impact force at the bottom of the cylindrical container is 4.67 kN. When the layer thickness of the granular bed is 0.8, 0.5, 2.5, 4.5, and 9.5 cm, the impact force at the bottom of the container varies with time as shown in Fig. 7.3a–c. The results show that spherical, cylindrical, and cubic particles have similar buffering properties. As the layer thickness of the granular bed increases, the peak value of the impact force decreases and the duration of impact is prolonged. The peak forces of the differently shaped particles with different layer thicknesses are counted, as shown in Fig. 7.3d.

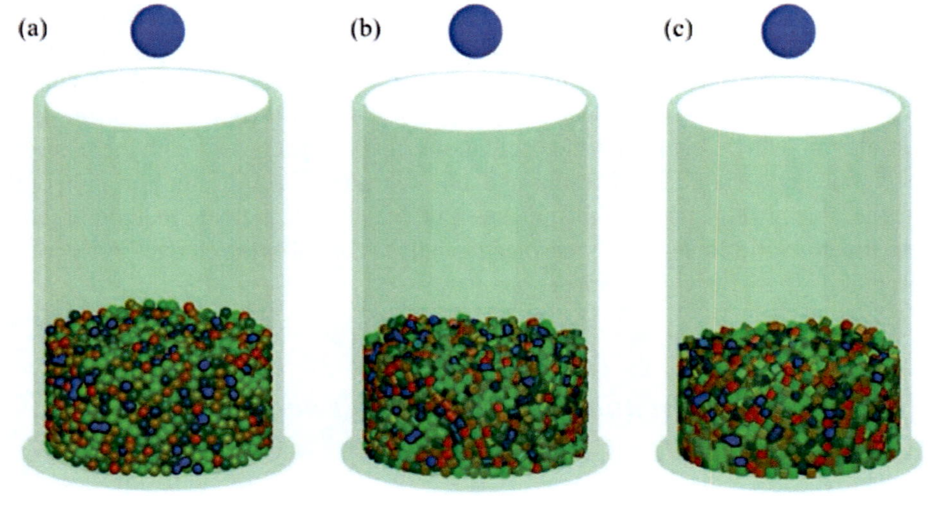

Fig. 7.2 DEM models of granular materials with different shapes: **a** spheres; **b** cylinders; **c** cubes

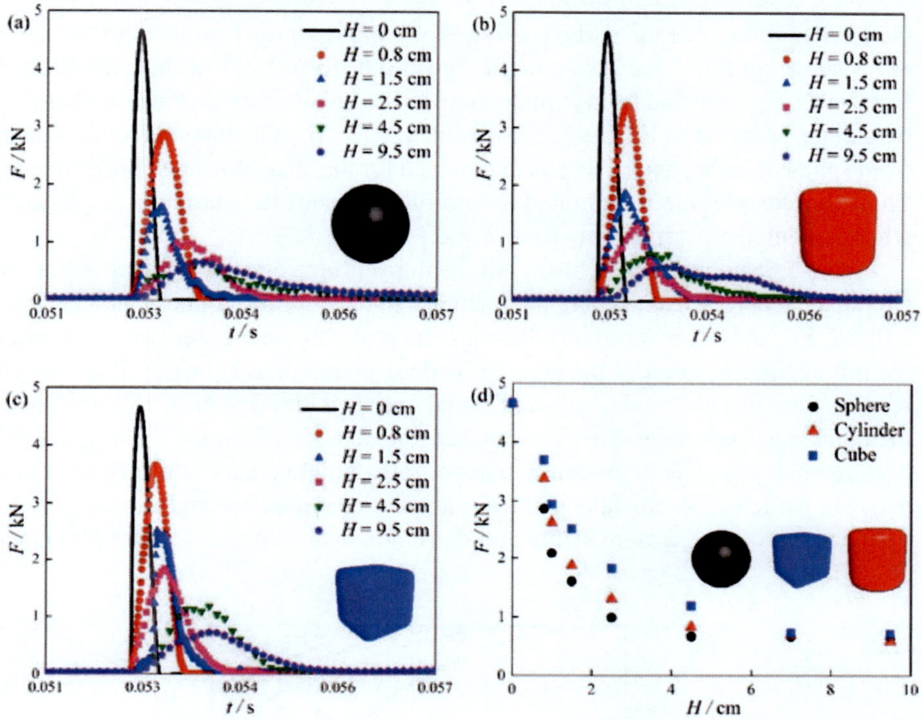

Fig. 7.3 Impact forces on the container bottom versus time under layer thickness of differently shaped particles: **a** spheres; **b** cylinders; **c** cubes; **d** comparison of the peak value of impact forces between differently shaped particles

When the layer thickness is less than 7.0 cm, spheres have the best buffering performance compared to cylindrical and cubic particles, while cylindrical particles have better buffering performance than cubic particles. When the layer thickness is greater than 7.0 cm, the layer thickness and particle shape have insignificant effects on the buffering properties of granular materials. This thickness is referred to as the critical layer thickness H_c, i.e., the critical layer thickness is 7.0 cm. However, the buffering properties of differently shaped particles are affected by different factors, such as material parameters, scale effects, impact energy. Thus, the critical layer thickness of granular materials usually varies under different conditions.

7.1.3 Effect of Particle Shapes on the Impact Force

In order to study the effect of surface sharpness of particles on the buffering properties of granular materials, differently shaped particles have the same volume so as to obtain the buffering properties of granular materials of different shapes. The particles constructed by the superquadric equation have geometric symmetry, the variation of the quarter contour of the particle surface with the sharpness parameter from 2.0 to 8.0 is shown in Fig. 7.4a. Two morphological transformations of the particle shape from sphere to cube and from cylinder to cube are obtained by varying the sharpness parameter, as shown in Fig. 7.4b. The parameter $n_1 = n_2$ is maintained for the change from sphere to cube, and $n_1 = 8$ is maintained for the change from cylinder to cube. Thus, thirteen particle morphologies with different surface sharpness are obtained when varying the parameter n_2 from 2 to 8.

Figure 7.5 shows the time history of the impact force on the container bottom for two shape changes from sphere and cylinder to cube with sharpness parameters of 2.0, 3.0, 5.0, and 8.0, respectively. Besides, the peak force at the container bottom and the initial concentration of the granular bed are counted, as shown in Fig. 7.6. The results show that the peak force and initial concentration increase with increasing the blockiness parameter. The fundamental role of surface sharpness during particle packing and impact is to gradually transform the point contact into surface contact between particles. Meanwhile, surface sharpness increases the contact area between particles, producing a more stable and dense granular system. A larger blockiness

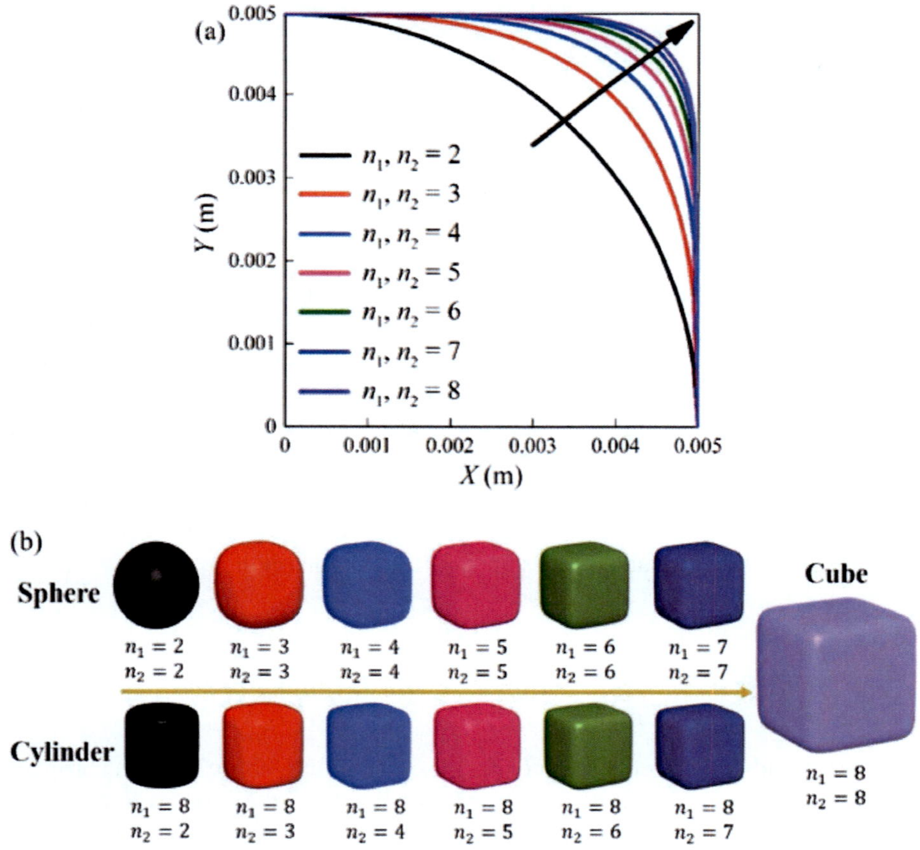

Fig. 7.4 The influence of blockiness parameters on the particle shapes

parameter limits the relative sliding and rolling of the particles, which allows the impactor to move relatively less distance within the granular system and produces a greater impact force on the bottom. Thus, spherical particles with smooth surfaces have better buffering performance than particles with sharp vertices and flat planes. Meanwhile, face-to-face contact between particles increases the concentration of the granular system, which reduces the buffering performance of the granular material.

To further investigate the effect of the aspect ratio of the particles on the buffering characteristics of the granular material, different aspect ratios are obtained by varying the functional parameters. The aspect ratio (σ) is defined as: $\sigma = c/a(= b)$. Here, a, b, and c are function parameters of the superquadric equation. Cylindrical and cubic particles with different aspect ratios are obtained by varying the aspect ratios, and the aspect ratios are set to 0.4, 0.6, 0.8, 1.0, 1.5, 2.0, 2.5, and 3.0, as shown in Fig. 7.7.

Figure 7.8 shows the influence of cylindrical and cubic particles with aspect ratios of 0.4, 1.0, 1.5, and 2.5 on the time history of the impact force at the container bottom.

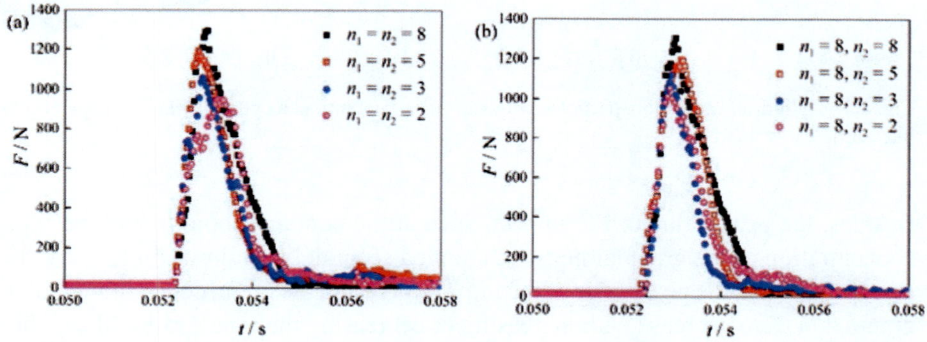

Fig. 7.5 Variation of impact force at the bottom of the with time during different shape changes: **a** sphere-cube; **b** cylinder-cube

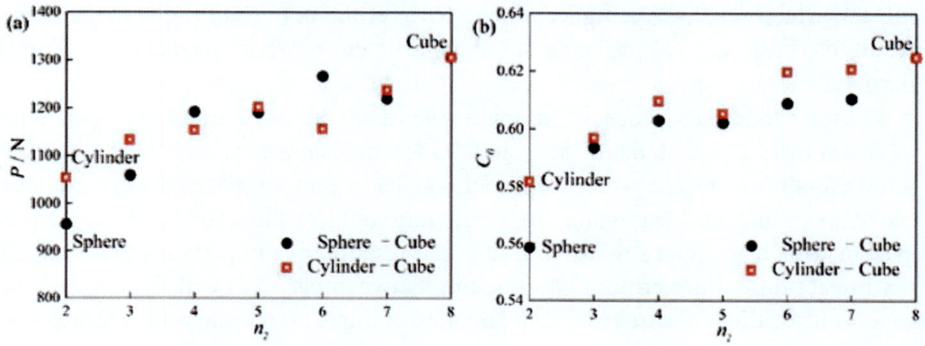

Fig. 7.6 The peak value of the impact force at the container bottom and the initial concentration of the granular system for different blockiness parameters: **a** peak value of the impact force at the container bottom; **b** initial concentration of the granular material

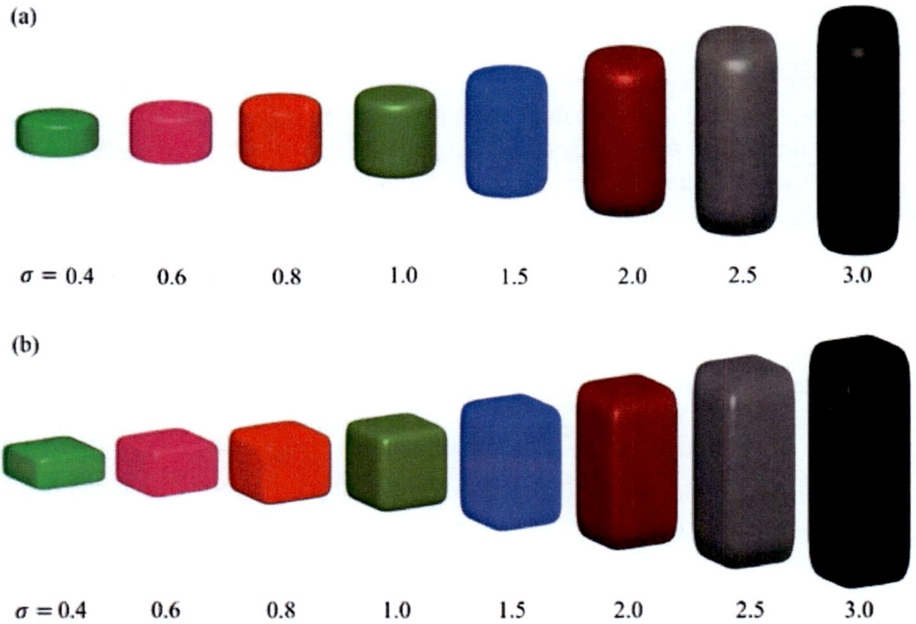

Fig. 7.7 Cylindrical and cubic particles with various aspect ratios: **a** cylinder-like particles; **b** cube-like particles

Besides, the peak value of the impact force at the container bottom and the initial concentration of the granular beds are counted separately, as shown in Fig. 7.9. The results show that the peak value of the impact force and the initial concentration of the granular system decrease with increasing or decreasing the aspect ratio. Meanwhile, the peak value of the impact force of cubic particles is higher than that of cylindrical particles for the same aspect ratio. This is because the primary contact mode between cubic particles is face-to-face contact, which results in a denser contact mode and a more stable granular system. The denser contact mode reduces the duration of the impact, which results in a higher impact force at the bottom of the container. As a result, the buffering performance of cubic particles is inferior to that of cylindrical particles.

During the impact process, the main role of the aspect ratio of the particles is to adjust the contact mode of the granular system. The aspect ratio of the particles allows for more pores between the particles and a larger free space for the particles to move under impact. Meanwhile, the aspect ratio reduces the stability of the granular system, which facilitates relative sliding and rotation of the particles and prolongs the impact time. Furthermore, interlocking between particles with higher or lower aspect ratios adjusts the transmission direction of impact force and achieves dispersed transmission. Localized impact forces are extended into three-dimensional space for the buffering performance.

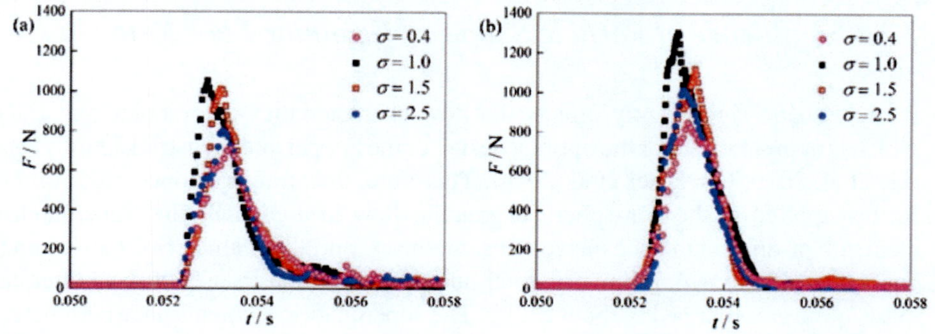

Fig. 7.8 The influence of aspect ratios on the impact force on the container bottom versus time for different particle shapes: **a** cylinder-like particles; **b** cube-like particles

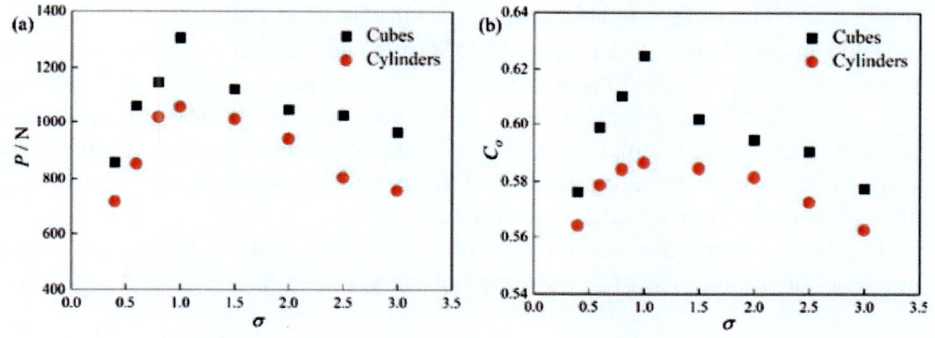

Fig. 7.9 The peak value of the impact force on the container bottom and the initial concentration of the granular system for different aspect ratios: **a** peak value of the impact force; **b** initial concentration of the granular material

7.2 Flow Characterization of Irregular Particles Driven by Gravity

In this section, the flow processes of granular materials in a conical silo were simulated by using the DEM, and the non-spherical particle shapes were constructed by superquadric equations. In the gravity-driven granular flow, the effects of the aspect ratios, blockiness parameters, and hopper angles on the discharge rate were studied. Then, the flow pattern transition between the mass and funnel flows and the distribution characteristics of the particle velocities were analyzed. Finally, the effects of particle shape on average coordination number, packing fraction, and contact force between particles were discussed to understand the basic flow characteristics of non-spherical granular materials.

7.2.1 Influence of Particle Shape on Granular Flow Rate

Particle shape significantly affects the flow characteristics of granular materials, which is of great value for the optimal design of the hopper and industrial applications (Ge et al. 2019; Govender et al. 2018). Therefore, the aforementioned algorithm is further applied to the non-spherical granular flow in a conical silo. Superquadric elements of different shapes have the same mass, and the diameter of the volume-equivalent sphere is 4.5 mm. The total number of particles is 17,600, and the total mass of the granular bed is about 2.1 kg. The superquadric elements have parameters of $a = b$ and $n_2 = 2$. The aspect ratio is defined as $\sigma = c/a$. When the aspect ratio (σ) is equal to 1, spheres and cylindrical particles with different blockiness parameters are shown in Fig. 7.10. When the blockiness parameters satisfy $n_1 = 6$ and $n_2 = 2$, cylindrical particles with aspect ratio varying from 1.0 to 3.0 are shown in Fig. 7.11. The main simulation parameters are listed in Table 7.1.

The conical silo has a diameter (D_c) of 90 mm and a height (H_c) of 240 mm. The opening diameter (D_0) is 30 mm, and the hopper angles $\theta = 15^{\circ}$, 30°, 45°, 60°, and 75°, respectively, as shown in Fig. 7.12. The spherical or cylindrical particles have a random position and orientation at the initial moment, and they are dropped into the conical container to form a granular bed under gravity. Then, the orifice is opened after all particles have no relative movement.

Figure 7.13 shows the mass flow rate of cylindrical particles in the conical hopper as a function of time. The mass flow rate and average mass flow rate are denoted as Q

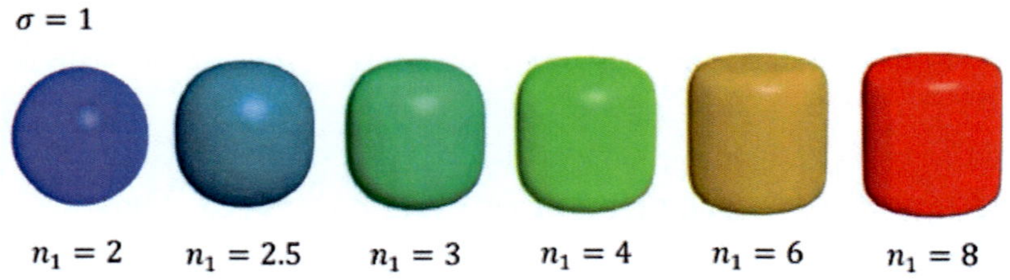

Fig. 7.10 Spheres and cylinder-like particles with different blockiness parameters

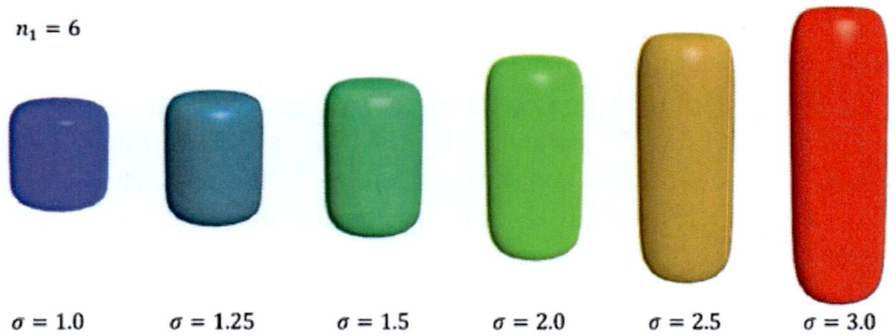

Fig. 7.11 Cylinder-like particles with different aspect ratios

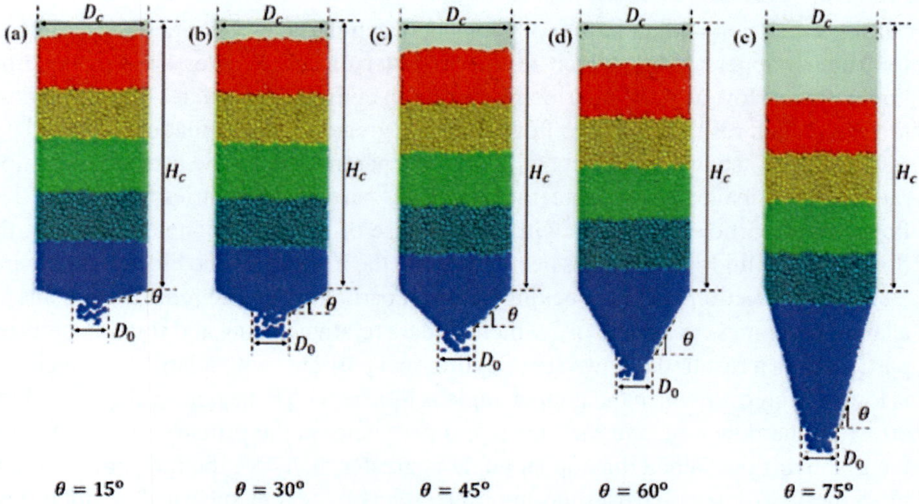

Fig. 7.12 Snapshots of the discharging processes of cylindrical particles for different base angles

and $\tilde{Q}$, respectively. The entire discharging process consists of three parts, namely, the phase (I) of increasing the flow rate, the phase (II) of steady granular flow, and the phase (III) of reducing the flow rate. The steady flow processes are commonly used to quantify the flowability of the granular materials and are critical to the design and application of conical hoppers. Therefore, the steady granular flow is the focus of this study.

Figure 7.14 shows the mass flow rate of spherical and cylindrical particles as a function of time during the steady flow. Here, the hopper angles $\theta = 30^o$ and 60°, respectively. The discharge rate increases and the flow fluctuation decreases as the hopper angle increases. The discharge rate of spheres is faster than that of cylindrical particles, while the flow fluctuation is smaller than that of cylindrical

Fig. 7.13 Mass flow rate of cylindrical particles as a function of time

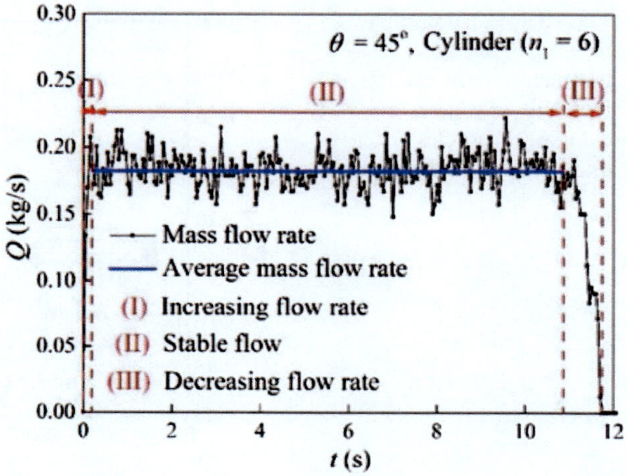

particles. Friction and inelastic collisions between particles become less significant for larger hopper angles, which results in faster discharge rates and smaller flow fluctuations. However, the interlocking between cylindrical particles hinders relative motion, which results in slower flow rates and greater flow fluctuations.

Figure 7.15 shows the effects of blockiness parameters on the average mass flow rate of cylindrical particles. Here, the blockiness parameter n_1 varies from 2 to 8. The flow rate of particles increases with the increase of the hopper angle. However, the flow rate of cylindrical particles decreases with the increase of the blockiness parameter n_1. The friction and interlocking between particles become more significant for a larger blockiness parameter n_1, which hinders relative sliding and rotation between particles. As a result, the flow rate of cylindrical particles with a larger parameter n_1 is lower. Moreover, when the hopper angle is less than 45°, the hopper angle has little effect on the flow rate, and the blockiness parameter is the primary factor affecting the granular flow. When the hopper angle is greater than 45°, the hopper angles and blockiness parameter have a superimposed effect on the non-spherical granular flow.

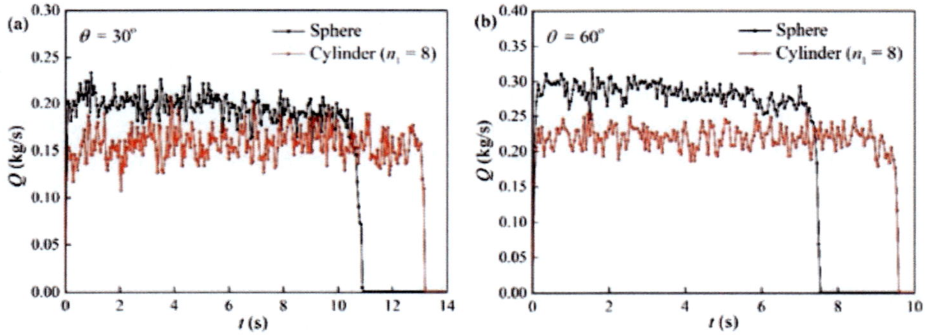

Fig. 7.14 Mass flow rate of spheres and cylindrical particles as a function of time at different hopper angles: **a** $\theta = 30^{\circ}$ and **b** $\theta = 60^{\circ}$

Fig. 7.15 Relationship between blockiness parameter n_1 and average flow rates of spheres and cylindrical particles under different hopper angles

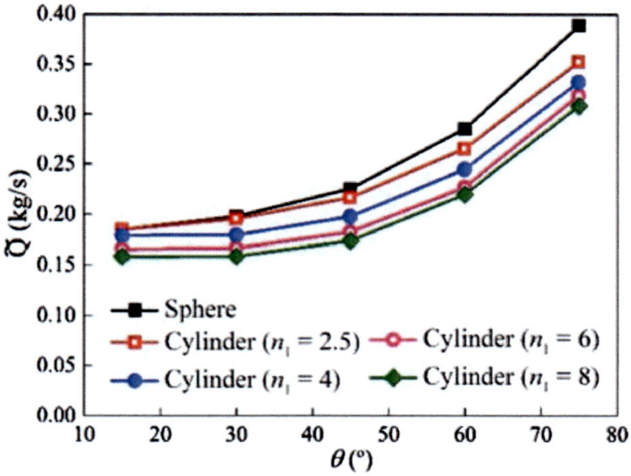

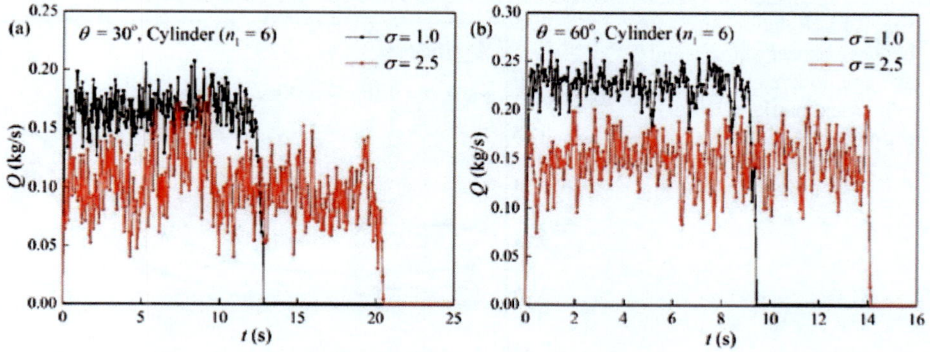

Fig. 7.16 Mass flow rate of cylindrical particles with different aspect ratios as a function of time at different hopper angles: **a** $\theta = 30^{\text{o}}$ and **b** $\theta = 60^{\text{o}}$

Figure 7.16 shows the relationship between the mass flow rate of cylindrical particles with different aspect ratios and time during the steady flow. Here, the aspect ratios are 1.0 and 2.5, respectively. The discharge rate of cylindrical particles decreases and the flow fluctuation increases with the increase of the aspect ratios. However, the discharge rate of cylindrical particles increases and the flow fluctuation decreases with the increase of the hopper angles. The aspect ratio of the particles causes more significant interlocking and local arching structure during the discharging process, which results in a more discontinuous granular flow and greater flow fluctuations.

Figure 7.17 shows the effects of aspect ratios on the average mass flow rate of cylindrical particles. Here, the aspect ratio σ varies from 1.0 to 3.0. The average flow rate of cylindrical particles decreases as the aspect ratio increases or the hopper angle decreases. Elongated cylindrical particles tend to align horizontally with their longer axis and form local clusters, which limits the relative motion between particles and reduces the vertical velocity of particles. Moreover, when the hopper angle is less than 45°, the hopper angle has no significant effect on the flow rate, and the aspect ratio appears to be the primary influencing factor. When the hopper angle is greater than 45°, the hopper angle and aspect ratios have a superposed effect on the flow rate.

7.2.2 *Transition of Granular Flow Pattern*

In industrial applications, the flow patterns of granular materials are critical to the design and optimization of hoppers. It has been reported that the porosity of the granular system decreases as the ratio of the particle diameter to the opening diameter decreases. As a result, more particles have a uniform velocity, and the number of particles in the mass flow increases while the number of particles in the funnel flow decreases (Zhang et al. 2018). Moreover, as the silo diameter increases, more particles may be affected by the wall, and the movement of the particles is further

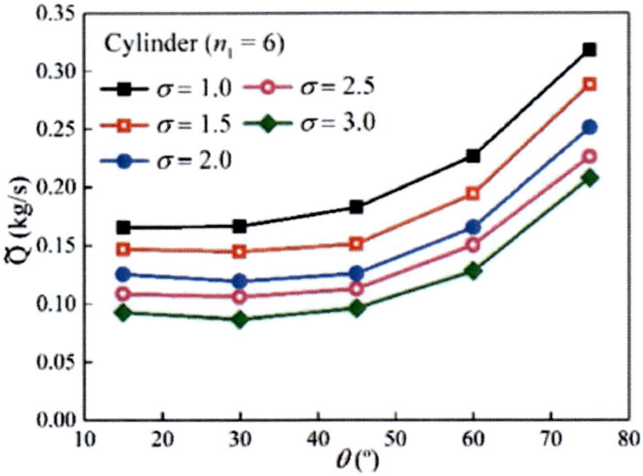

Fig. 7.17 Relationship between aspect ratios σ and average flow rates of cylindrical particles under different hopper angles

hindered by the wall. The non-uniformity of the velocity distribution of the granular system becomes more pronounced, and more particles are in the funnel flow (Zhang et al. 2018). It is worth noting that the particle shape is a crucial factor affecting the flow pattern transition of the granular materials. The aspect ratio and blockiness parameters of the particles limit the relative motion between the non-spherical particles and enhance interlocking, which results in discontinuous granular flow and reduces the flowability of granular materials. Therefore, the non-spherical granular flow in the conical hopper needs to be further analyzed by the flow pattern transition and the particle velocity.

To illustrate the effect of the particle shape on the discharging process of granular materials, the cross sections of flow patterns of the spherical and cylindrical particles at different moments are compared, as shown in Fig. 7.18. Here, the hopper angle $\theta = 15^{\circ}$. In the gravity-driven granular flow, a V-shaped flow pattern gradually appears. Eventually, this pattern disappears as more particles flow out of the container. However, the flow pattern of the granular systems is governed by the particle shapes. The vertical velocity of the cylindrical particles with an aspect ratio of 1 at the center is obviously higher than that of particles near the wall of the container, which results in a more pronounced V-shaped pattern. Spheres and cylindrical particles with an aspect ratio of 2.5 have a V-shaped pattern in the lower part of the granular bed and have a uniform vertical velocity in the upper part of the granular bed. In addition, compared with spherical particles, cylindrical particles near the wall of the container are more likely to form a pile during the discharging process.

Figure 7.19 shows the cross sections of flow patterns of the spherical and cylindrical particles at different moments. Here, the hopper angle $\theta = 60^{\circ}$. Particles of different shapes have uniform vertical velocities, and the vertical velocities of the particles located at the center of the container are substantially equal to those of the particles near the wall of the container. As a result, V-shaped flow patterns are not observed for differently shaped particles. Meanwhile, it is difficult for cylindrical particles near the wall of the container to form a pile during the discharging process.

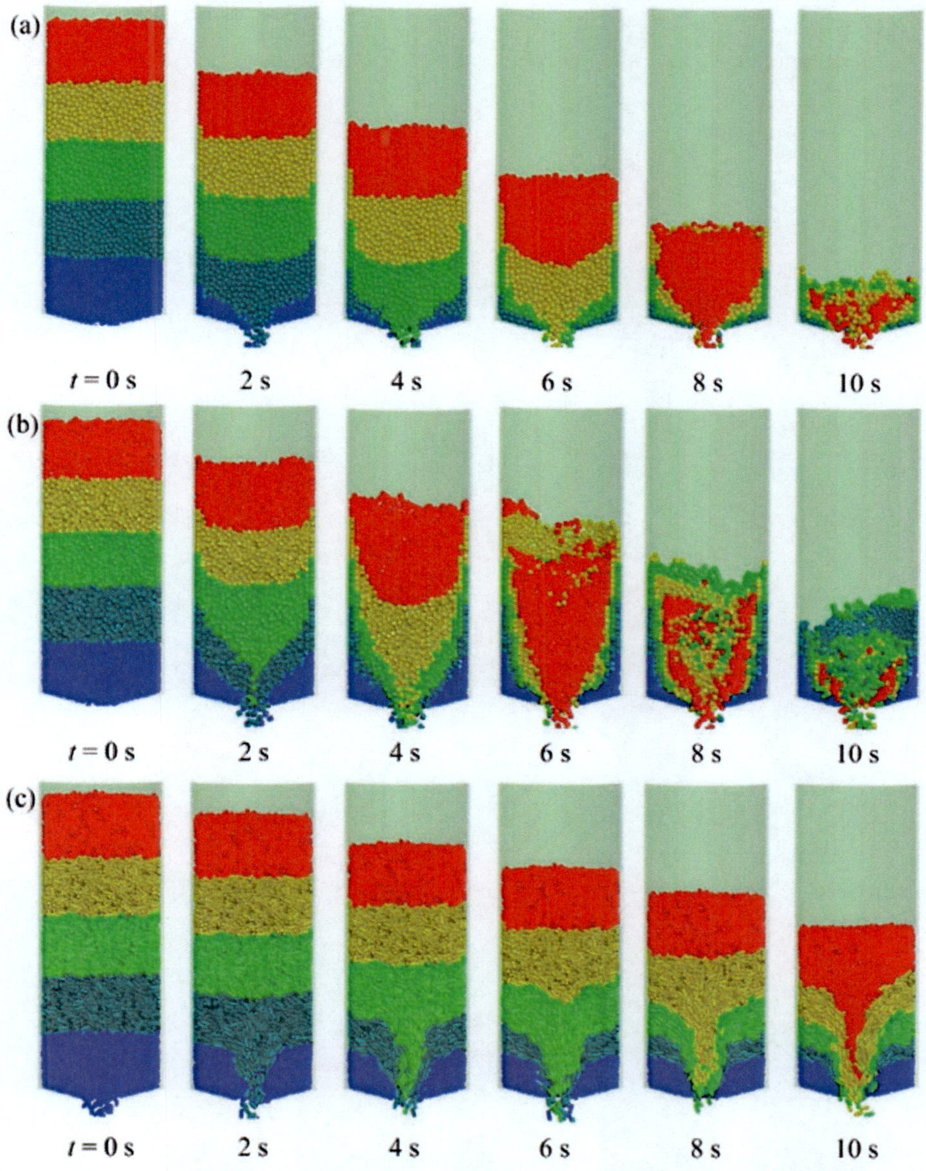

Fig. 7.18 Snapshots of the discharging process of the spherical and cylindrical particles when the hopper angle $\theta = 15^{\circ}$: **a** spherical particles ($n_1 = 2$ and $\sigma = 1$), **b** cylindrical particles ($n_1 = 6$ and $\sigma = 1$), and **c** cylindrical particles ($n_1 = 6$ and $\sigma = 2.5$)

As the hopper angle increases, the particles slide or rotate more easily and the flowability of the particles increases. Therefore, the particle shape has no effect on the flow pattern of the granular systems for larger hopper angles.

The flow pattern transition of granular materials is closely related to the vertical velocity distribution of particles in the container, which is crucial for the transition

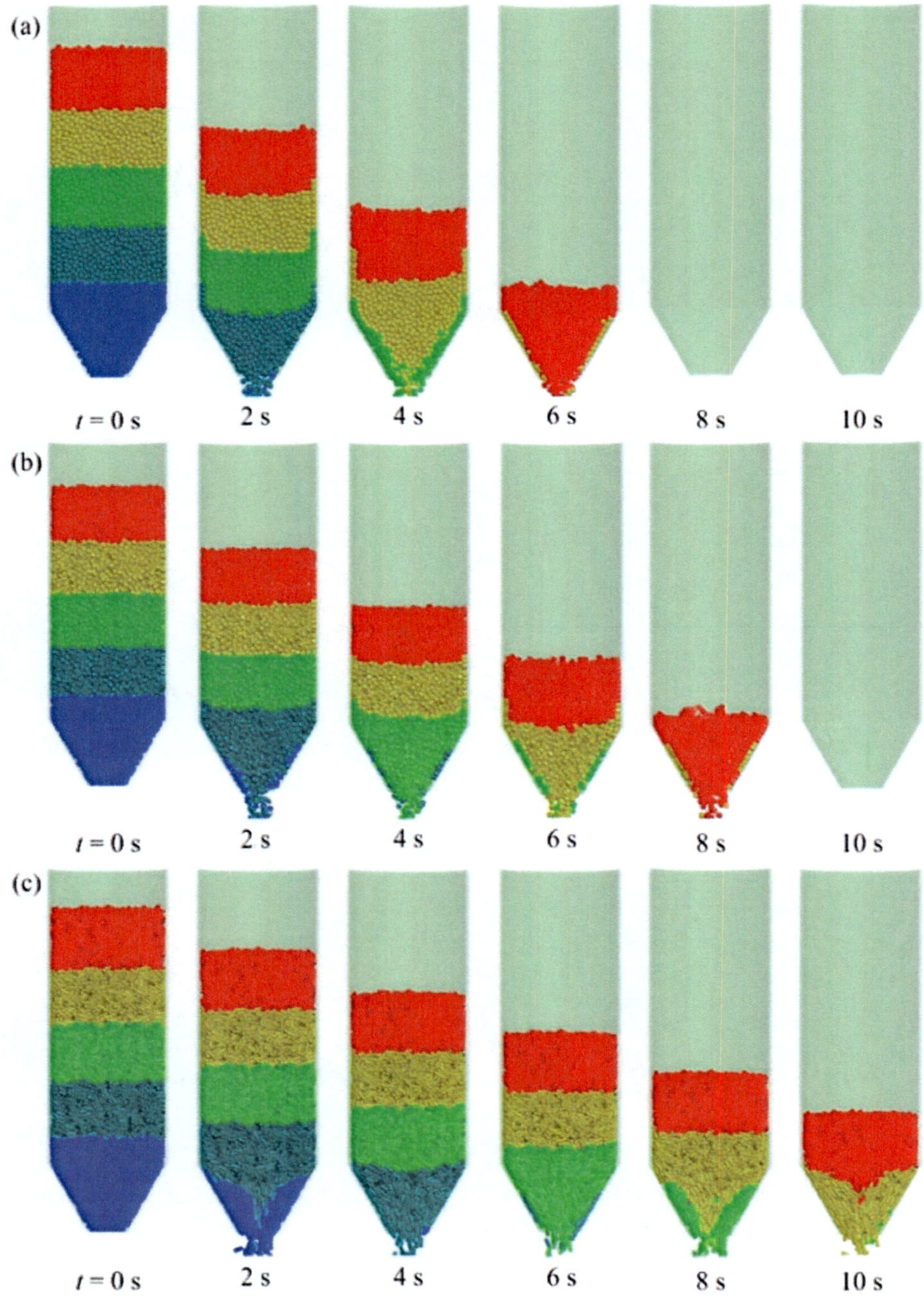

Fig. 7.19 Snapshots of the discharging process of the spherical and cylindrical particles when the hopper angle $\theta = 60^{\circ}$: **a** spherical particles ($n_1 = 2$ and $\sigma = 1$), **b** cylindrical particles ($n_1 = 6$ and $\sigma = 1$), and **c** cylindrical particles ($n_1 = 6$ and $\sigma = 2.5$)

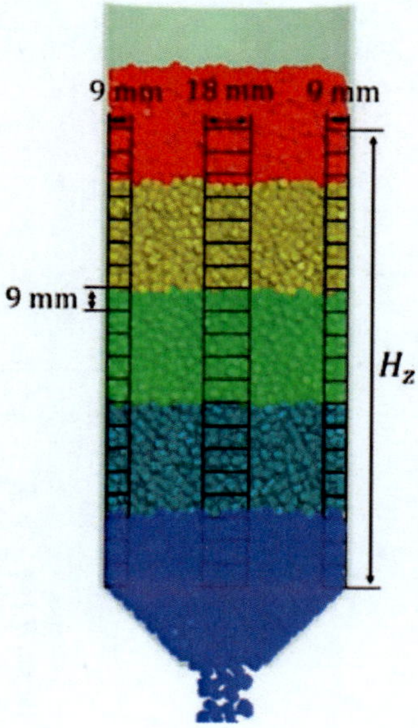

Fig. 7.20 Cross-sectional view of the calculated region of particle velocity

between the mass flow and the funnel flow of granular systems. The vertical velocities of the particles located in the center of the container and near the wall are measured, as shown in Fig. 7.20. The black cell has a length of 9 mm and a total height of 180 mm. The centroid positions of the particles are used to determine in which cell they are located. Moreover, the vertical mean velocities of the particles in each cell are obtained at each DEM step, and the statistical time is the time taken from the initial height of the granular systems to 180 mm.

Figure 7.21 shows the height of the granular bed as a function of the vertical velocity of differently shaped particles. Here, the hopper angle $\theta = 15^{\circ}$. The vertical velocities of particles located in the center of the container and near the wall are denoted as V_c and V_w, respectively. V_w increases and V_c decreases with the increasing height of the granular bed. When the height (H_z) is less than the critical height (H_c), V_c is greater than V_w and the flow state of the granular materials is funnel flow. When H_z is equal to or greater than H_c, V_c is approximately equal to V_w and the flow state of the granular systems is mass flow. The cylindrical granular systems with an aspect ratio of 1 appear in the funnel flow, while spherical systems and cylindrical granular systems with an aspect ratio of 2.5 are in both the funnel and mass flows. This means that the cylindrical granular systems with an aspect ratio of 1 have a more pronounced flow pattern. In addition, the critical height of the spherical systems is higher than that of the cylindrical granular systems with an aspect ratio of 2.5. A larger number of spherical particles are in the funnel flow than cylindrical particles with an aspect ratio of 2.5.

Figure 7.22 shows the relationship between the height of the granular bed and the vertical velocity of differently shaped particles when the hopper angle $\theta = 60^{\circ}$. Particle shape has no significant effect on the transition between the funnel and mass flows. V_w increases and the difference between V_c and V_w decreases for a larger

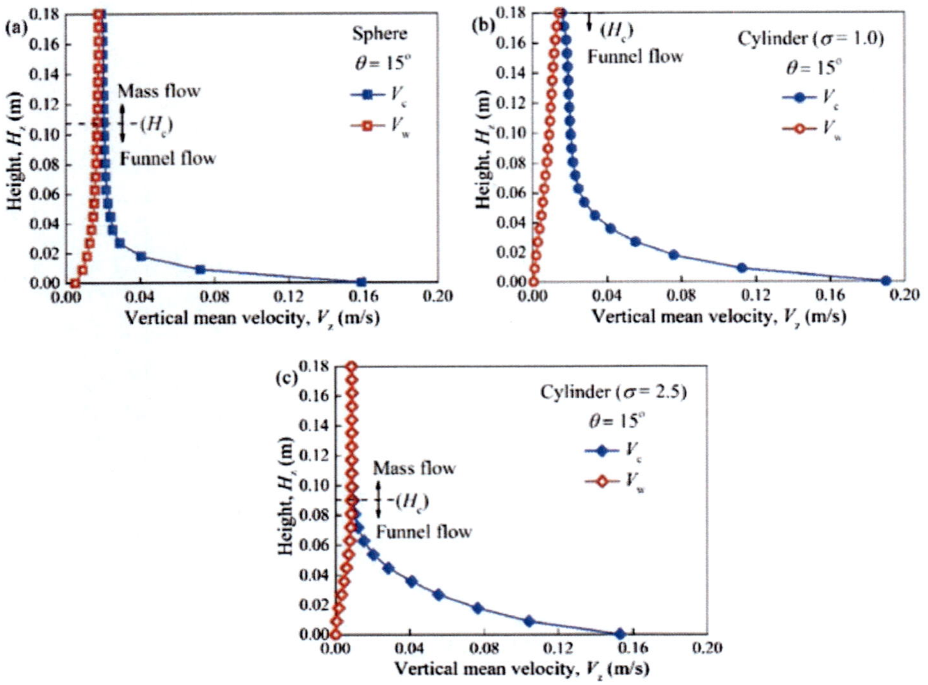

Fig. 7.21 Relationship between the heights of spherical and cylindrical granular beds and the vertical mean velocities of particles when the hopper angle $\theta = 15^{\mathrm{o}}$: **a** Sphere ($n_1 = 2$ and $\sigma = 1$), **b** cylindrical particles ($n_1 = 6$ and $\sigma = 1$), and **c** cylindrical particles ($n_1 = 6$ and $\sigma = 2.5$)

hopper angle. Therefore, granular systems composed of differently shaped particles basically have a uniform vertical velocity and are in the mass flow. Moreover, particles are easier to slide or rotate, and it is difficult to form interlocking and partially arching structures, which reduces the difference in velocity distribution due to particle shape and forms a uniform flow pattern.

In addition, the critical heights of the differently shaped granular systems under different hopper angles are investigated, as shown in Fig. 7.23. The critical height of differently shaped granular systems decreases as the hopper angle increases and reaches a steady state. This indicates that fewer particles are in the funnel flow. When the hopper angle is greater than 60°, the granular system generally has a uniform vertical velocity and all particles are in the mass flow.

Furthermore, the aspect ratio and blockiness parameter of particles significantly affect the critical height of the granular system; however, this effect becomes insignificant as the hopper angle increases. When the blockiness parameter (n_1) is less than 3, the critical height (H_c) increases as the blockiness parameter increases. When the blockiness parameter (n_1) is greater than 3, the blockiness parameter has no effect on the critical height, and the flow patterns of the granular materials are not changed. When the aspect ratio (σ) is less than 1.5, the critical height (H_c) decreases as the aspect ratio increases. A greater number of particles have a uniform vertical velocity

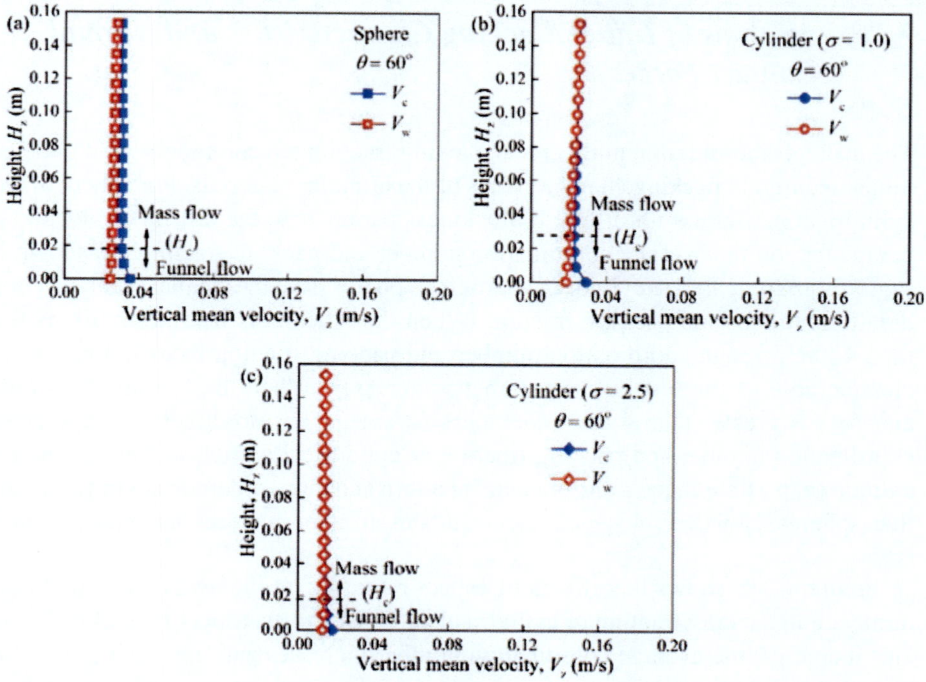

Fig. 7.22 Relationship between the heights of spherical and cylindrical granular beds and the vertical mean velocities of particles when the hopper angle $\theta = 60°$: **a** Sphere ($n_1 = 2$ and $\sigma = 1$), **b** cylindrical particles ($n_1 = 6$ and $\sigma = 1$), and **c** cylindrical particles ($n_1 = 6$ and $\sigma = 2.5$)

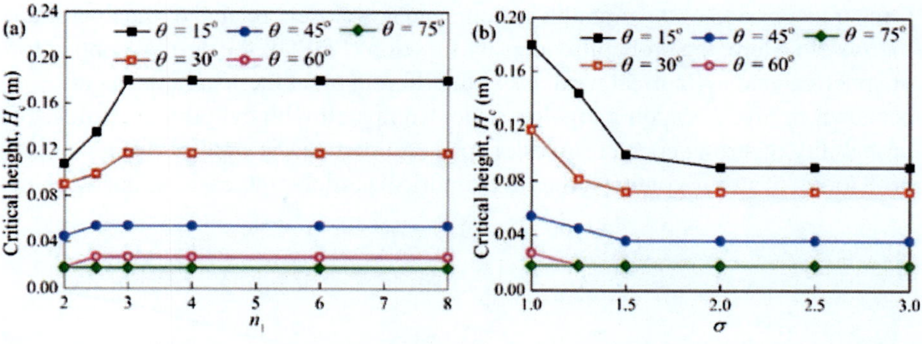

Fig. 7.23 Effects of the blockiness parameter (n_1) and aspect ratio (σ) on the critical height of granular systems at different hopper angles: **a** The blockiness parameter (n_1) and **b** the aspect ratio (σ)

and are in the mass flow. When the aspect ratio is greater than 1.5, the aspect ratio has no effect on the critical height and the flow patterns of the granular materials are not changed.

7.2.3 *Analysis of Initial Stacking Characteristics and Normal Contact Forces*

The average coordination number and packing fraction are commonly used to determine the internal packing characteristics of the granular materials. For spherical and cylindrical particles with different blockiness parameters, the effects of blockiness parameters on the average coordination number and packing fraction are shown in Fig. 7.24a. Compared to cylindrical particles, spheres have the smallest average coordination number and packing fraction. When the blockiness parameter (n_1) is less than 4, the average coordination number and packing fraction of cylindrical particles increase as the blockiness parameter increases. When the blockiness parameter (n_1) is greater than 4, the blockiness parameter has no effect on the average coordination number and packing fraction of cylindrical particles. This means that cylindrical particles have a closer contact pattern and more ordered packing structure than spheres. However, closer contact results in stronger interlocking and lower flow rates.

Figure 7.24b shows the effects of aspect ratio (σ) on the average coordination number and packing fraction of cylindrical particles. As the aspect ratio of the particles increases, the average coordination number increases and the packing fraction decreases. The aspect ratio increases the number of contacts between particles, which results in more local clusters and stronger interlocking. Moreover, cylindrical granular systems with large aspect ratios have more voids, which reduces the stability of the system.

The contact force between particles can be used to characterize the stability and force transmission of the granular system, and it is closely related to the flow state of the system. Here, the probability density function (PDF) of the normal contact force of spherical and cylindrical particles with different blockiness parameters at $t = 1$ s is shown in Fig. 7.25a on a log–log scale. Compared with cylindrical particles, the probability of strong contact forces of spherical particles is smaller. Meanwhile, the probability of strong contact forces of cylindrical particles increases as the blockiness

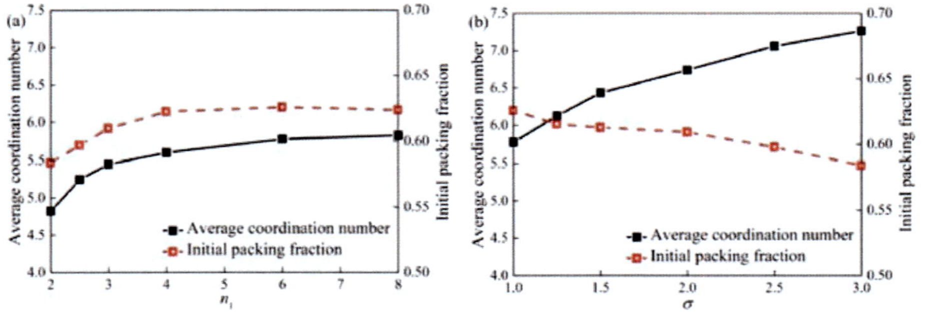

Fig. 7.24 Effects of the blockiness parameter (n_1) and aspect ratio (σ) of the particles on the initial packing fraction and average coordination number: **a** The blockiness parameter (n_1) and **b** the aspect ratio (σ)

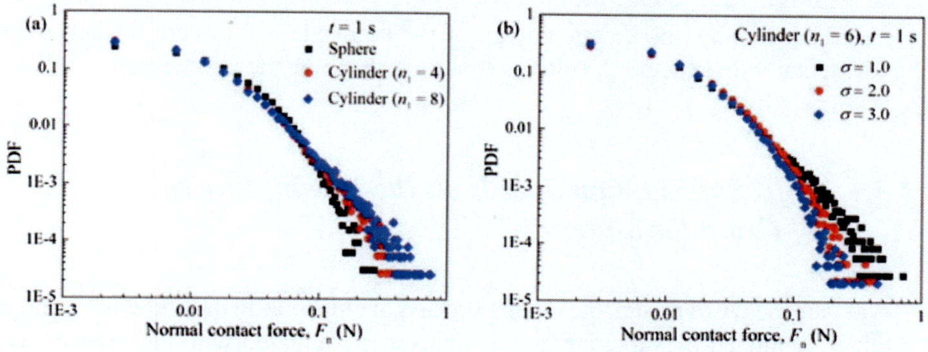

Fig. 7.25 Effects of the blockiness parameter (n_1) and aspect ratio (σ) of the particles on the probability density function of the normal contact force between particles when the hopper angle $\theta = 15^{\text{o}}$: **a** The blockiness parameter (n_1) and **b** the aspect ratio (σ)

parameter increases. This is mainly because the flow state of the cylindrical particles is funnel flow, while spherical particles are in both mass and funnel flow. The non-uniformity of the velocity distribution of the granular system is more pronounced in the funnel flow than in the mass flow. Therefore, the collisions between cylindrical particles are more violent, which leads to greater flow fluctuations and stronger contact forces.

In addition, the effect of the aspect ratio on the probability density function of the normal contact force of cylindrical particles is shown in Fig. 7.25b. As the aspect ratio increases, more cylindrical particles have uniform velocity, and the flow state of the granular system becomes mass flow and funnel flow. As a result, the collision between the cylindrical particles of high aspect ratio becomes weak, and the probability of strong contact forces between cylindrical particles decreases. Furthermore, cylindrical particles with a high aspect ratio have stronger interlocking, resulting in more local cluster structures. The relative movement between the particles is hindered by the clusters, which makes more particles have a uniform velocity and the contact force between particles become smaller. Therefore, the particle shape causes changes in the contact force between particles at the microscopic scale, which may further cause the flow pattern transition of the granular materials at the macroscopic scale.

7.3 Mixing Characteristics of Granular Materials in a Horizontal Rotating Drum

In this section, the discrete element method based on superquadric particles is used to simulate the mixing processes of spherical and non-spherical particles within a horizontal rotating drum. Then, the DEM model is verified based on good agreement between the numerical results and previous experiments (You and Zhao 2018). Moreover, the influences of rotating speeds and particle shapes on the mixing rate are studied. Finally, the kinetic energy of the non-spherical system is analyzed to fundamentally understand the effect of particle shape on the mixing rate.

7.3.1 *Effect of Rotation Speeds on the Mixing Process of Granular Materials*

To validate the DEM model, the mixing process of ellipsoids in the horizontal rotating drum is simulated by superquadric elements and compared with the experimental results (You and Zhao 2018). The oblate ellipsoidal particles used in the experiments are produced by 3D printing of the Plexiglas and have parameters of $2a = 2b = 8\text{mm}$, $2c = 4\text{mm}$, and $n_1 = n_2 = 2$. In the DEM simulation, the diameter and length of the rotating drum are 200 and 20 mm, respectively. The three rotating speeds are used: 20, 40, and 60 rpm. The key calculation parameters are listed in Table 7.2. A total of 1000 ellipsoids with random orientations and initial positions are dropped into the horizontal drum, and then the drum does not begin to rotate until all the particles remain motionless in the container. To further validate the accuracy of the DEM models, the mixing degree of granular system is quantified by the Lacey mixing index, which can be expressed as Lacey (1954):

$$M = \frac{S_0^2 - S^2}{S_0^2 - S_r^2}, \tag{7.1}$$

where S_0^2 is the variance corresponding to the completely separated state and expressed as $S_0^2 = p(1 - p)$. S_r^2 corresponds to the completely mixed state and expressed as $S_r^2 = p(1 - p)/N$. p is the volume ratio of one type of particle in the system and N is the average number of particles in a sample. S^2 is the variance of the current mixed state. The horizontal drum is separated into a number of cubes with a fixed sample size ($3D_s \times 3D_s \times 3D_s$). D_s is the diameter of the volume-equivalent sphere. Then, the position of the particle center is used to determine which cell the particle is in. Considering that there are no particles in the upper part of the drum, a weighting scheme is used to calculate the mixing index (Jiang et al. 2011). Therefore, a sample containing a larger number of particles corresponds to a larger weight, while a sample containing a smaller number of particles corresponds to a smaller weight. If there are no particles in a sample, the corresponding weight is 0. Moreover, the variance (S^2) of the current mixed state based on the weighting concept is obtained by $S^2 = \frac{1}{k}\sum_{i=1}^{N_c} k_i (a_i - \overline{a})^2$. N_c is the total number of samples in the drum. a_i is the volume ratio of particles of the reference type in a sample, and $\overline{a}$ is the volume ratio of particles of the reference type in the drum. k is the sum of the weights of all samples, expressed as $k = \sum_{i=1}^{N_c} k_i$. k_i is the weight of sample i, obtained by $k_i = N_i/N_t$. N_i is the number of particles in sample i, and N_t is the total number of particles in all samples.

Table 7.2 Major simulation parameters of granular systems

Definition	Value	Definition	Value
Young's modulus E (GPa)	1.0	Sliding friction coefficient μ_s (–)	0.3
Poisson's ratio ν (–)	0.3	Normal damping coefficient C_n (–)	0.05
Particle density ρ (kg/m^3)	1150	Tangential damping coefficient C_t (–)	0.05

Figure 7.26 shows the experimental (You and Zhao 2018) and simulation snapshots of mixing process at different revolutions. According to this figure, a good agreement between the experiment and the simulation is observed for the mixing processes. In order to quantitatively compare the mixing degree between the experiment and the simulation, the Lacey mixing index at different rotating speeds is calculated, as shown in Fig. 7.27. The mixing degree between the ellipsoids increases as the number of revolutions increases, and basically reaches a quasi-static state. The simulation results of superquadric DEM agree well with the experimental results, which verifies the applicability of DEM model for the mixing processes of non-spherical particles.

It is well known that the effect of particle shape on the mixing degree is critical to industrial production and design. Therefore, the mixing process of the non-spherical particles in a horizontal rotating drum is further simulated by the aforementioned method. The diameter and length of the drum are 200 and 50 mm, respectively, and the periodic boundary conditions are applied in the z-direction to eliminate the end wall effect, as shown in Fig. 7.28. The rotating speeds of σ = 10, 20, 30, 40, 60, 80 rpm are used for non-spherical particles to show the superposed effect of the drum operation. Particles of different shapes have the same mass, and the total number of particles is 35,000. The diameter of the volume-equivalent sphere is 2.9 mm, and the aspect ratio is obtained by $\alpha = c/a(= b)$. The cube-like ($n_2 = n_1$) and cylinder-like ($n_2 = 2$) particles with different blockiness are obtained by

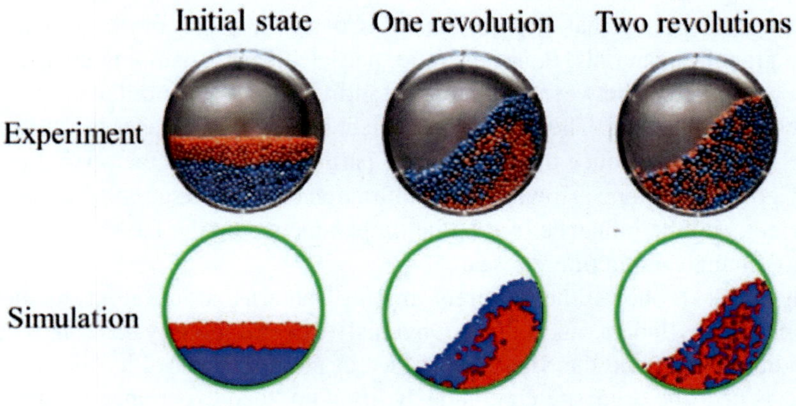

Fig. 7.26 Comparison of simulated mixing processes with experimental results (You and Zhao 2018) at the rotating speed of 20 rpm for oblate ellipsoids

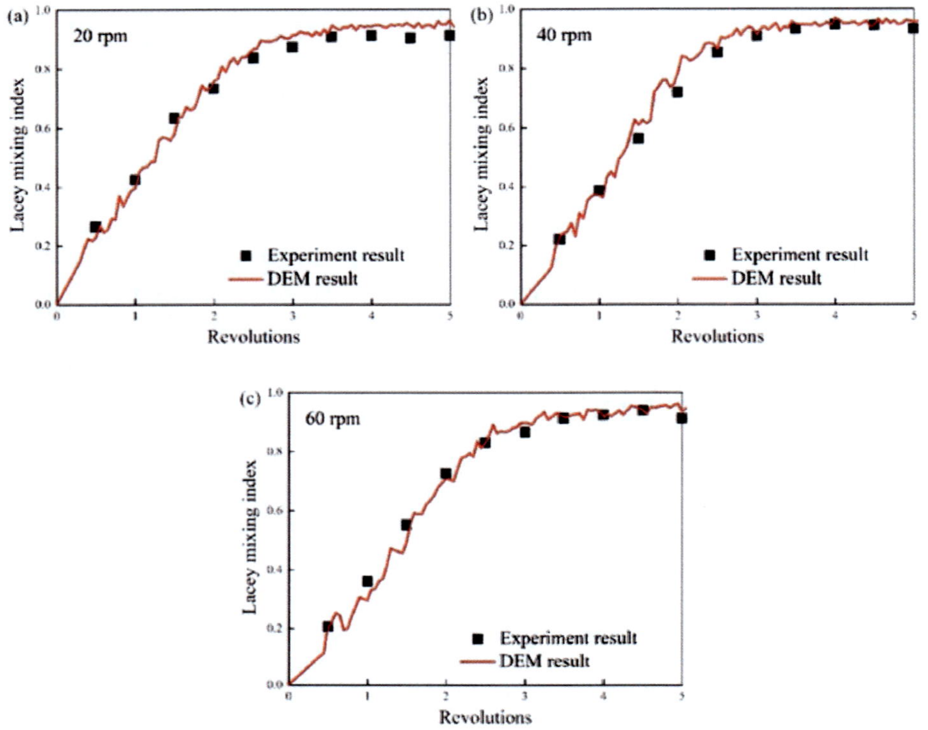

Fig. 7.27 The experimental (You and Zhao 2018) and simulated results of the Lacey mixing index for the different rotating speeds: **a** 20 rpm, **b** 40 rpm, and **c** 60 rpm

changing the parameter n_1, as shown in Fig. 7.29. Aspect ratios for cylinders and cubes vary from 0.25 to 3.0, as shown in Fig. 7.30. The main DEM parameters of the granular systems are listed in Table 7.3. The low Young's modulus is to reduce the running time and improve the calculation efficiency of the DEM simulation. It has been reported that Young's modulus has little effect on the mixing results of the granular materials. Poisson's ratio, particle density, damping coefficient, and friction coefficient between the particles and the drum are similar to the material properties of the glass. The friction coefficient between particles is less than the true value, which is to reduce the influence of particle friction on the mixing processes of the granular systems. However, the calibration of the simulation parameters is not sufficient, and the influence of calculation parameters on the DEM results needs to be further analyzed in future research.

Figure 7.31 shows the different mixing patterns at different moments for the spheres, cylinders, and cubes. Obviously, the granular systems have reached cascading regime and the S-shaped surface of the granular bed has been observed. Particles near the drum are continuously lifted up, then experience the avalanches, and flow downward the free surfaces. As the granular bed is repeatedly rotated, the red and blue particles exhibit a spiral shape and a hierarchical pattern gradually appears. When the rotating time is greater than 6 s, this pattern disappears, and the

Fig. 7.28 Schematic diagram of three-dimensional horizontal drum simulated by DEM

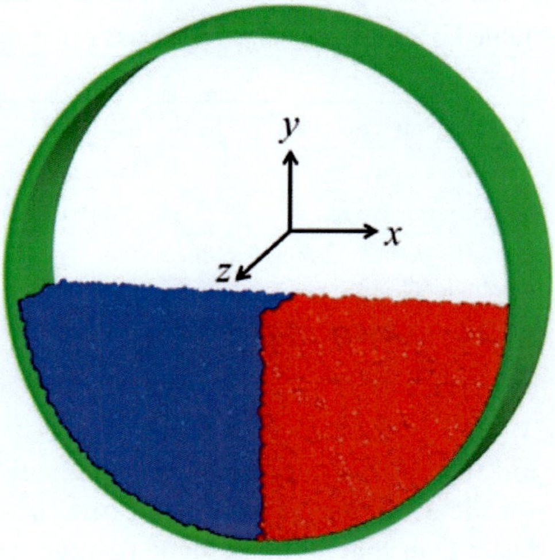

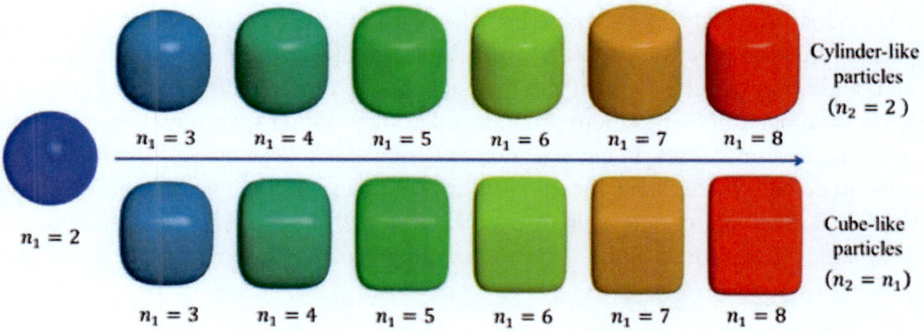

Fig. 7.29 Influence of blockiness parameters on the particle shape

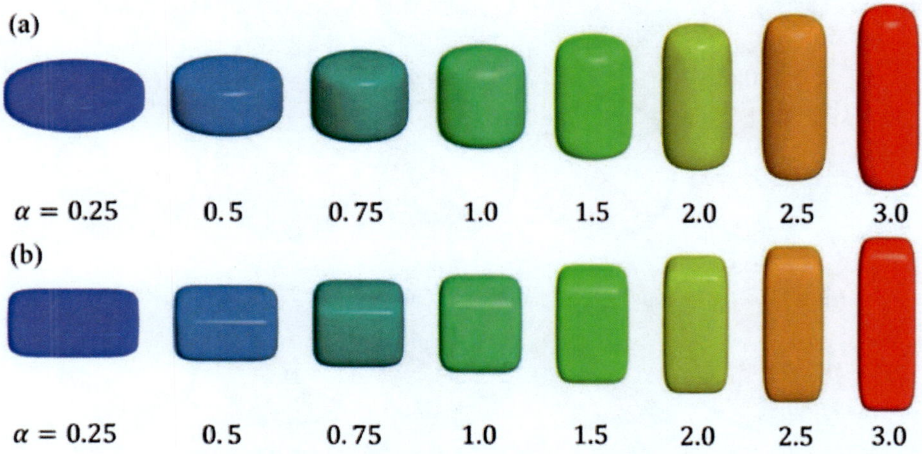

Fig. 7.30 Examples of superquadric particles with different aspect ratios: **a** cylinder-like particles and **b** cube-like particles

Table 7.3 DEM simulation parameters of the segregation process

Definitions	Values	Definitions	Values
Young's modulus (Pa)	1.0×10^7	Normal damping coefficient	0.3
Poisson's ratio	0.3	Tangential damping coefficient	0.3
Density (kg/m^3)	2500.0	Rolling friction coefficient	0.002
Particle–particle sliding friction coefficient	0.4	Filling level (%)	40
Particle–wall sliding friction coefficient	0.4	Time step (s)	1×10^{-6}

red and blue particles are substantially uniformly distributed in the drum. However, for the evolution of the mixing patterns, non-spherical particles are clearly mixed faster than spheres at the same rotating speed.

Figure 7.32 shows the velocity distribution of the granular bed at 10, 20, 40, 60, 80 rpm for differently shaped particles. The flow field can be clearly divided into three parts: a flowing layer close to the drum, a static base in the middle of the granular bed, and an avalanche layer of the free surface. Meanwhile, this layered pattern of particle velocity becomes more pronounced as the rotating speed and the irregularity of the particle increases. In general, non-spherical particles that are close to the drum are more easily lifted than spheres, resulting in greater particle velocities. Therefore, there is a relatively greater velocity at the surface of the non-spherical granular bed compared to the surface of the spherical granular bed. As the rotating speed increases,

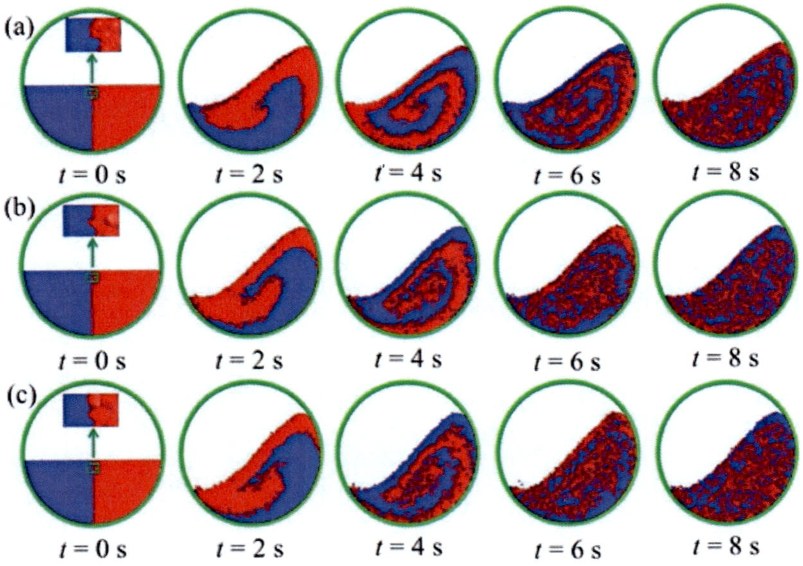

Fig. 7.31 Temporal evolutions of mixing patterns for differently shaped particles at 30 rpm: **a** spheres, **b** cylinders, and **c** cubes

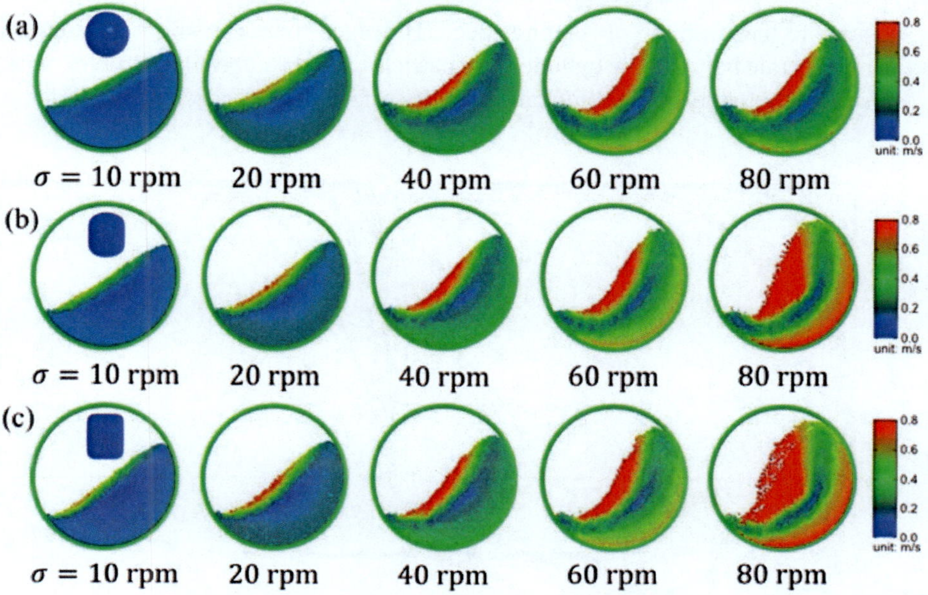

Fig. 7.32 The velocity profiles of granular bed at 10, 20, 40, 60, 80 rpm for differently shaped particles: **a** spheres, **b** cylinders, and **c** cubes

the areas of the flowing layer and the avalanche layer are enlarged and the area of the static base is reduced. As a result, the average velocity of the granular systems increases significantly. When the rotation speed is 80 rpm, the non-spherical particles are upcasting and cataracting. The particles are not in close contact with each other as in the cascading regime, and the flow state of the non-spherical granular systems has been transformed into the cateracting regime.

At the initial moment, the two types of particles are completely separated, i.e., the initial mixing index M_0 is close to 0. After sufficient revolution of the drum, the granular system eventually reaches a well-mixed state, i.e., the final mixing index M_f is close to 1. For spherical or non-spherical particles, the evolution of the mixing index can be fitted by an error function, which is given by He et al. (2017):

$$M(t) = M_0 + \left(M_f - M_0\right)\mathrm{erf}(\mathrm{Rt}), \tag{7.2}$$

where $\mathrm{erf}(\mathrm{Rt})$ is an error function, given by $\mathrm{erf}(x) = 2/\sqrt{\pi}\int_0^x e^{-x^2}\mathrm{d}x$. R is the mixing rate, obtained by the fitting of simulation results. Figure 7.33 shows the temporal evolutions of Lacey mixing index for spheres, cylinders, and cubes. Here, the solid curves represent the results of fitting the data using Eq. (7.2). For differently shaped particles, the mixing index eventually becomes stable and close to 1. Therefore, the mixing rate of the particles at different rotating speeds is quantitatively compared, as shown in Fig. 7.34. Clearly, as the rotating speed increases, the particles mix faster. Meanwhile, the non-spherical particles are mixed faster than spheres for the same rotating speed. This means that the mixing rate of the non-spherical particles is more sensitive to the variation of rotating speed. However, there is a slight difference in the mixing rate between the cylinder-like particles and the cube-like particles, which is mainly due to the similar blockiness of the surface of the particles.

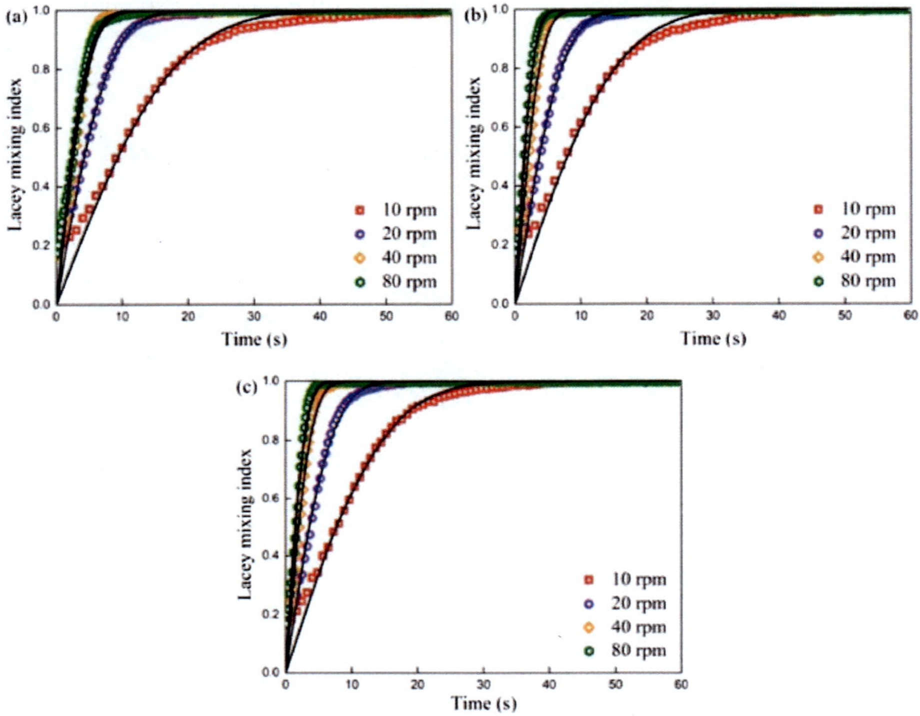

Fig. 7.33 The Lacey mixing index as a function of time for differently shaped particles: **a** Spheres, **b** cylinders, and **c** cubes. The solid curves are the fitting results with Eq. (7.2)

Fig. 7.34 The influence of rotating speed on the mixing rate for various particle shapes

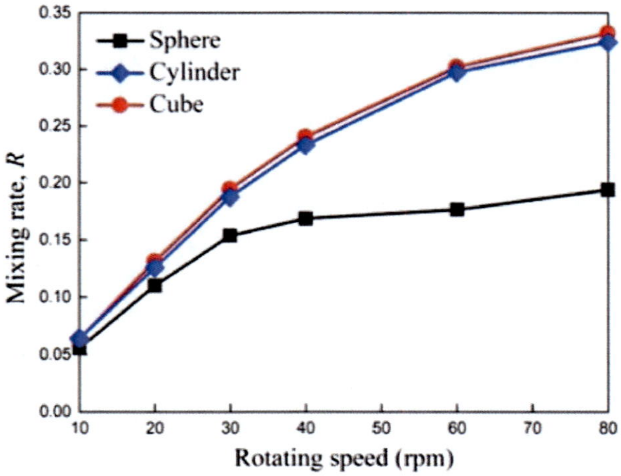

7.3.2 Effect of Particle Shapes on the Mixing Process of Granular Materials

Figure 7.35a shows the mixing rate of particles with different blockiness parameters. The mixing rate increases significantly as the blockiness parameter n_1 is changed from 2 to 3, mainly because of the direct change of particle shape from spherical to non-spherical, as shown in Fig. 7.29. When the parameter n_1 is greater than 3, the particle blockiness has no significant effect on the mixing rate. It is mainly because the mixing process of non-spherical particles is closely related to the packing state, while the dense face-to-face contact and the ordered packing structure result in a better quality of mixing. However, the blockiness of particle surface has little effect on the packing state of the granular bed when the parameter n_1 is changed from 3 to 8, as shown in Fig. 7.35b. As a result, the blockiness parameters do not significantly affect the mixing rate. Therefore, the rotating speed has a primary effect on the mixing rate, whereas the effect of the blockiness on the mixing rate is a secondary factor for non-spherical granular systems.

Figure 7.36a shows the mixing rate for particles with different aspect ratios. The mixing rate decreases as the aspect ratio deviates from 1.0. This is mainly because the increase or decrease of the aspect ratio results in the granular bed changing from a densely packing state to a loosely packing state, as shown in Fig. 7.36b. Particles with higher or lower aspect ratios have more voids and interlocking structures, resulting in lower quality of mixing. For the same aspect ratio, the mixing rate of cube-like particles is slightly higher than cylinder-like particles. More voids are formed within the cylindrical granular systems than in the cubic granular systems. It has been reported that the voids in the granular system reduce the particle–particle collision and driving force transfer efficiency, which results in lower particle motion in the drum (Gui et al. 2018).

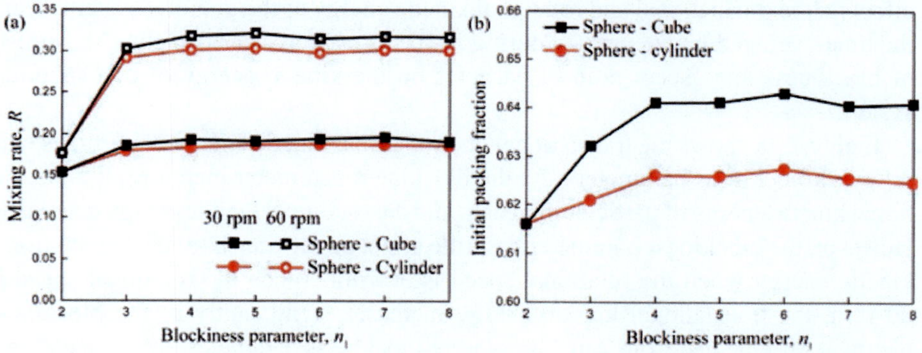

Fig. 7.35 The influence of the blockiness parameter on the mixing rate (**a**) and the initial packing fraction (**b**)

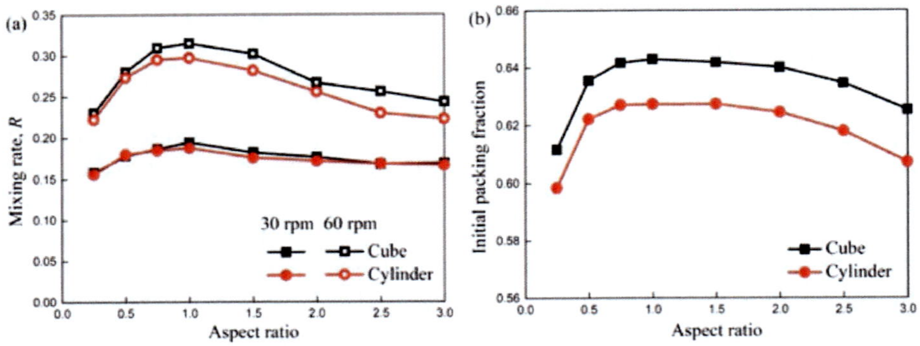

Fig. 7.36 The influence of the aspect ratio on the mixing rate (**a**) and the initial packing fraction (**b**)

7.3.3 Analysis of Translational and Rotational Kinetic Energies of Particles

As a pure energy input, the rotating drum plays a vital role in accelerating or driving particle motion and mixing. The external energy is transformed into the kinetic energy of the particle assembly throughout the friction and inter-locking between particles. Figure 7.37 shows the translational and rotational kinetic energy at different rotating speeds for spheres and cubes. When the drum starts to rotate, non-spherical particle systems require more energy to be driven from a stationary packing state to a motion state. This is mainly because non-spherical particles have a tighter face-to-face contact and a more stable packing structure. When the rotating time is longer than 3 s, the granular system reaches a stable flow and mixing state. As the rotational speed increases, both the translational and rotational kinetic energy of particles increases. Moreover, the particle shape and rotational speed have a superposed effect on the translational and rotational kinetic energy of the granular system. Thus, the time-averaged kinetic energy during 3–10 s is used to clearly analyze the effects of blockiness and aspect ratio of particles on the kinetic energy of non-spherical systems.

Figure 7.38 shows the translational and rotational kinetic energy of particles with different blockiness parameters. As the blockiness parameter increases, the translational kinetic energy of particles increases and the rotational kinetic energy decreases. However, the blockiness parameter has little effect on the translational and rotational kinetic energy when the rotational speed is 30 rpm. When the rotational speed is 60 rpm, the translational kinetic energy increases significantly as the blockiness parameter is changed from 2 to 3. For larger blockiness parameters, the translational kinetic energy does not change significantly. Moreover, the rotational kinetic energy is clearly reduced as the blockiness parameter increases. This is mainly because the less round particles (namely with larger n_1) are almost one-by-one arranged and have more tight surface-to-surface contacts with their neighbors. As a result, the cubic particles have larger internal interactions and are more difficult to roll. In other

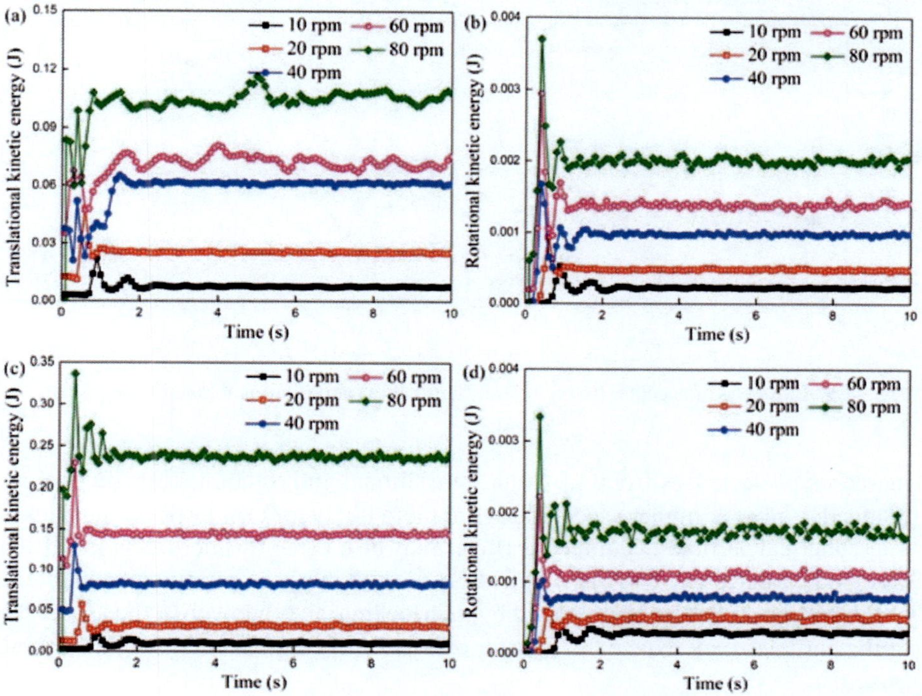

Fig. 7.37 Translational (**a**, **c**) and rotational kinetic energy (**b**, **d**) at different rotating speeds for differently shaped particles: **a**, **b** Spheres and **c**, **d** cubes

words, the surface blockiness of particles may enhance the transfer efficiency of external energy to the non-spherical granular systems.

Figure 7.39 shows the translational and rotational kinetic energy of particles with different aspect ratios. The aspect ratio has less influence on the translational kinetic energy and has a greater influence on the rotational kinetic energy. As the aspect ratio

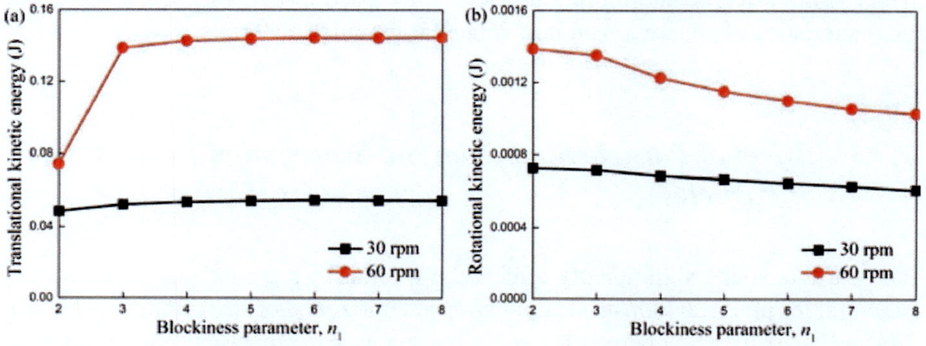

Fig. 7.38 Influence of blockiness parameter on the translational (**a**) and rotational kinetic energy (**b**)

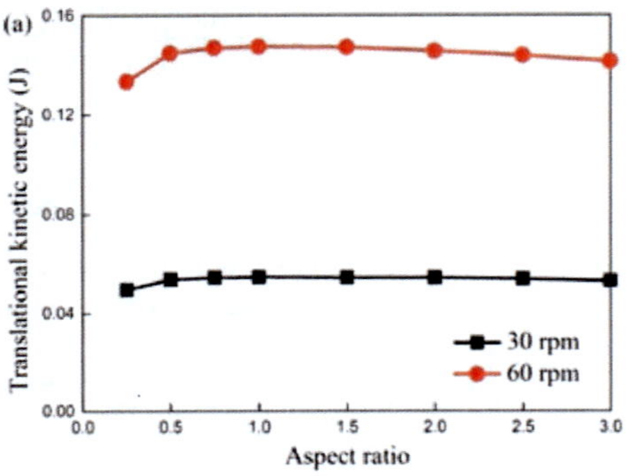

Fig. 7.39 Influence of aspect ratio on the translational (**a**) and rotational kinetic energy (**b**)

increases or decreases from 1.0, both translational and rotational kinetic energy is reduced. Lower or higher aspect ratios are more likely to form interlocking between non-spherical particles. Clumped particles may be created by interlocking, and all of clumped particles move as one element. As a result, both the movement and rotation of the particles seem to be restricted by interlocking. In other words, the aspect ratio of the particles may reduce the transfer efficiency of external energy to the granular systems.

7.4 Radial Segregation Characteristics of Particles in a Horizontal Rotating Drum

In this section, the superquadric equation is used to construct non-spherical particles, and the DEM is used to simulate the radial segregation behaviors of the Gaussian-dispersed mixture. Then, the effects of particle shape, the mixture standard deviation, and rotating speed on the radial segregation of granular materials are analyzed. The gyration degree and mixing index are discussed to understand the segregation characteristics of spherical and non-spherical granular materials.

7.4.1 Effect of Particle Shapes on the Segregation Process of Particles

The particle shape significantly affects the motion of granular materials, which is essential for practical industrial applications in horizontal drums (Höhner et al. 2014; Ma and Zhao 2018). Superquadric equations are used to construct spheres, ellipsoids with aspect ratios of 0.5 and 2.0, and cube-like particles with aspect ratios of 0.5 and 2.0, as shown in Fig. 7.40. Sphericity is used to quantify the change in the shape of the superquadric element, which is determined based on the ratio of the surface area

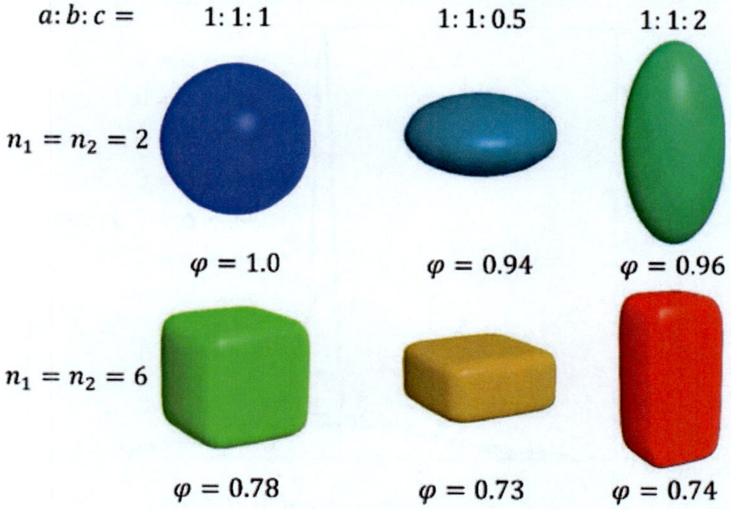

Fig. 7.40 Particles of different shapes obtained by the superquadric equations

of a sphere with the same volume as a given particle (S_s) to the actual surface area of the particle (S_p). The equation of sphericity (φ) is expressed as Zhao and Chew (2020).

$$\varphi = \frac{S_s}{S_p} = \frac{\pi^{1/3}(6V_p)^{2/3}}{S_p}, \tag{7.3}$$

where V_p is the volume of the particle, and S_p is the surface area of the particle.

The equivalent diameter is defined as the diameter of a sphere with the same volume as the superquadric element, expressed as $d_e = 2\sqrt[3]{3V_s/(4\pi)}$. Here, d_e is the equivalent diameter, and V_s is the volume of the superquadric element. The number of particles of different sizes satisfies the Gaussian distribution, and its range is $[\mu - 2\sigma, \mu + 2\sigma]$, as shown in Fig. 7.41a. Here, σ and μ are the mixture standard deviation and the mixture mean size, respectively. Note that particles with sizes out of the range $[\mu - 2\sigma, \mu + 2\sigma]$ are not included in the simulation. Meanwhile, in Fig. 7.41b, the color of particles is used to indicate the equivalent diameter. Blue indicates that the particle has the smallest equivalent diameter, and red indicates that the particle has the largest equivalent diameter. The standard deviations of $\sigma = 0.1$, 0.25, 0.5, 0.75, 1.0, 1.25 mm are used for granular flow to show the effect of the Gaussian distribution of particle sizes on the segregation characteristics. Moreover, the total mass of the granular bed is 2.8 kg, and the DEM simulation parameters are similar to those used in the previous study (He et al. 2019), which are listed in Table 7.3.

The drum has a diameter of 280 mm and a length of 70 mm, and the periodic boundary conditions are applied in the z direction to eliminate the end wall effect. The range of rotation speed (ω) is 7.5~130 rpm to show the superimposed effect

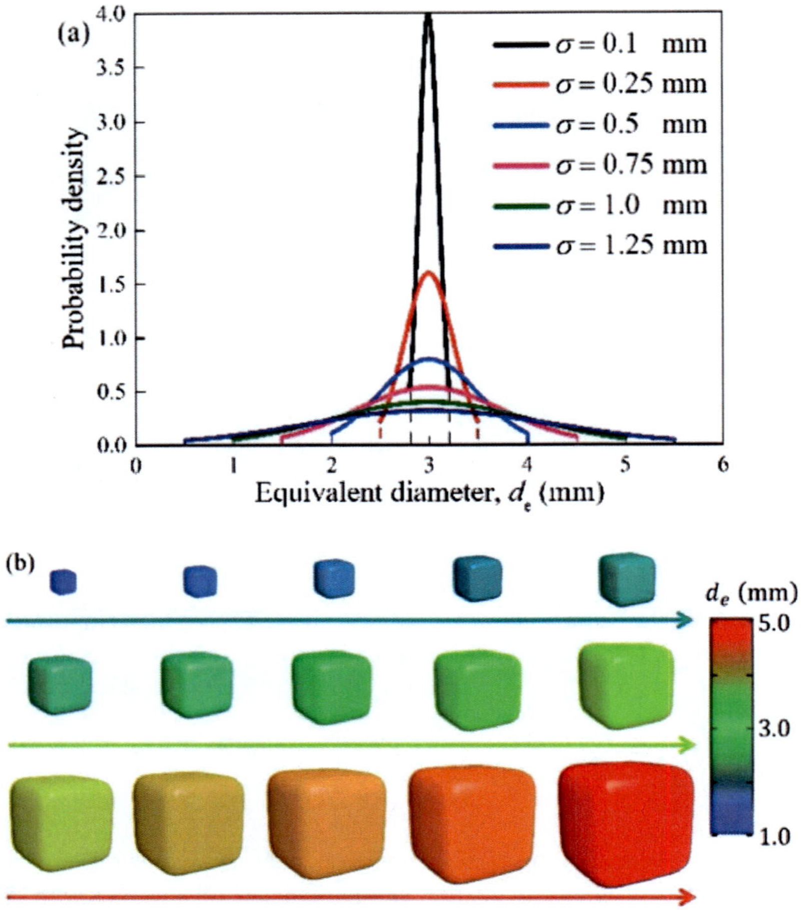

Fig. 7.41 **a** Particle size distribution (PSD) used in the simulation where the mean size $\mu = 3$ mm, and the standard deviation σ varies from 0.1 to 1.25 mm. **b** Particle sizes used in PSD of cubic particles ranging from 1 to 5mm where the color also indicates the magnitude of particle size

of the cylinder operation on granular flow. The dispersion behavior of particles is considered as the main factor that counterbalances the particle segregation and leads to the mixing of various types of particles in the rotating drum (Peng et al. 2016). The gyration radius of the granular materials is defined as the distance of the particles in the mixture from the core of the granular bed. A small gyration radius means that the particles are mainly located in the core of the granular bed, while a large gyration radius means that the particles are mainly located on the periphery of the granular bed. In the rotating drum, if the particles basically have no radial movement relative to the core of the granular bed, these particles are in a relative equilibrium state. Therefore, a small gyration radius also means that the particles reach a relative equilibrium state. The gyration radius is expressed as Yang et al. (2017):

$$R_d = \sum_{i=1}^{N_{\text{st}}} \frac{\sqrt{(x_i - x_c)^2 + (y_i - y_c)^2}}{N_{\text{st}}}, \tag{7.4}$$

where R_d represents the average gyration radius of all particles in a sample. x_i and y_i are the coordinates of particle i in the x and y directions, respectively. x_c and y_c are the coordinates of the core of the granular bed in the x and y directions, respectively. N_{st} is the total number of particles in a sample. Furthermore, ψ_r is defined as the gyration degree of a component in the mixture, obtained by $\psi_r = \left|R_{\text{d}f} - R_{\text{d}0}\right|$. R_{df} and $R_{\text{d}0}$ are the average gyration radius of the components in the mixture at 120 and 0 s, respectively. A large value of ψ_r means that the gyration radius of the particles at the final moment is significantly different from the gyration radius of the particles at the initial moment. Therefore, particles with a large gyration degree are considered to have strong radial flowability and dispersibility, while particles with a small gyration degree are considered to have a weaker radial flowability. To evaluate the segregation behavior of the entire granular system, the gyration coefficient ψ_r^*, which is defined as the average value of the dimensionless gyration degree of the various components in the mixture and expressed as

$$\psi_r^* = \frac{1}{N_{\text{type}}} \sum_{i=1}^{N_{\text{type}}} \left(\frac{R_{\text{df},i} - R_{\text{d}0,i}}{R_{\text{d}0,i}} \right)^2, \tag{7.5}$$

where N_{type} is the total number of particle types, and $N_{\text{type}} = 6$. Considering that the entire granular system has a continuous size distribution, the entire granular system is equally divided into six ranges according to the equivalent diameter, which is helpful for further analysis of the radial segregation characteristics of particles of different sizes. A larger value of ψ_r^* indicates that particles of different sizes have a larger difference in the positions from the core of the granular bed, and also indicates that the granular system has a greater segregation degree.

Besides, the mixing index of the particles in the rotating drum is calculated using Boltzmann's expression, which is more suitable for polydisperse granular systems. A smaller mixing index indicates a greater segregation degree of particles. The mixing index is expressed as Alchikh-Sulaiman et al. (2016):

$$M_s = \frac{S_0 - S_t}{S_0 - S_r}, \tag{7.6}$$

where S_0 is the average global entropy when the granular system is completely separated, and S_r is the average global entropy when the granular system is completely mixed. S_t is the global entropy of the current state, expressed as $S_t = \frac{-1}{\text{In} N_{\text{type}}} \sum_{j=1}^{N_{\text{cell}}} \sum_{i=1}^{N_{\text{type}}} k_j p_{ij} \text{In} p_{ij}$. Here, N_{cell} and N_{type} are the total number of cubic cells and the total number of particle types, respectively. The horizontal drum is divided into a series of cubes with a fixed sample size ($4\mu \times 4\mu \times 4\mu$), and μ is the mean size of the Gaussian-dispersed size distribution. Considering that there are no particles in upper cubes within the drum, a weighting strategy is used to calculate the current mixed state (S_t). k_j is the weight of cell j, obtained by $k_j = N_j/N_s$. N_j is the number of particles in cell j, and N_s is the total number of particles in the drum. p_{ij} is the number fraction of particles of type i in cell j. However, it is necessary to further analyze the influence of grid size and division on the calculation results in future work.

Figure 7.42 shows the radial segregation process of spheres and cube-like particles at different moments. Here, the rotation speed is 15 rpm and the total simulated time is 120 s. The average size μ and the standard deviation σ are 3 and 1 mm, respectively. Initially, particles of different sizes are uniformly distributed in the drum ($t = 0$ s). As the granular bed is repeatedly rotated, the granular system forms a dynamic angle of repose and gradually reaches the cascading and rolling regime. The small blue particles gradually move to the core of the granular bed, and the large red particles move to the periphery of the granular bed. Meanwhile, the large red particles near the drum are constantly lifted, then avalanche and flow down along the free surface. When the rotating time is > 40 s, a stable radial segregation pattern forms. However, spherical particles have a more slightly pronounced segregation pattern than cube-like particles at the same rotational speed.

To further demonstrate the distribution of Gaussian-dispersed particle mixtures in the drum, the entire particle sizes are equally divided into 6 ranges according to the equivalent diameter, as shown in Fig. 7.43 where $\mu = 3$ mm and $\sigma = 1$ mm. It can be observed that for spheres (Fig. 7.43a), the particles with small diameters (e.g., less than 2.33 mm) are located at the core of the granular bed, while the particles

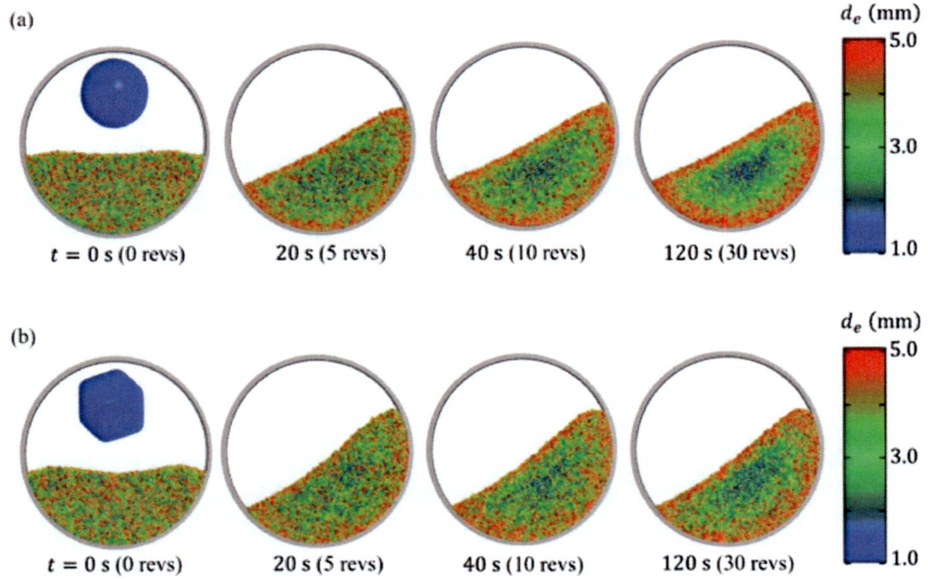

Fig. 7.42 Temporal evolutions of segregation patterns for spherical (**a**) and cube-like (**b**) particles at 15 rpm

with the large diameters (e.g., larger than 3.67 mm) are located at the periphery of the granular bed. Particles with sizes in the range of 2.33~3.67 mm are more evenly distributed in the drum. It is worth noting that the radial segregation behavior of spherical particles is slightly different from cube-like particles. Cube-like particles have a larger dynamic angle of repose and a smaller degree of segregation in a drum (Fig. 7.43b).

To further analyze the radial distribution characteristics of particles of different sizes, the probability distribution function of the spherical particles in the rectangular area along the mid-chord is counted, as shown in Fig. 7.44a. The mid-chord is the line connecting the center of the drum and the mass center of the granular bed, and the width of the rectangular area is approximately one-third of the length of the bed surface. A dimensionless radial distance of zero indicates the center of the drum, and a dimensionless radial distance of 1 indicates the wall of the drum. At the initial moment, the probability of particles of different sizes in the radial direction is basically the same, which means that these particles are evenly distributed in the granular bed, as shown in Fig. 7.44b–g. As the drum is repeatedly rotated, the

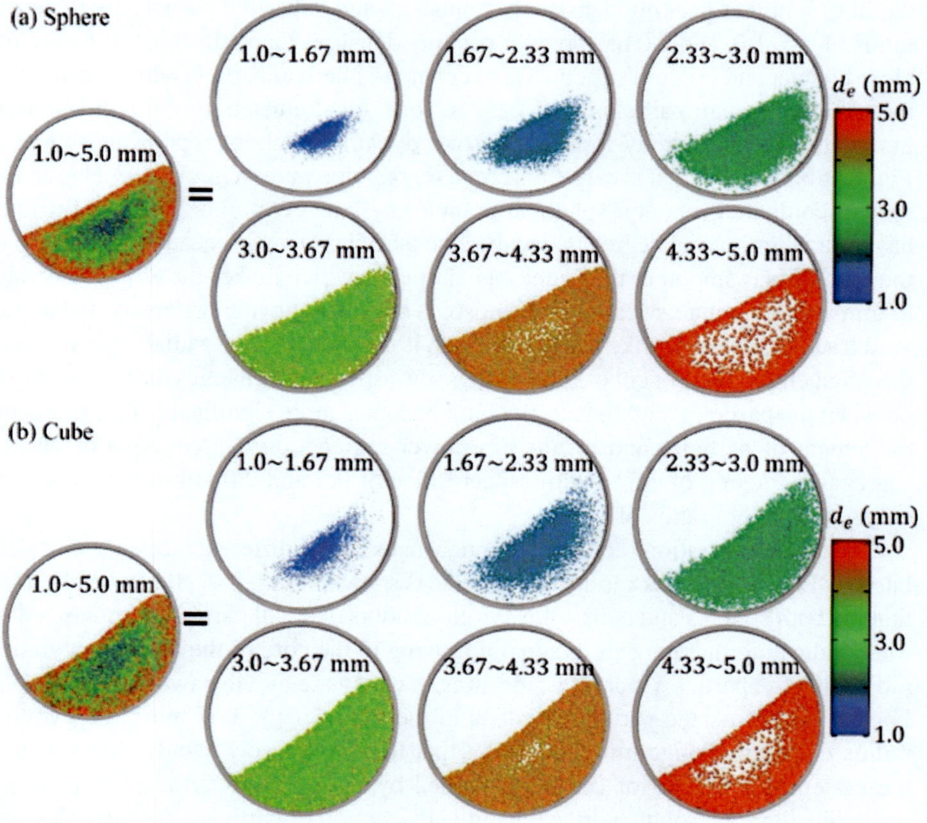

Fig. 7.43 Radial distribution of Gaussian-dispersed particles in the drum: **a** spheres, and **b** cube-like particles

probability distribution curve with particle sizes of 1~2.33 mm has a peak, which indicates that the particles are mainly located in the core of the granular bed; the probability distribution curve with particle sizes of 2.33~5 mm has double peaks, which indicates that the particles are close to the bed surface and drum. As the particle size increases, the value of the single peak gradually decreases and becomes a double peak, and then the value of the double peak increases. The probability distribution curves are consistent with the segregation patterns shown in Fig. 7.43a.

Moreover, the temporal variation of the mixing index for different shapes of particles is shown in Fig. 7.45a, b shows the final mixing index at the steady state together with the segregation patterns (inset figures). It can be observed that spheres have a smaller mixing degree (or a greater segregation degree) than non-spherical particles, and the mixing degree of the granular system basically decreases as the sphericity of the particles increases. This is mainly because the formation of the segregation pattern depends on the penetration process of small particles. The larger void ratio facilitates the movement of small particles through the voids to the center of the granular bed. Figure 7.46a shows the effect of particle shape on the volume fraction of granular materials at $t = 0$s. Ellipsoids with an aspect ratio of 0.5 or 2.0 have a higher packing density than spheres, and cube-like particles with aspect ratios of 0.5, 1.0, and 2.0 have higher packing densities than ellipsoids. Besides, the close contact and ordered structure between non-spherical particles limit the relative movement between particles. It is easy to form local clusters in the non-spherical granular systems, thereby forming a larger dynamic angle of repose, as shown in Fig. 7.46b. As a result, it is easier to form a segregation pattern in a spherical granular system compared to a non-spherical granular system. Also note that the ellipsoids have a smaller mixing degree than cube-like particles. This is because the surface of the ellipsoids is smoother than cube-like particles, which makes the ellipsoids easier to translate and rotate. Meanwhile, ellipsoids have lower packing density and higher void fraction than cube-like particles, which is conducive to the radial penetration of the small ellipsoids. For cube-like particles, the aspect ratio changes the contact mode between the particles, and the interlocking becomes more significant. This limits the movement of particles and results in a lower segregation degree. As a result, the segregation degree of cubes with aspect ratios of 0.5 and 2.0 is lower than that of cubes with aspect ratios of 1.

The average gyration radius of granular materials of different sizes is also calculated by Eq. (7.4) and its variation with time is shown in Fig. 7.47. It can be observed that for both spheres and cubes, the gyration radius of small particles decreases with time, indicating that they are gradually moving to the core of the bed. The gyration radius of large particles increases, meaning that they are moving away from the core. For the medium-sized particles such as in the range of [3, 3.67 mm], the gyration radius does not change much, showing that they are largely evenly distributed in the system. This behavior can be explained by the particle–particle and particle–wall collision forces which drive the radial dispersion of particles, counterbalancing the convective behavior of particles and reaching the equilibrium state (Peng et al. 2016). Small particles are generally located in the voids between large particles, which makes it difficult for small particles to obtain external energy directly through

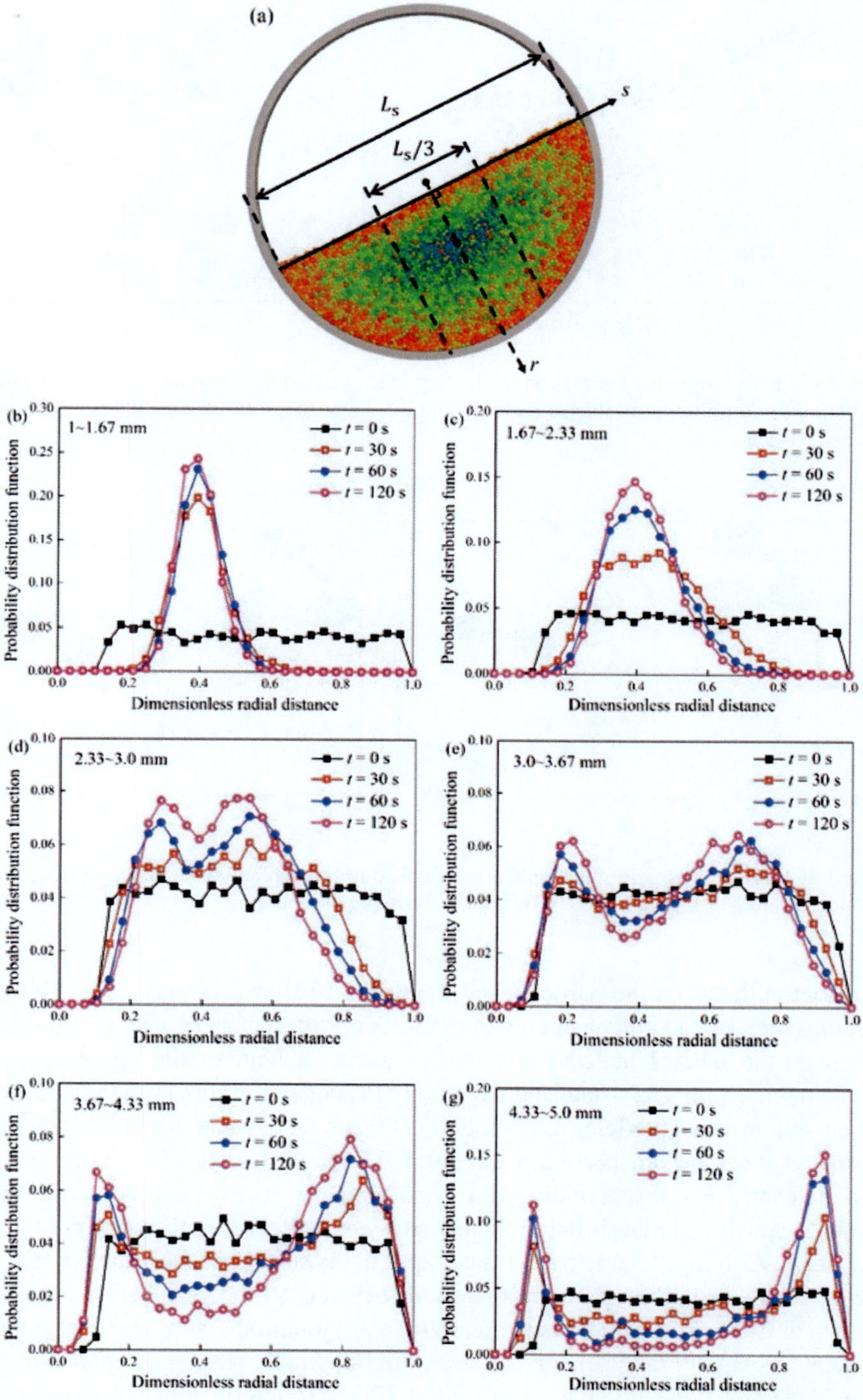

Fig. 7.44 The probability distribution of particles of different sizes in spherical granular systems at 15 rpm: **a** a snapshot showing the extraction area of the particles, **b** $d_e \in [1, 1.67\text{mm})$, **c** $d_e \in [1.67, 2.33\text{mm})$, **d** $d_e \in [2.33, 3\text{mm})$, **e** $d_e \in [3, 3.67\text{mm})$, **f** $d_e \in [3.67, 4.33\text{mm})$, **g** $d_e \in [4.33, 5\text{mm}]$

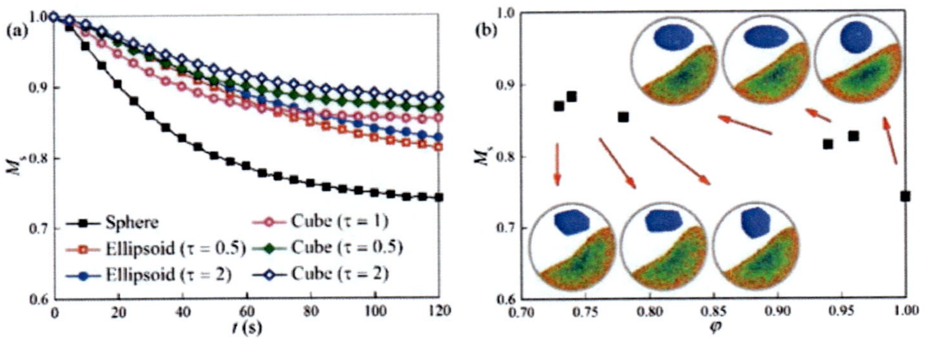

Fig. 7.45 Effect of particle shapes on the mixing index of granular materials: **a** time evolution of mixing index for differently shaped particles and **b** comparison of the mixing index of differently shaped particles at 120 s

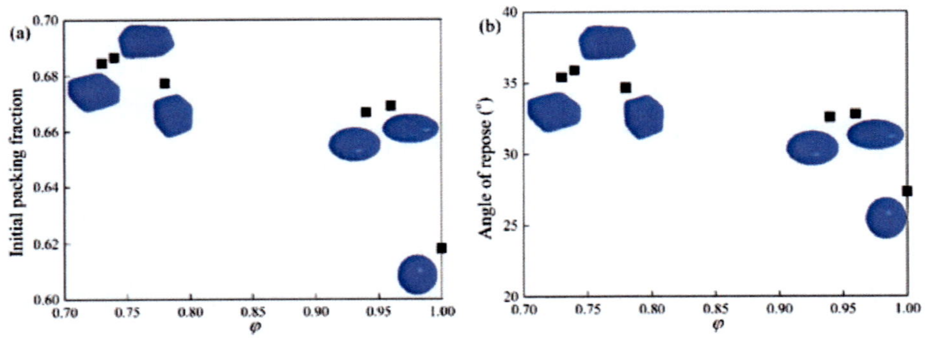

Fig. 7.46 Effect of particle shape on the initial packing fraction and dynamic angle of repose of granular materials: **a** initial packing fraction and **b** angle of repose

the friction between the particles and the drum, as shown in Fig. 7.48a. Here, $\overline{E}_k$ denotes the average kinetic energy of particles of different sizes. The smallest particles have the lowest kinetic energy, which indicates that the smallest particles have a more stable equilibrium state and are mainly located in the core of the granular bed. The medium-sized particles have moderate kinetic energy and dispersion, and these particles are evenly dispersed in the drum. The largest particles have the greatest kinetic energy and dispersibility, which makes these particles in a non-equilibrium state. Meanwhile, large particles obtain driving energy directly through friction with the drum. As a result, large particles collapse to the surface of the granular bed after being lifted, which makes them mainly distributed on the periphery of the granular bed. Compared with the spherical granular system, the non-spherical granular system has larger kinetic energy, as shown in Fig. 7.48b. Here, E_k denotes the total kinetic energy of the granular system at $t = 120$s. The interlocking and close contact between non-spherical particles is beneficial to improve the conversion efficiency of external energy to the granular system. However, particles with higher kinetic energy are more likely to move around the core of the granular bed, which limits the radial penetration of small particles. Therefore, the spherical granular system with the smallest kinetic energy has the best segregation behavior. The kinetic energy of the cube-like granular system is greater than that of the ellipsoids, which causes its segregation behavior to be less obvious than that of the ellipsoids.

In addition, the gyration degree of different particle sizes for different shapes is also plotted and shown in Fig. 7.49. It is observed that with the particle diameter increasing, the gyration degree first decreases and then increases (Fig. 7.49a). Both small and large particles have a large gyration degree, illustrating that they move from the core to the periphery or vice versa. Medium-sized particles have the lowest gyration degree, which indicates that these particles neither move to the core of the granular bed nor the periphery. Figure 7.49b shows the gyration coefficients of differently shaped particles, indicating that spheres have the largest gyration coefficient and cube-like particles have a low gyration coefficient. This is because particle shape changes the contact mode and penetration ability of particles, thereby restricting the movement of small and large particles. Note that cube-like particles with different aspect ratios have similar sphericity, and their gyration coefficients do not change

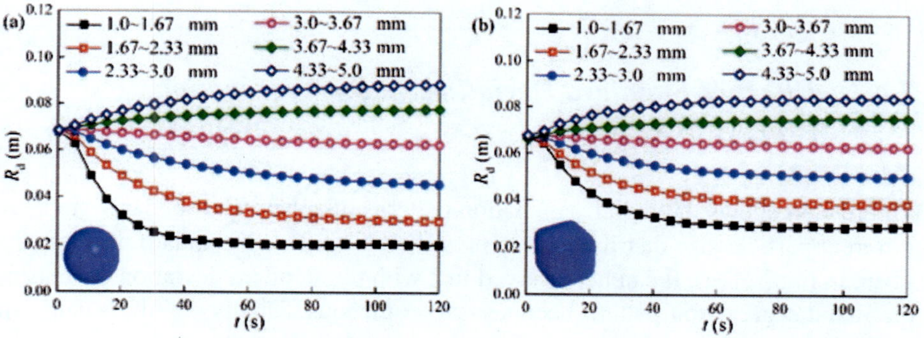

Fig. 7.47 Effect of particle diameters on the average gyration radius of granular materials: **a** spheres and **b** cubes

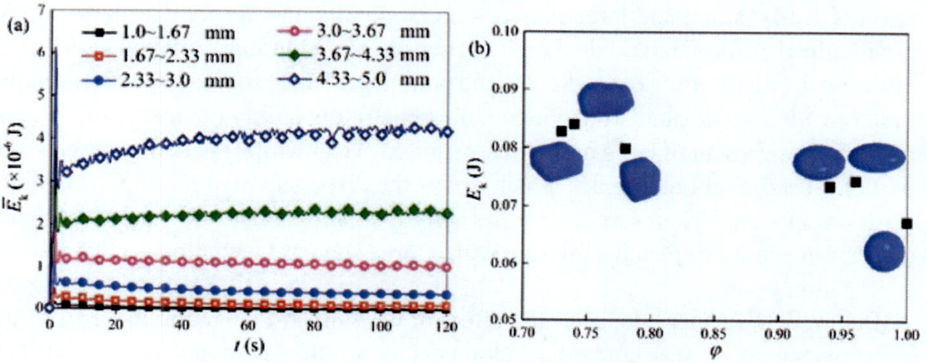

Fig. 7.48 **a** The average kinetic energy of spherical particles with different sizes as a function of time. **b** Effect of particle shape on the total kinetic energy of granular systems

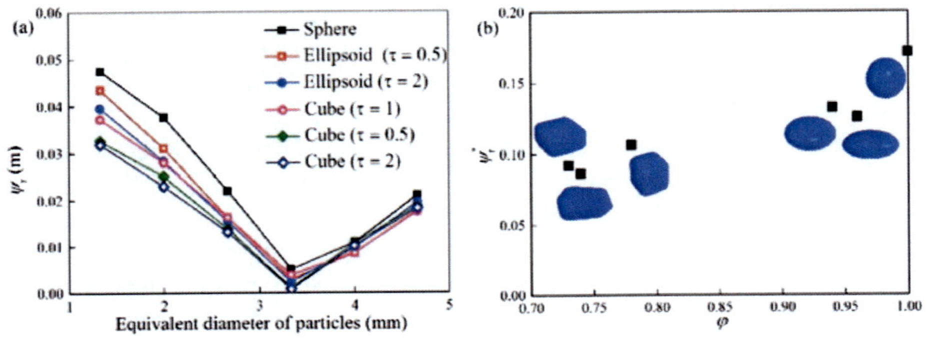

Fig. 7.49 **a** Effect of particle diameters on the gyration degree of granular materials. **b** Effect of particle shape on the gyration coefficient of the mixture

much with the aspect ratio. This is consistent with the results of the mixing index shown in Fig. 7.45b. Therefore, in the following sections, only the results for spheres and cube-like particles at an aspect ratio of 1 are presented.

7.4.2 *Effect of Standard Deviation on the Segregation Process of Particles*

Figure 7.50 shows the radial segregation patterns of spherical granular systems for different particle size distributions (represented by different standard deviations as given in Fig. 7.41a). It can be observed that with the standard deviation decreasing, the radial segregation pattern becomes inconspicuous. This is mainly because the size difference of particles in the system becomes smaller as the standard deviation decreases. As a result, it is difficult for small particles to penetrate through the voids between large particles and move to the core of the granular bed. Moreover, each granular bed in Fig. 7.50 is separated into six figures according to the particle size range. Clearly, small and large particles are gradually and evenly distributed in the granular bed as the standard deviation decreases, while the standard deviation has no effect on the distribution of medium-sized particles. In a Gaussian-dispersed granular system with a small standard deviation, the penetrating ability of small particles and the radial movement of large particles are limited. Meanwhile, the collisions between particles become more intense, which drives the dispersion and mixing of particles of different sizes. Therefore, a mixture with a small standard deviation facilitates dispersion and mixing, and a mixture with a large standard deviation is conducive to segregation.

Figure 7.51 further quantifies the effect of the standard deviation in terms of the gyration degree. As the standard deviation decreases, the gyration degree of particles of different sizes decreases. When the standard deviation is equal to 0.1 mm, particles of different sizes have a similar gyration degree. It means that the distance between

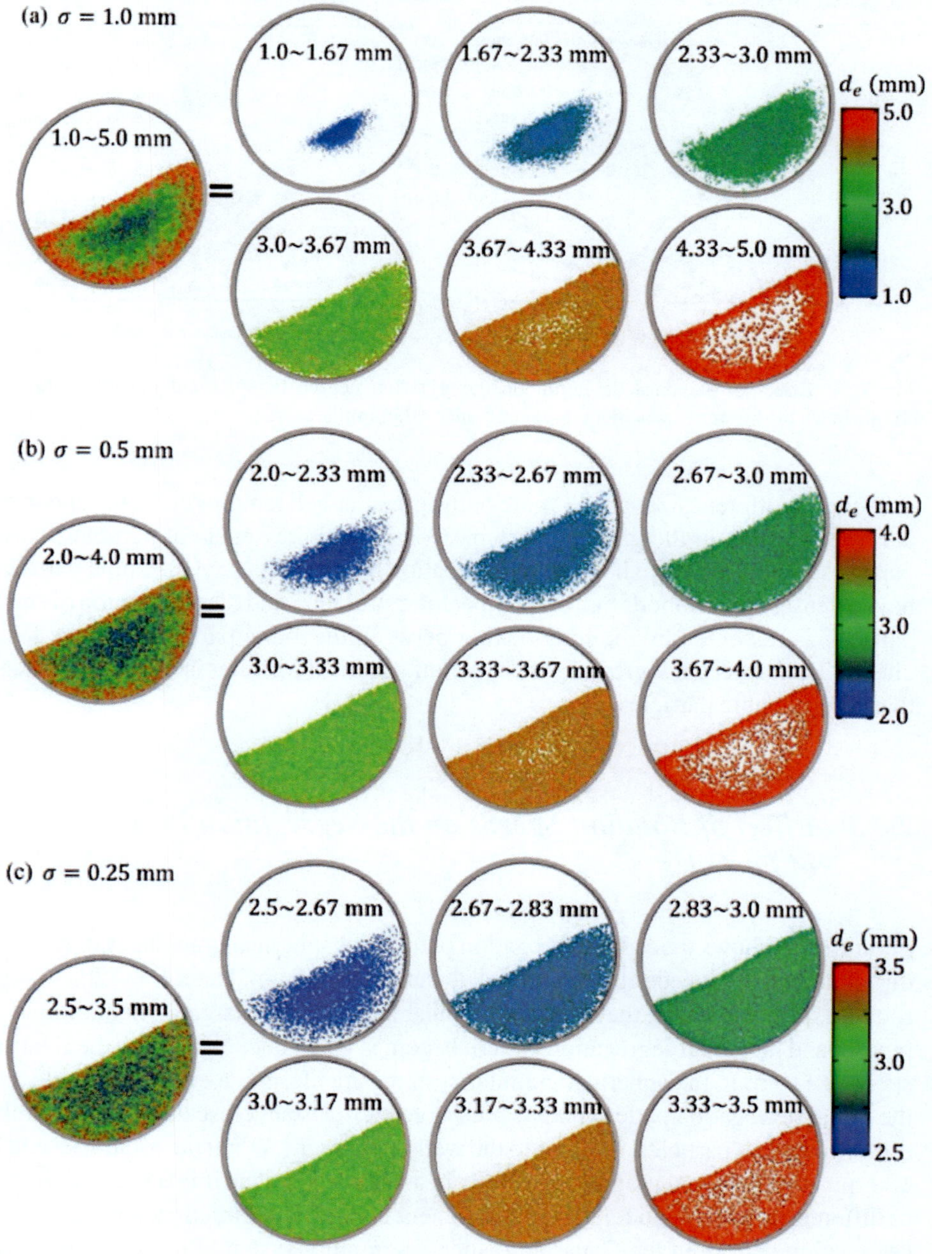

Fig. 7.50 Effect of standard deviation on the radial distribution of spherical particles of different sizes in a Gaussian-dispersed granular system: **a** $\sigma = 1.0$mm, **b** $\sigma = 0.5$mm, and **c** $\sigma = 0.25$mm

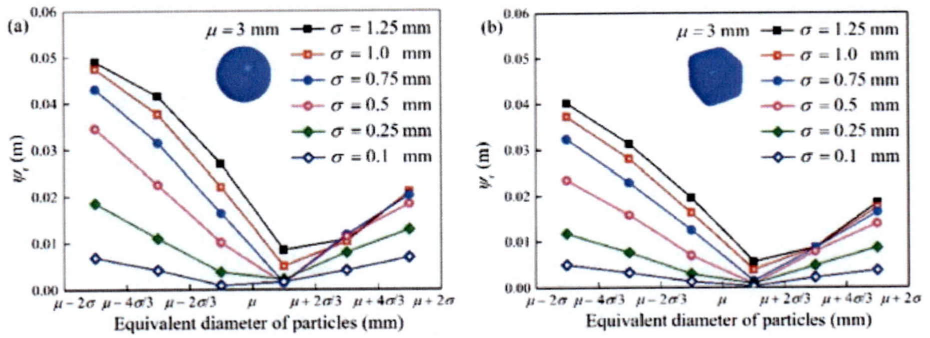

Fig. 7.51 Effect of standard deviation on the gyration degree of spherical (**a**) and cube-like (**b**) particles of different sizes in a Gaussian-dispersed granular system

particles of different sizes and the core of the granular bed remains constant, and there is no segregation in the granular system. Compared with spherical particles, cubic particles of the same size have a lower gyration degree (Fig. 7.51b). Cubic particles have a denser contact mode and orderly packing structure, and the penetration of small particles is restricted by the interlocking between the cubic particles and the local cluster. Therefore, the segregation behavior of spherical particles is more pronounced than that of cubic particles.

7.4.3 Effect of Rotation Speeds on the Segregation Process of Particles

Figure 7.52a shows the radial segregation patterns of spherical granular systems with different rotating speeds at the standard deviation $\sigma = 1\text{mm}$. Here, $t = 120\text{s}$. As the rotating speed increases, the spherical granular material has a larger dynamic angle of repose, and the radial segregation pattern becomes inconspicuous. When the rotation speed is 110 rpm, the spherical granular systems are close to a centrifugal state and the radial segregation pattern reappears. Large particles are close to the center of the drum, and small particles are close to the wall of the drum. When the rotation speed is 130 rpm, the spherical granular system is in a completely centrifuged state. Particles of different sizes have no relative displacement in the radial direction, and there is no segregation pattern in the Gaussian-dispersed granular system. Figure 7.52b shows the patterns for cubic particles at different rotating speeds. The close contact mode between cubic particles facilitates energy transfer through the particle–wall friction and inter-particle interlocking, thereby accelerating the dispersion and mixing of cube-like particles. As a result, cubic particles reach a centrifugal state at a lower rotating speed than spherical particles. Although spherical and cubic particles have different flow characteristics, the radial distribution of particles of different sizes in

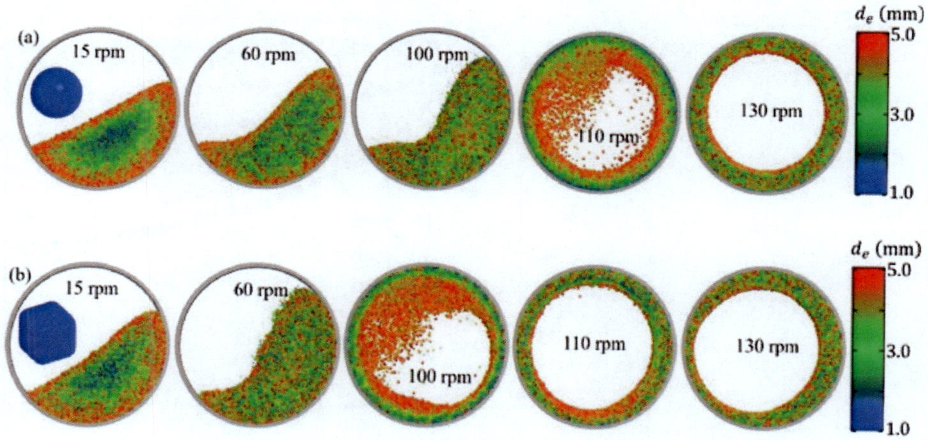

Fig. 7.52 Influence of rotating speed on the radial distribution patterns of spherical (**a**) and cube-like (**b**) granular materials

a Gaussian-dispersed mixture is similar. Therefore, the radial segregation characteristics of spherical granular materials at different rotation speeds are further analyzed through the mixing index and the positional relationship of particles of different sizes.

In the cascading and rolling regime (e.g., rotating speed in the range of 7.5~100 rpm), the temporal variation of the mixing index of spherical granular materials for different rotating speeds is shown in Fig. 7.53. The segregation state of the granular materials and the time to reach the segregation state are different at different rotating speeds. When the rotating speed is 7.5 rpm, the granular materials have the lowest mixing index and the best segregation state. As the rotating speed increases, the mixing index gradually increases, which indicates that the segregation behavior of granular materials is not obvious. To further compare the radial positions of particles of different sizes in the granular system, the relationship between the radial distribution of spherical particles of different sizes and the rotation speed is shown in Fig. 7.54a–c. When the rotation speed is 15 rpm, small particles are located in the core of the granular bed, and the largest particles are distributed on the periphery of the granular bed. As the rotation speed increases, small particles gradually diffuse toward the periphery of the granular bed, while large particles gradually diffuse toward the core of the granular bed. When the rotation speed is 100 rpm, small particles are distributed on the periphery of the granular bed, and large particles are evenly distributed in the granular system. It is worth noting that the rotation speed has no obvious effect on the radial distribution of medium-sized particles, and these particles are uniformly distributed in the granular system regardless of the rotation speed. When the rotation speed is greater than 100 rpm, the granular system is close to a centrifuged state. Figure 7.54d, e show the radial distribution of particles of different sizes under high-speed centrifugal rotation. When the rotation speed is 110 rpm, small particles are close to the wall of the drum, while large particles are

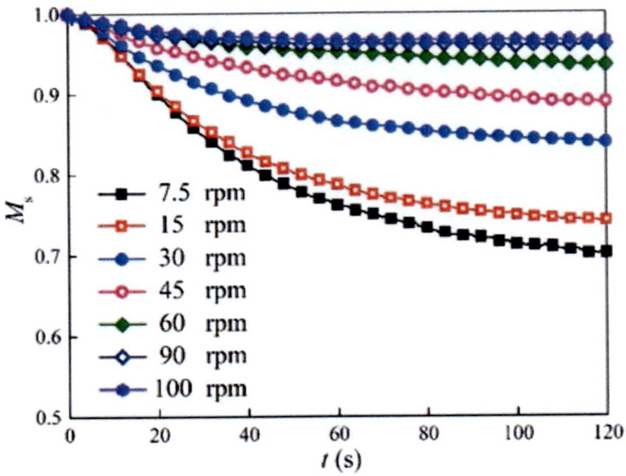

Fig. 7.53 Effect of rotating speeds on the mixing index of spherical granular materials

close to the center of the drum. As the particle size increases, the particles gradually move toward the center of the drum, which makes it more difficult for larger particles to obtain enough energy to reach the centrifugal state. As the rotation speed is further increased, the granular system reaches a completely centrifugal state, and particles of different sizes basically have the same velocities. As a result, all particles are evenly distributed in the granular system.

The rotating drum as a pure energy input causes violent collisions between particles, thereby driving the mixing and segregation of the granular system. Figure 7.55 shows the average normal contact force and coordination number of spherical granular systems at different rotation speeds. Here, $t = 120$s. As the rotation speed increases, the average normal contact force of the granular system first increases and then decreases, while the average coordination number first decreases and then increases. When the rotation speed is less than 100 rpm, the entire granular system is in a cascading and rolling regime. The particles collide more violently and obtain more driving forces in the drum with a higher rotation speed. Although the lower coordination number and looser contact mode between particles are conducive to the penetration behavior of small particles, the collision force is considered as the main mechanism driving the particle dispersion and mixing, which counterbalances the convective motion of the particles (Peng et al. 2016). Therefore, a lower rotation speed is conducive to achieving a better segregation state of the granular system.

When the rotation speed is equal to 110 rpm, the entire granular system is in a centrifugal state and the segregation pattern reappears (Fig. 7.54d). It is worth noting that the segregation pattern in the centrifugal state is mainly caused by the centrifugal movement of particles of different sizes, which is significantly different from the segregation mechanism of the granular system in the cascading regime. The collision force between the particles acts as a centripetal force to make the particles move in a circle. Driven by the same centripetal force, particles with small masses are farther from the center of the drum than particles with large masses. Besides, the granular system has a loosely packing structure and a low average coordination

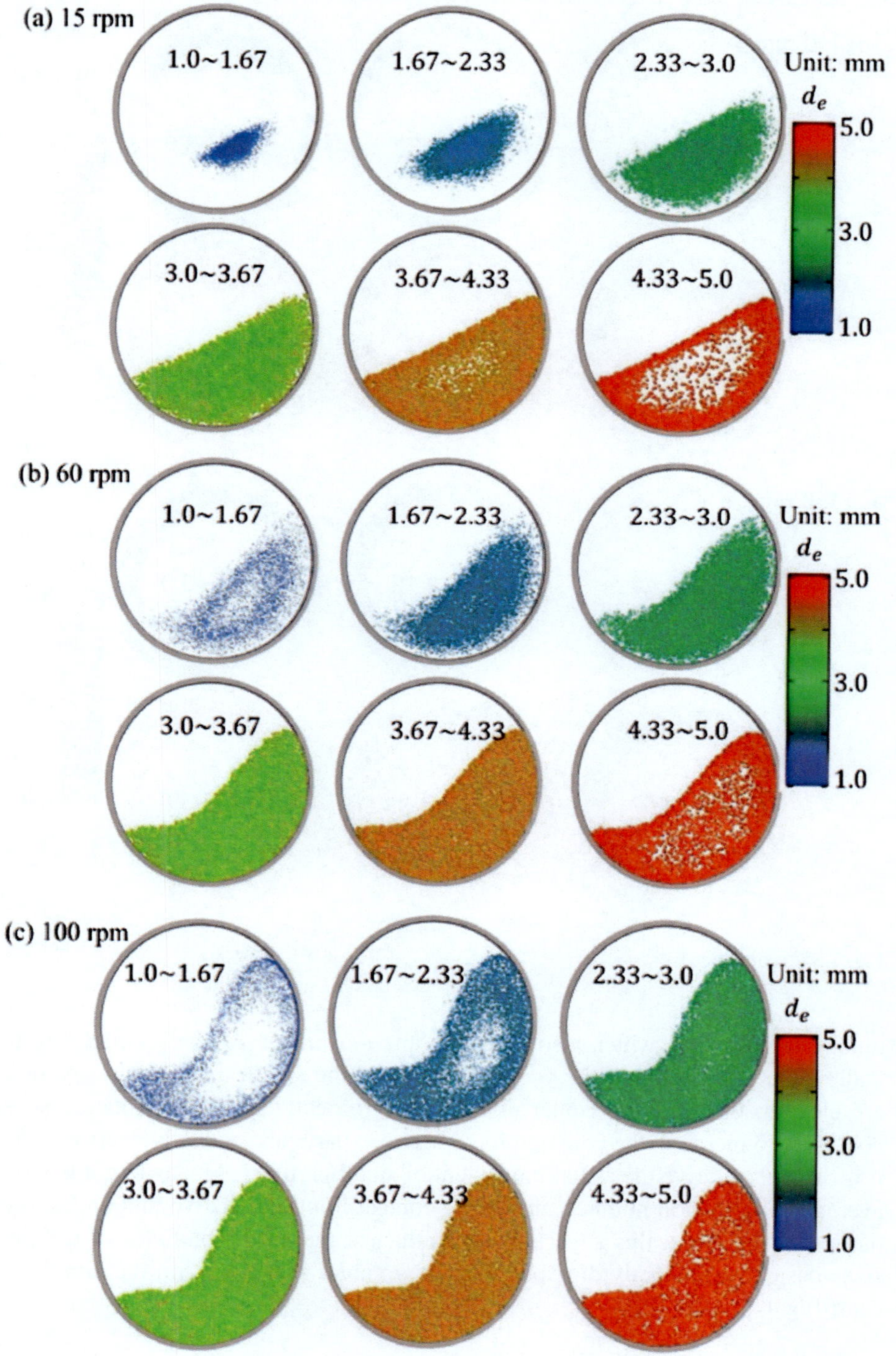

Fig. 7.54 Effect of rotating speed on the radial distribution of spherical particles of different sizes: **a** 15 rpm, **b** 60 rpm, **c** 100 rpm, **d** 110 rpm, and **e** 130 rpm

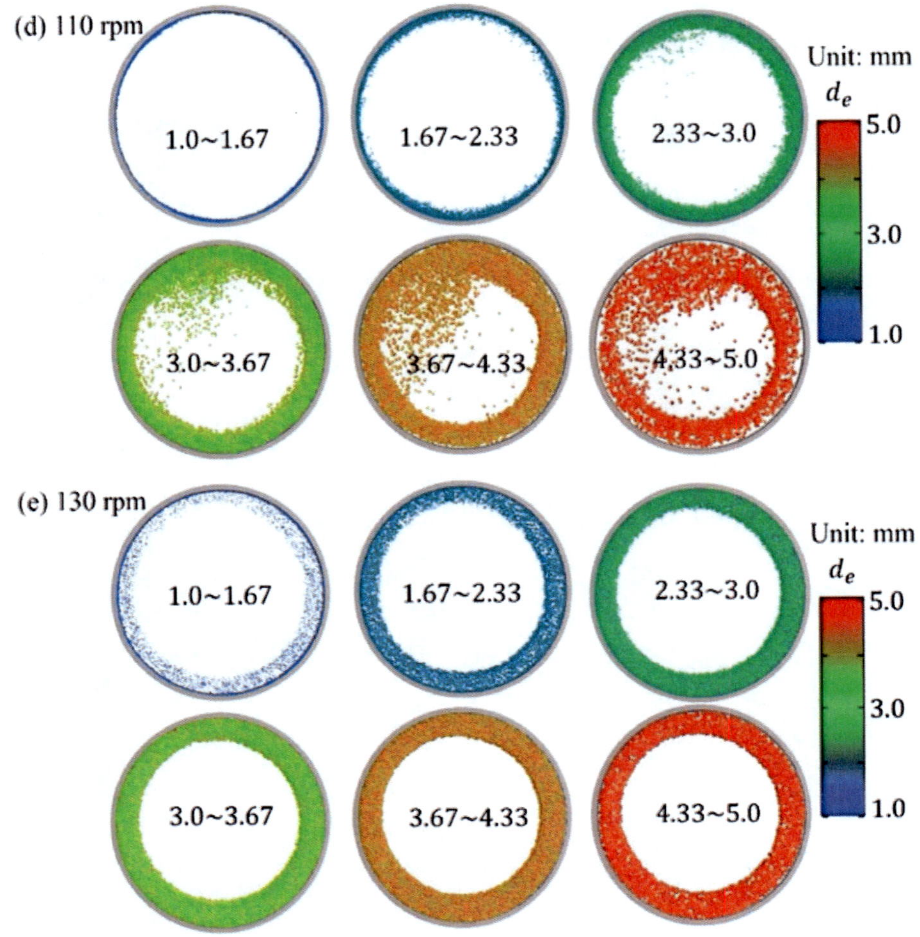

Fig. 7.54 (continued)

number at 110 rpm, which also facilitates the movement of small particles to the wall of the drum through the voids. As a result, the smallest and largest particles are close to the wall and center of the drum, respectively. As the rotation speed continues to increase, the collision force between particles gradually decreases, and it is difficult to drive the radial movement of small particles. Meanwhile, the larger average coordination number and denser contact limit the relative motion between particles. Therefore, this superimposed influence makes the particles of different sizes basically uniformly distributed in the granular system within the high-speed centrifugal drum.

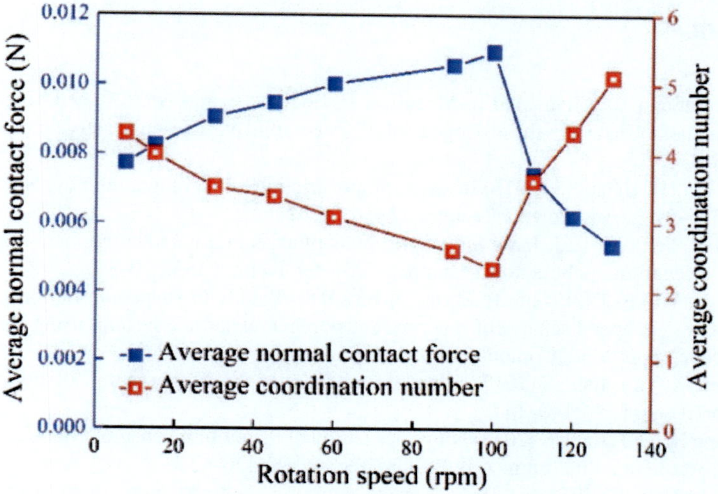

Fig. 7.55 Effect of rotation speed on average normal contact force and coordination number of spherical granular systems

7.5 Summary

In this chapter, the DEM simulation of the flow process of spherical and irregular particles based on superquadric equations and the related experimental validation are presented, mainly including:

1. In gravity-driven granular flows, the particle shape and the hopper angle jointly determine the flow pattern of granular materials. The critical height between the mass flow and the funnel flow decreases and reaches a steady state with the decrease of the sharpness parameter or the deviation of the aspect ratio from 1. Meanwhile, the critical height decreases with the increase of the hopper angle, and it is no longer sensitive to the change of the particle shape. In addition, the particle shape at the micro-scale causes changes in the contact pattern and force between particles, and further causes the flow pattern transition of the granular material at the macroscopic scale.
2. The particle shape and the rotating speed jointly determine the mixing and segregation mode of the granular material in the horizontal rotating drum. In a granular bed with the same particle size, the particle shape changes the conversion efficiency of the external driving energy to the granular bed. The close contact pattern of particles and higher driving energy are conducive to the mixing behavior of the granular material. Within the granular beds with particle sizes satisfying Gaussian distribution, loose contact patterns of particles and lower driving energies favor the segregation behavior of the granular materials.

References

Alchikh-Sulaiman B, Alian M, Ein-Mozaffari F, Lohi A, Upreti SR (2016) Using the discrete element method to assess the mixing of polydisperse solid particles in a rotary drum. Particuology 25:133–142

Chou SH, Hu HJ, Hsiau SS (2016) Investigation of friction effect on granular dynamic behavior in a rotating drum. Adv Powder Technol 27:1912–1921

Ge L, Gui N, Yang X, Tu J, Jiang S (2019) Effects of aspect ratio and component ratio on binary-mixed discharging pebble flow in hoppers. Powder Technol 355:320–332

Govender N, Wilke DN, Pizette P, Abriak N-E (2018) A study of shape non-uniformity and poly-dispersity in hopper discharge of spherical and polyhedral particle systems using the Blaze-DEM GPU code. Appl Math Comput 319:318–336

Gui N, Yang X, Tu J, Jiang S (2017) Effect of roundness on the discharge flow of granular particles. Powder Technol 314:140–147

Gui N, Yang X, Tu J, Jiang S (2018) Numerical study of the motion behaviour of three-dimensional cubic particle in a thin drum. Adv Powder Technol 29:426–437

He S, Gan J, Pinson D, Zhou Z (2017) Transverse mixing of ellipsoidal particles in a rotating drum. EPJ Web Confer 140:06018

He SY, Gan JQ, Pinson D, Zhou ZY (2019) Particle shape-induced radial segregation of binary mixtures in a rotating drum. Powder Technol 341:157–166

Höhner D, Wirtz S, Scherer V (2014) A study on the influence of particle shape and shape approximation on particle mechanics in a rotating drum using the discrete element method. Powder Technol 253:256–265

Huang W, Huang L, Sheng D, Sloan SW (2014) DEM modelling of shear localization in a plane couette shear test of granular materials. Acta Geotech 10:389–397

Ji SY, Li PF, Chen XD (2012) Experiments on shock-absorbing capacity of granular matter under impact load. Acta Phys Sin 61:184703

Jiang M, Zhao Y, Liu G, Zheng J (2011) Enhancing mixing of particles by baffles in a rotating drum mixer. Particuology 9:270–278

Lacey PMC (1954) Developments in the theory of particle mixing. Russ J Appl Chem 4:257–268

Liu AJ, Nagel SR (1998) Jamming is not just cool any more. Nature 396:21–22

Ma H, Zhao Y (2018) Investigating the flow of rod-like particles in a horizontal rotating drum using DEM simulation. Granular Matter 20:41

Peng Z, Joshi JB, Moghtaderi B, Khan MS, Evans GM, Doroodchi E (2016) Segregation and dispersion of binary solids in liquid fluidised beds: a CFD-DEM study. Chem Eng Sci 152:65–83

Yang S, Sun Y, Zhang L, Chew JW (2017) Segregation dynamics of a binary-size mixture in a three-dimensional rotating drum. Chem Eng Sci 172:652–666

You Y, Zhao Y (2018) Discrete element modelling of ellipsoidal particles using super-ellipsoids and multi-spheres: a comparative study. Powder Technol 331:179–191

Zeng Y, Jia F, Zhang Y, Meng X, Han Y, Wang H (2017) DEM study to determine the relationship between particle velocity fluctuations and contact force disappearance. Powder Technol 313:112–121

Zhang Y, Jia F, Zeng Y, Han Y, Xiao Y (2018) DEM study in the critical height of flow mechanism transition in a conical silo. Powder Technol 331:98–106

Zhao Y, Chew JW (2020) Discrete element method study on hopper discharge behaviors of binary mixtures of non-spherical particles. AIChE J 66:16254

Zhong W, Yu A, Liu X, Tong Z, Zhang H (2016) DEM/CFD-DEM modelling of non-spherical particulate systems: theoretical developments and applications. Powder Technol 302:108–152